Subsurface CO_2 Sequestration

Ebrahim Fathi · Fatemeh Belyadi

Subsurface CO_2 Sequestration

A Thorough Analysis of Sustainable Emission Solutions and Regulatory Frameworks

Springer

Ebrahim Fathi
The Department of Engineering
in the College of Science & Engineering
Texas Christian University
Fort Worth, TX, USA

Fatemeh Belyadi
The Department of Engineering
in the College of Science & Engineering
Texas Christian University
Fort Worth, TX, USA

ISBN 978-3-032-28295-8 ISBN 978-3-032-28296-5 (eBook)
https://doi.org/10.1007/978-3-032-28296-5

This Springer imprint is published by the registered company Springer Nature Switzerland AG
The registered company address is: Gewerbestrasse 11, 6330 Cham, Switzerland

Competing Interests The authors have no competing interests to declare that are relevant to the content of this manuscript.

About This Book

In this book, we will discuss multiple aspects of Carbon Capture and Storage (CCS) technologies, investigating their critical role in addressing the environmental challenges associated with excess carbon dioxide emissions. Our discussion not only highlights CCS as a viable long-term solution but also emphasizes its significance in the broader framework of sustainable energy development. We will analyze different CCS techniques, economic implications, and regulatory frameworks, to provide a thorough understanding of how CCS can shape a low-carbon future. Brief discussion on carbon utilization will also be covered to introduce the Carbon Capture Utilization and Storage (CCUS).

We will start our discussion with fundamentals of climate change and the global energy transition. In the opening chapters, we will investigate the environmental implications of CCS technologies, evaluating their effectiveness in mitigating greenhouse gas emissions. Through case studies and real-world examples, we will evaluate the potential benefits of CCS implementation and its alignment with international climate agreements and summits, such as the Paris Agreement, COP28 and COP29. Understanding CCS within the context of these agreements will reveal CCS technology's role in supporting national and international commitments to mitigate climate change. We then move to an in-depth look at carbon capture and storage technologies, facilities, regulatory and economic frameworks. Covering methods like post- and pre-combustion capture, oxyfuel combustion, and direct air capture, these chapters provide insights into both well-established and emerging technologies. We will examine each technique's applications, challenges, and contributions to the evolving CCS landscape. Further chapters address the essential steps involved in CO_2 transport, from pipeline infrastructure to leak detection and monitoring technologies, emphasizing the importance of safe and efficient CO_2 movement from capture sites to storage facilities.

A key element of CCS involves site selection and characterization for geological storage. We will explore the methodologies used to evaluate geological viability and long-term stability. We detail the site screening and characterization process that ensure CO_2 securely stored underground. Next, we explore the concept of carbon storage modeling and simulation, which enables us to predict and simulate the behavior of stored carbon dioxide over time specially CO_2 saturation and pressure during injection (i.e., area of review, AoR) and after injection stops (i.e., post injection site care, PISC). We will also discuss regulatory aspects, including the complex permitting requirements for Class VI wells in the U.S., potential policy developments such as pore space ownership, and the strategic importance of regional partnerships in advancing carbon storage projects. As we are encouraging discussions on the analysis of current and ongoing research and development efforts on CCS projects such as those applied in the Illinois Basin-Decatur Project (IBDP), we will gain insight into the evolving landscape of CCS, including emerging trends and technological advancements, Figs. 1 and 2.

Fig. 1 Photo by Ronak Naik on Unsplash

Fig. 2 Archer Daniels Midland's Decatur facility: CO_2 capture from biofuel production and storage in Decatur, Illinois. NETL.doe.gov.

Finally, we extend our scope to consider carbon-negative solutions and alternative storage options. This closing chapter covers innovative approaches such as bioenergy with carbon capture and storage (BECCS), direct air capture (DAC), and other carbon-negative technologies that go beyond traditional CCS methods. We will explore the potential for these solutions to complement existing CCS efforts, as well as their feasibility in different industrial sectors. By the end of this book, you will have gained a comprehensive understanding of the environmental, regulatory, and technological dimensions of CCS. Equipped with this knowledge, you will be prepared to:

- Evaluate and select carbon storage sites,
- Model CO_2 sequestration behavior,
- Analyze the social and economic aspects of CCS projects, and
- Quantify the impactful CCS solutions with respect to global climate goals.

About Chap. 1: Foundations of Climate Change and Energy Transition for CO_2 Sequestration

This chapter explores the climate change and the urgent need for an energy transition/energy mix policies to manage CO_2 emissions effectively. Beginning with an overview of the carbon cycle, it examines the impact of fossil fuel emissions on global warming. Key mitigation strategies across sectors are introduced, focusing on renewable energy adoption, energy mix, and the environmental challenges linked to these technologies. The chapter also discusses energy demand forecasting, emphasizing the role of reliable and scalable energy sources.

International climate agreements, from the Paris Agreement onward, are reviewed to understand global climate governance. Additionally, the chapter explores climate adaptation strategies, emphasizing nature-based and community-driven approaches. Practical applications of these mitigation and adaptation efforts also will be discussed. The chapter concludes by illustrating how integrated, sustainable energy mix and carbon management can help create a robust, low-carbon future.

About Chap. 2: Carbon Capture and Storage (CCS) Technologies, and Challenges

During this chapter, we present discussions regarding different classification of CCS facilities, including Carbon Negative, Carbon Neutral or Near Carbon Neutral, and Carbon Positive. This classification serves as a framework for understanding the challenges in scaling CCS technologies. We also examine the CCS landscape's evolution, illustrating how advancements in technology are driving more efficient carbon capture and storage processes. Through real-world examples, we highlight

CCS applications across hard-to-decarbonize industries, such as steel, cement, and chemical manufacturing. The chapter also looks ahead, considering future CCS technologies and industry trends intended at further reducing emissions. Understanding the complexities and opportunities in these sectors helps us better strategize for a sustainable, low-carbon future.

About Chap. 3: Regulatory and Economic Frameworks for CCS

In this chapter, we discuss the regulatory and economic frameworks that shape carbon management and CCS adoption. We explore key regulatory approaches, including the Carbon Tax System, Cap-and-Trade System, and Voluntary Carbon Markets. We investigate how each mechanism incentivizes and regulates CCS initiatives with unique benefits and trade-offs. The chapter then covers economic mechanisms aimed at reducing carbon emissions, such as carbon pricing, subsidies, and tax credits. We discuss policy and incentive structures, highlighting how effective policies can attract investments and enhance public–private partnerships in carbon management. Key integrations and justifications underscore the importance of aligning regulatory and economic strategies to support CCS within broader environmental goals. The chapter concludes with a real-world case study that exemplifies successful CCS implementation.

About Chap. 4: Fundamentals of Combustion and CO_2 Capture Technologies

This chapter provides combustion principles, and the technologies developed to capture CO_2 emissions from combustion processes, which are a major source of greenhouse gases in industries and power generation. We begin with an introduction to the fundamentals of combustion and its role in CO_2 emissions. Following this, we detail various CO_2 capture technologies: post-combustion capture (PCC), pre-combustion capture, and oxyfuel combustion. Each technology is analyzed in terms of its operational mechanisms, efficiency, scalability, and technical challenges. The chapter also discusses the concept of CCS-readiness in combustion technologies, focusing on innovations that facilitate CO_2 capture from existing combustion systems.

Contents

About the Authors

Dr. Ebrahim Fathi is a Full Professor and Holder of the Ross B. Matthews Professorship in Petroleum Engineering in the Department of Engineering at Texas Christian University. His research focuses on carbon capture, utilization, and storage (CCUS), geothermal energy, data analytics and machine learning applications in subsurface systems, hydraulic fracturing, and unconventional reservoir development.

Throughout his academic career, Dr. Fathi has led and contributed to research projects supported by organizations including the U.S. Department of Energy (DOE), NETL, RPSEA, Leidos, the U.S. Department of State, and industry partners. His work has included developing reduced-order models for CO_2 sequestration at Citronelle Dome in Alabama, developing machine learning tools for reservoir characterization in the Bioenergy Carbon Capture and Storage (BECCS) initiative at the Illinois Basin-Decatur Project (SMART Phase I and II), and characterization of stacked reservoir, caprock systems across West Virginia, Ohio, and Pennsylvania, the Tri-State CO_2 Storage Hub project, to assess their suitability and integrity for long-term CO_2 storage. He has also conducted a feasibility analysis of Geothermal Deep Direct-Use combined with Reservoir Thermal Energy Storage at West Virginia University's campus, and developing of Marcellus shale gas reservoir through Marcellus Shale Energy and Environment Laboratory (MSEEL) project.

Dr. Fathi has authored several journal articles, technical papers, and books on unconventional reservoir hydraulic fracturing. He has received several honors for his work, including the Outstanding Technical Editor Award from the SPE Journal (2013 and 2018) and the prestigious AIME Rossiter W. Raymond Memorial Award in 2015. He contributes to the academic community as Associate Editor for Frontiers in Earth Science and as Guest Editor for Energies and Geofluids.

Dr. Fatemeh Belyadi is a Research Professor in the Department of Engineering at Texas Christian University. She is an accomplished reservoir engineer, author, entrepreneur, and educator with extensive expertise in Carbon Capture, Utilization, and Storage (CCUS), geothermal energy, and integrated field development. She previously served as Senior Reservoir Engineer at Pangea Geoscience, Assistant Professor at West Virginia University, and Process Engineer at Exterran Energy Solutions. She also founded Integrated Smart Data Solutions, LLC, where she served as Chief Executive Officer.

Her research has focused on the integration of subsurface characterization, reservoir modeling, and data-driven optimization to enhance the efficiency and safety of CO_2 storage operations. She has contributed to projects on systematic Uncertainty and Contingency Planning for subsurface storage, machine learning-driven monitoring of CO_2 plume dynamics, and the evaluation of geologic formations. In addition, she has explored synergies between geothermal energy systems and CCUS.

Dr. Belyadi has authored several peer-reviewed publications, conference papers, and books, including Hydraulic Fracturing in Unconventional Reservoirs (Gulf Professional Publishing, 2016/2019). Her academic service includes roles as Review Editor for Economic Geology of Frontiers in Earth Science and Guest Editor of a special issue on subsurface CO_2 sequestration in Energies.

List of Figures

List of Tables

Chapter 1
Foundations of Climate Change and Energy Transition for CO_2 Sequestration

1.1 Chapter Overview and Objectives

1.1.1 Overview

This chapter serves as an introduction into the global energy systems, climate change, and sustainable development. It begins with an overview of global climate change, especially carbon dioxide (CO_2) emissions from fossil fuel use. The chapter outlines the scientific agreement on climate change, highlighting how carbon dioxide emissions primarily from coal, oil, and natural gas are changing Earth's climate systems. It presents data and trends to illustrate the scale and urgency of the problem. Next, the importance of using fossil fuels will be discussed by closer look to forecast global energy production and demand, breaking down current and projected consumption by region and energy type. Cost analyses of different energy sources, including fossil fuels and renewables, are presented to assess feasibility and affordability. In evaluating energy sources, the chapter introduces a multi-criteria decision-making framework. It covers essential factors such as cost, energy storage capability, scalability, and environmental footprint. A detailed investigation of environmental concerns related to renewable technologies follows. Solar, wind, hydropower, geothermal, and biomass are analyzed not only for their potential benefits but also for their environmental trade-offs. The chapter then addresses climate change mitigation strategies, including carbon capture and storage (CCS). International cooperation is emphasized through a review of key climate agreements and summits, including the Paris Agreement and the COP28, 29 summits. These discussions outline global collaborative efforts aimed at limiting temperature rise beyond 2 degrees Celsius as outlined in Paris Agreement by defining clear goals and policy developments. The chapter concludes with a case study that discusses the strategy planned by USA government to cut methane emissions.

E. Fathi and F. Belyadi, *Subsurface CO_2 Sequestration*,
https://doi.org/10.1007/978-3-032-28296-5_1

1.1.2 Learning Objectives

Upon completion of this chapter, you will be able to:

1. Define the relationship between global energy systems and climate change.
2. Analyze the economic implications of different energy sources.
3. Evaluate various renewable and non-renewable energy technologies.
4. Assess international climate agreements and policy tools.

1.2 Global Climate Change

1.2.1 Overview

The evolution of systematic weather data collection is critical in our understanding of global climate change. The earliest systematic weather recordings began in the eighteenth century, such as efforts by Thomas Jefferson recording temperature, wind, and precipitation (NOAA 2025a, b). In the mid-nineteenth century, the formal meteorological organizations created such as the UK Met Office in 1854 and the U.S. Weather Bureau in 1870 (Craft 2023). Collections of reliable global temperature records began around 1880 by agencies like NASA's and the NOAA's Global Historical Climatology Network (GHCN). The modern satellite weather data collection started with the launch of TIROS-N in 1978, providing high-resolution global climate monitoring (Wentz and Schabel 1998). Today weather information is collected from over 100,000 weather stations across 180 countries globally (https://climate.nasa.gov). These stations gather data from various sources like weather stations, balloons, ships, radar, and satellites. The (GHCNd) is responsible for collecting and compiling this data. The National Oceanic and Atmospheric Administration (NOAA), a U.S. organization, utilizes the GHCNd data and oversees global climate data management.

The Earth's temperature has exhibited a consistent upward trend, increasing by 0.14 °F (0.08 °C) per decade since 1880. Notably, the rate of warming has more than doubled since 1981, reaching 0.32 °F (0.18 °C) per decade, as depicted in Fig. 1.1. In 2021, the average temperature across land and sea was 1.51 degrees Fahrenheit higher than the twentieth-century average of 57 degrees Fahrenheit and 1.87 degrees Fahrenheit warmer than the pre-industrial period (1880–1900) (NOAA 2022). Additionally, the years spanning from 2013 to 2025 have consistently ranked among the warmest on record, highlighting the persistent trend of rising temperatures on our planet. 2025 declared the hottest year globally at the time, with temperatures average above 1.5 °C above pre-industrial levels (Fig. 1.2). Figure 1.2 shows the upper and lower bounds of temperature variations along with average temperature change compared to pre-industrial era before 1850.

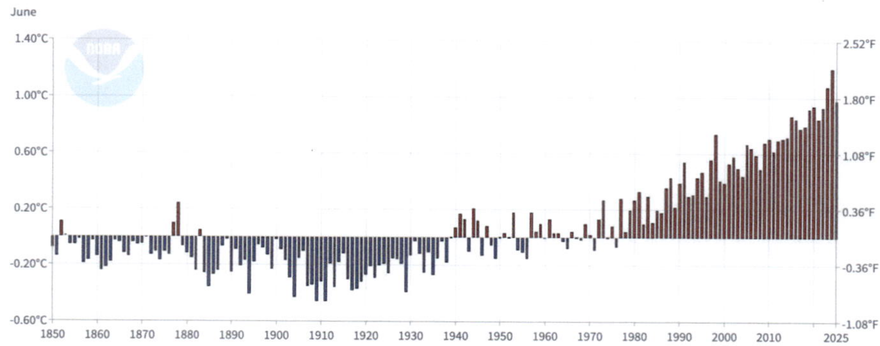

Fig. 1.1 Global average surface temperature NOAA https://Climate.gov graph, based on data from the national centers for environmental information

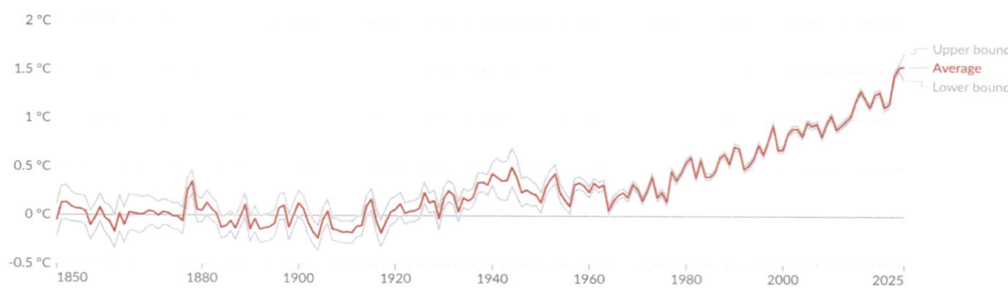

Fig. 1.2 Global temperature trend from 1850 to 2025. *Data Source* met office Hadley centre-HadCRUT5 (2025)

Table 1.1 compares recorded temperatures to the 1901–2000 baseline, ranking them as "Coolest" or "Warmest" historically (since 1850). All periods (18–60 months) are ranked as "first Warmest," indicating exceptionally high temperatures. Temperature deviations range from +1.06 °C to +1.28 °C above the baseline with consistent warming trend, with recent years (2024–2025) being the warmest on record for all measured periods.

The extensive volume and heat capacity of the world's oceans requires a considerable amount of heat energy to create even a minor increase in the Earth's annual average surface temperature. Despite the modest increase of about 2 degrees Fahrenheit or 1 degree Celsius in the global average surface temperature since the pre-industrial era, it represents a significant accumulation of heat. This excess heat has significant consequences, contributing to the occurrence of extreme temperatures in different regions and seasons. Moreover, they are responsible in shrinking snow cover and sea ice, increasing the number of heavy rainfall, and causing shifts in habitat ranges for various plants and animals. The ongoing changes lead to the expansion of certain habitats and the reduction of others in response to these environmental shifts (*Source* https://www.climate.gov). Tangible examples of these changes include shrinking of sea ice cover in the Antarctic between 1980 and 2020, and the rise in

Table 1.1 Record warmest days of all time

Record Coolest	Bottom 1/10	Bottom 1/3	Normal	Top 1/3	Top 1/10	Record warmest
Period		Anomaly (1901–2000)	Rank (1850–2025)	Coolest/ warmest since	Record	
Jan 2024–Jun 2025 18-month		1.26 °C (2.27 °F)	175th coolest	Coolest since: 2024	1909	
			1st warmest	Warmest to date	2025	
Jul 2023–Jun 2025 24-month		1.28 °C (2.30 °F)	174th coolest	Coolest since: 2024	1918	
			1st warmest	Warmest to date	2025	
Jul 2022–Jun 2025 36-month		1.17 °C (2.11 °F)	173rd coolest	Coolest since: 2024	1911	
			1st warmest	Warmest to date	2025	
Jul 2021–Jun 2025 48-month		1.11 °C (2.00 °F)	172nd coolest	Coolest since: 2024	1911	
			1st warmest	Warmest to date	2025	
Jul 2020–Jun 2025 60-month		1.06 °C (1.91 °F)	171st coolest	Coolest since: 2024	1913	
			1st warmest	Warmest to date	2025	

mean sea level since 1993. Since the late 1990s, satellite observations have shown a persistent decline in Arctic Sea ice extent (Parkinson 2019; Meier et al. 2014). This loss has intensified since the 1990s and is part of a broader pattern of Arctic changes, including rising air temperatures, melting glaciers and ice sheets, and warming ocean temperatures (Walsh 2013; Jeffries et al. 2013). The rise in mean sea level is recorded to be 10.10 cm since 1992 (Mizanuzzaman and Kabir 2025).

The consequences of accumulated heat energy become evident as we understand the magnitude of the temperature increase. These changes have significant consequences for various aspects of our planet's climate system, impacting ecosystems, and weather patterns. Several factors contribute to the rise in temperature and global warming. These include carbon dioxide emissions; changes in land use like deforestation; and an increase in particulate matter such as black carbon from incomplete combustion, methane, nitrous oxide, and human-made compounds like chlorofluorocarbons (CFCs) and hydrochlorofluorocarbons (HCFCs). Specially, carbon dioxide (CO_2) stands out as the most prevalent greenhouse gas, contributing significantly to the overall emissions.

Figure 1.3 shows there is a direct and almost linear relationship existing between the increase in cumulative CO_2 emission and global surface temperature increase. It also illustrates annual carbon emissions and corresponding temperature changes. The left figure depicts historical annual carbon emissions and a range of future emissions in gigatons of carbon (GtC) per year. The right figure shows observed historical temperature changes and projected future changes relative to the 1901–1960 average. These projections include both central estimates (lines) and uncertainty ranges (shaded areas, representing two standard deviations), based on simulations from the full set of Coupled Model Intercomparison Project Phase 5 (CMIP5) global climate models. There is an almost linear relationship between the accumulation of carbon dioxide and global warming. To estimate future climate conditions, a variety of plausible scenarios are assumed by Intergovernmental Panel on Climate Change (IPCC). Among these scenarios, RCP8.5 represents a high-emissions future with minimal mitigation efforts, while the other RCPs reflect varying degrees of emission reduction.

The Representative Concentration Pathways (RCPs) describe possible future climate scenarios based on different levels of greenhouse gas emissions.

1. **RCP 2.6** is a *very low-emissions* scenario. It assumes strong actions to reduce emissions and remove carbon from the atmosphere, keeping global warming well below 2 °C.
2. **RCP 4.5** is a *low-emissions* pathway. Here, it assumes higher emissions around mid-century and then emissions decrease.

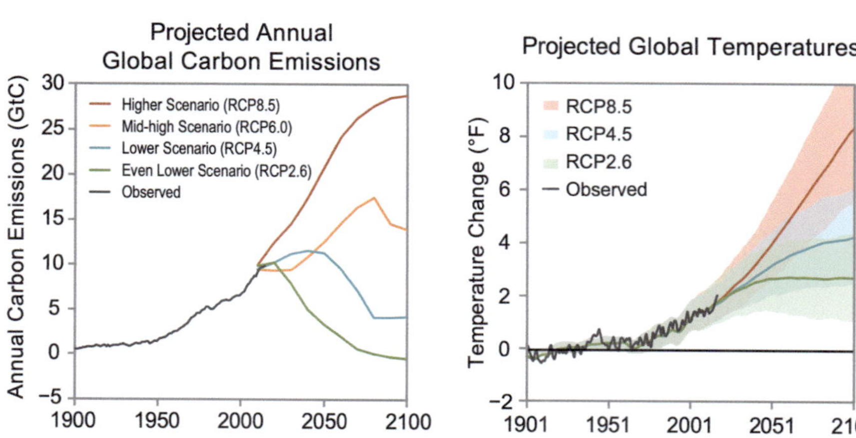

Fig. 1.3 Projected annual global emissions (left) and global temperature (right), USGCRP, Wuebbles et al. 2017

3. **RCP 6.0** is a *medium to high-emissions* scenario. Emissions continue to rise slowly before stabilizing later in the century.
4. **RCP 8.5** assumes emissions keep growing throughout the century, without significant reductions. This leads to very high warming scenario (Van Vuuren et al. 2011).

1.2.2 CO_2 Emission from Fossil Fuel

The concentration of carbon dioxide in the atmosphere is currently at a relatively low level, around 427 ppm. Figure 1.4 displays the last 5 full years of Mauna Loa CO_2 measurements, along with preliminary data from the current year (https://gml.noaa.gov/ccgg/trends). The monthly mean values are shown as red line with highlighted measured data points and the black lines with data points show this same graph smoothed to remove the average seasonal cycle.

Prior to the start of the industrial revolution, the atmospheric CO_2 concentration was even lower, at approximately 0.028% (Thoning and Komhyr 1989).

However, the use of carbon found in fossil fuels on an industrial scale and the subsequent combustion of these fuels have resulted in a marginal increase in CO_2 concentration. Although the increase may seem small, it has significant consequences for the equilibrium of the planet's carbon cycle.

The carbon cycle represents the neutral process by which carbon is exchanged between various carbon reservoirs, such as the atmosphere, oceans, land, and living organisms. This cycle maintains a balance in the movement of carbon, ensuring that carbon is continuously cycled and recycled between these reservoirs. However, the

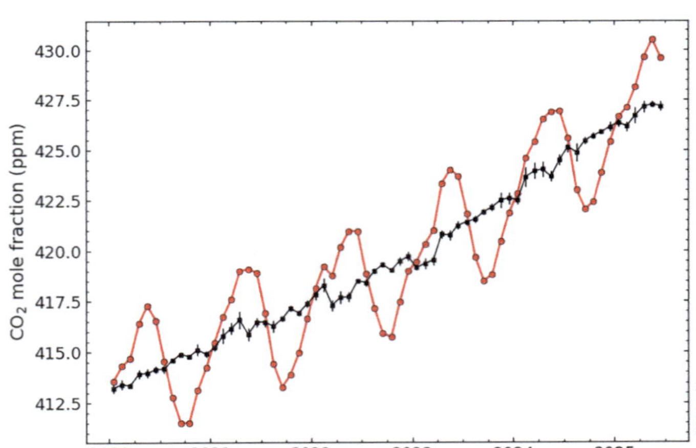

Fig. 1.4 Recent monthly mean CO_2 at Mauna Loa observatory (https://gml.noaa.gov/)

combustion of fossil fuels has disturbed this balance by introducing additional carbon dioxide into the atmosphere at an accelerated rate.

The higher levels of CO_2 in the air from burning fossil fuels have significant consequences. CO_2 is a greenhouse gas that traps heat, causing the greenhouse effect, global warming, and climate change. The disruption in the natural carbon cycle, caused by these emissions, has led to a general rise in greenhouse gases. This imbalance is altering the Earth's climate system. It is important to quantify and analyze the out-of-balance carbon cycle and the associated consequence of increased CO_2 emissions. Figure 1.5 visually depicts the primary carbon reservoirs on Earth. Earth's crust serves as a massive carbon store, containing over 100,000 trillion tons of carbon within sedimentary rocks beneath the surface. Additionally, fossil fuels like oil contribute 5–10 trillion tons, while soil contains 1.5 trillion tons of carbon. The oceans hold a substantial 38 trillion tons, mainly in submerged sedimentary rocks. The atmosphere contains 750 billion tons of carbon, primarily as carbon dioxide, along with some methane and other gases. Plants store 560 billion tons of carbon. Arrows in the diagram represent the flux or transfer between these carbon reservoirs. Processes such as soil decomposition, plant respiration, volcanic activities, deforestation, and changes in land use release a significant 208 billion tons of carbon annually into the atmosphere (U.S. DOE, Biological and Environmental Research Information System). This release is offset by the photosynthesis process and ocean uptake. However, human activities, particularly the burning of fossil fuels, introduce additional carbon into the atmosphere, disrupting the natural carbon cycle balance.

Burning fossil fuels releases carbon previously stored in the geological cycle which exceeds the emissions from natural processes, such as volcanic emissions. This disruption leads to an imbalance in the carbon cycle. The excess carbon dioxide release contributes to the accumulation of greenhouse gases, resulting in climate

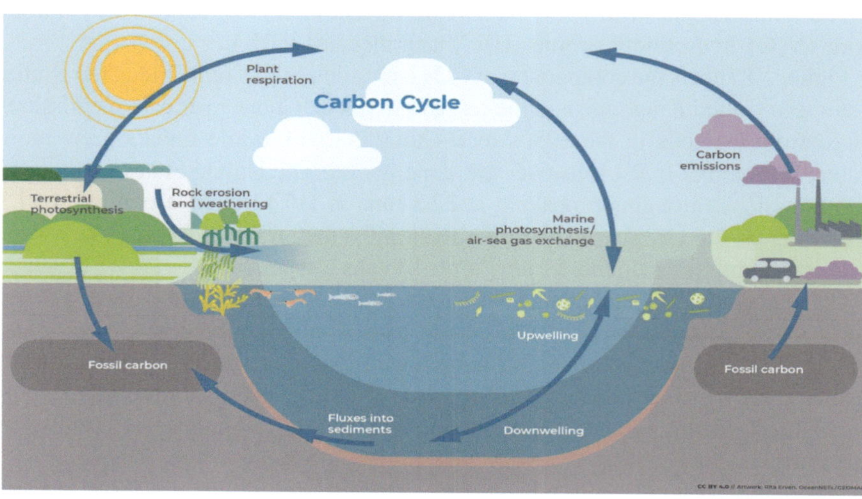

Fig. 1.5 Global carbon cycle image (Rita Erven, GEOMAR/OceanNETs)

change and related environmental impacts. Understanding the dynamics of the global carbon cycle and the disturbances caused by human activities is critical for developing strategies to restore balance and mitigate the consequences of carbon emissions. It is also important to be more specific on type of gases that contributes to greenhouse gas emissions in a specific time span.

The greenhouse gas emissions include F-gases, nitrous oxide (N_2O), methane (CH_4), CO_2 from land use change and forestry, and CO_2 from fossil fuels and industrial processes. Figure 1.6 illustrates the atmospheric concentrations of three major greenhouse gases, namely, carbon dioxide (CO_2), methane (CH_4), and nitrous oxide (N_2O) from 1850 to 2023. Carbon dioxide (CO_2) exhibits the highest percentage of gas emission in whole time. Methane emissions are the second significant gas emission and nitrous oxides are the third most important component. The graph indicates in average more than 75% of the emissions are due to CO_2, methane contributes 16%, nitrous oxide accounted for 7%, and F-gases much smaller portion (less than 2% of total emissions, not shown in this graph). F-gases or fluorinated gases are human-made gases in a variety of applications such as refrigeration, air conditioning, and electronic manufacturing. They include gases like hydrofluorocarbons (HFC), tetrafluoromethane PFC-14 (CF_4), and sulfur hexafluoride (SF_6). These gases have high global warming potentials, meaning they can trap heat in the atmosphere much more effectively than CO_2. While F-gases contribute a relatively small percentage to the overall greenhouse gas emissions, their high global warming potential makes them important to address. To standardize emissions across different gases, carbon dioxide equivalents (CO_2e) are used that is calculated by multiplying the emissions of a given greenhouse gas by its corresponding Global Warming Potential over 100 years (GWP100). The GWP100 metric quantifies how much heat a greenhouse gas traps in the atmosphere relative to CO_2 over a century. Figure 1.7 shows among F-gases sulfur hexafluoride (SF_6) has the highest impact 24,300 times more than CO_2, the next is tetrafluoromethane with 7380 times more impact than CO_2, followed by nitrous oxide (N_2O), hydrofluorocarbons (HFC), and methane (CH_4).

Figure 1.8 shows the comparison of CO_2 emissions from China, India, the United States, the United Kingdom, and the rest of the world. China's per capita emissions raised from 0.1 tons in the 1950s to 8.4 tons in 2023, while India's grew more gradually from 0.2 to 2.1 tons over the same period. The United States maintained high emissions throughout, peaking at 22.3 tons in 1973 before declining to 14.3 tons, whereas the UK peaked at 11.7 tons in 1973 then fell to 4.4 tons. The rest of the world's emissions doubled from 2.4 to 4.7 tons per person.

China now leads in total annual emissions, followed by the US and the rest of the world. These patterns reveal both historical responsibility and recent growth areas. The trend shows there is a need for coordinated global action to transition toward cleaner energy and sustainable development. The differences in national trajectories highlight the importance of differentiated yet collective approaches to emission reduction.

When comparing the emissions among these major contributors, China consistently has the highest annual CO_2 emissions, followed by the United States and the rest of the world, while per capita United States emits the highest emission followed

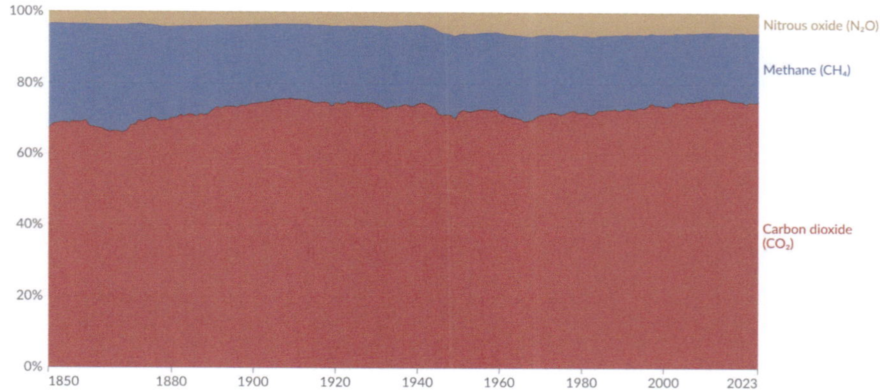

Fig. 1.6 Total greenhouse gas emissions by sources from 1850 to 223. *Data Source* Jones et al. (2025)—OurWorldinData.org/co2-and-greenhouse-gas-emissions

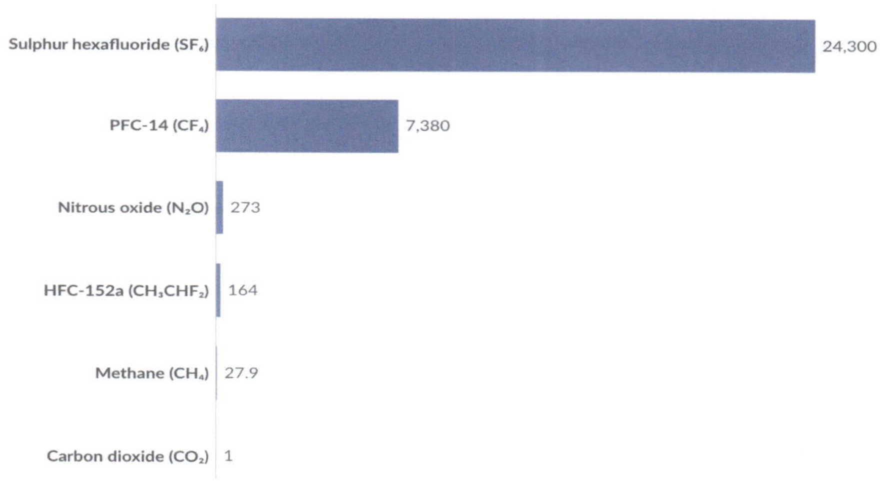

Fig. 1.7 Comparative warming effect of greenhouse gases relative to CO_2. *Data Source* IPCC (2021) OurWorldinData.org/co2-and-greenhouse-gas-emissions

by China. These countries together represent a significant portion of global emissions. Understanding the contribution of different countries to global CO_2 emissions is important as it helps in addressing climate change, since successful plan requires direct contribution from each one of these major contributors to global emissions.

Figure 1.9 presents a detailed perspective on global primary energy consumption, breaking down the contributions from various fossil fuel sources in terawatt-hours (10^{12} watt-hours). This presentation shows the correlations between energy demand and the sources that support global energy demand. It represents the global fossil fuel consumption, or energy production, based on Vaclav Smil data from (2017).

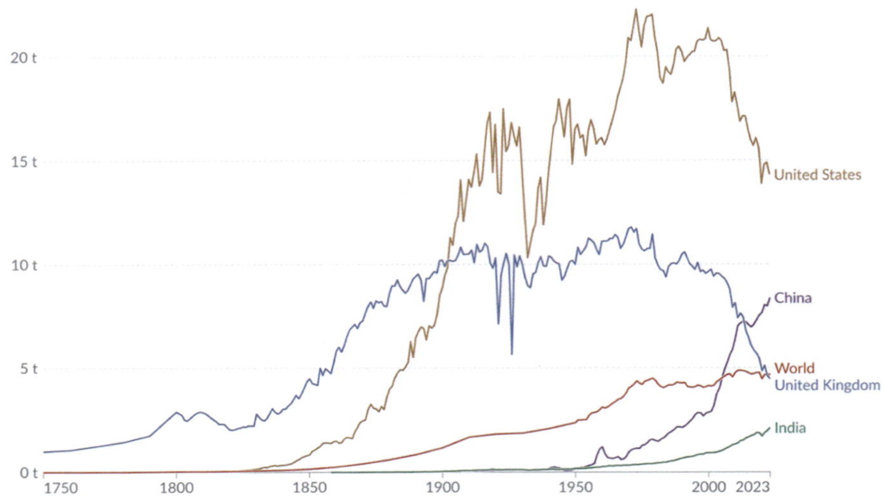

Fig. 1.8 Per capita CO_2 emissions from fossil fuels by region. *Data Source* Global Carbon Budget (2024); Population based various sources (2024) OurWorldinData.org/co2-and-greenhouse-gas-emissions

According to the data, oil stands out as the major source of energy, with the consumption exceeding 54,564 terawatt-hours (TWh) by 2023. Oil has been widely used for various purposes, including transportation, industrial processes, and power generation. Coal follows oil in terms of energy production, with a consumption of approximately 45,565 terawatt-hours. Coal has historically been a significant contributor to electricity generation, particularly in regions heavily reliant on this fossil fuel. Gas consumption ranks third, with a total consumption of around 40,102 terawatt-hours. Natural gas is commonly used for heating and electricity generation and as a cleaner alternative to coal in some cases. This breakdown indicates that oil remains the primary source of energy globally, followed by coal and gas, so called the "Kings of Energy."

Above discussions showed the current representation of the CO_2 status in the atmosphere, the source of that CO_2, the major contributors, the need for the energy, and where the energy is coming from. So, it tries to build that chain, but this is the current situation.

Now, What About the Future?

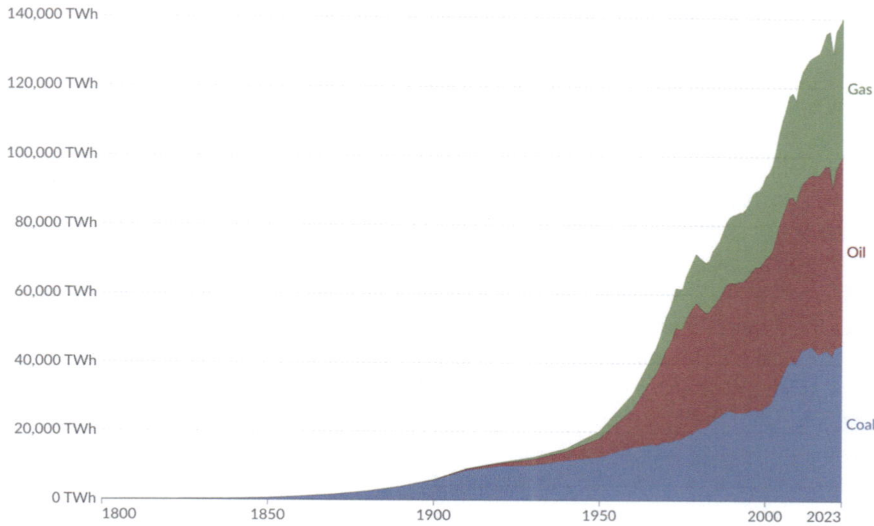

Fig. 1.9 Global primary energy consumption by fossil fuel source in terawatt-hours (TWH). *Source* Energy Institute—statistical review of world energy (2024); Smil (2017) OurWorldinData.org/fossil-fuels

1.3 Forecasting Energy Production and Demand

Energy production and demand forecasting is a key aspect of understanding and predicting future energy trends, which helps inform policy decisions, and investment strategies. One approach to energy forecasting involves seeking information and insights from reputable organizations such as the International Energy Agency (IEA) and the Energy Information Administration (EIA). The IEA and the EIA provide comprehensive analysis, research, and forecasts related to energy markets, consumption, production, and policy. These organizations employ a wide range of methodologies, data collection techniques, and models to develop their energy forecasts.

When utilizing forecasts from the IEA and EIA, it is important to recognize that energy systems are complex and influenced by many factors, including technological advancements, policy changes, economic conditions, and social shifts. As a result, energy forecasts often involve a degree of uncertainty. Factors like changes in energy prices, advancements in renewable energy technologies, government policies, and global energy demand are among major factors impacting these forecasts. It is important to remember that the forecasts are not crystal balls. Unexpected events or developments can impact the actual trajectory of energy systems, like changes in policies, changes in governments, war, or energy demand.

1.3.1 World Energy Consumption

Energy consumption reflects the total amount of energy used, including various sectors such as residential, commercial, industrial, and transportation. Understanding this consumption is essential for evaluating the demands on our energy resources and shaping sustainable practices. In Fig. 1.10, we are looking at the world's energy consumption in quadrillion British thermal units (10^{15} BTU) over a period of 2010–2050. The vertical axis represents the energy consumption measured in quadrillion BTUs, while the horizontal axis denotes the time. BTU stands for British Thermal Unit. It is a unit of energy commonly used in the United States and other countries that use the Imperial system. One BTU is approximately the amount of energy needed to heat one pound of water by one degree Fahrenheit. In the context of energy consumption, it is often used to quantify the amount of heat energy produced or consumed by various processes, including those related to heating, ventilation, air conditioning, and industrial processes. At the starting point of energy consumption forecast in 2020, the global energy consumption stands at approximately 600 quadrillion BTUs. This value provides a baseline for understanding the scale of energy demand at the beginning of the projection. As the progress along the time toward 2050, the graph shows a significant increase in energy consumption. The world's energy consumption is expected to reach a 50% increase compared to the initial value. This implies a substantial rise in the demand for energy resources and services to meet the needs of a growing population, expanding industries, and increasing urbanization.

The clear trend in energy consumption indicates the challenges associated with meeting the growing energy demand. It highlights the importance of implementing strategies for efficient energy use, exploring renewable energy sources, and adopting

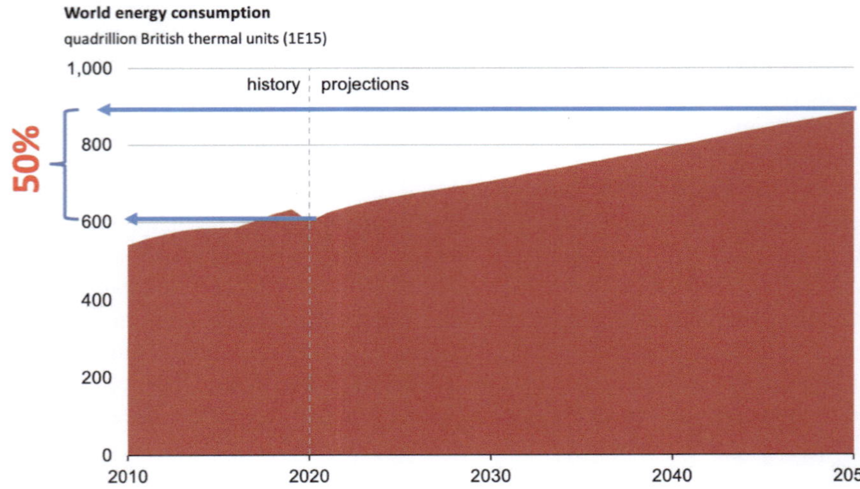

Fig. 1.10 World energy consumption. *Source* U.S. Energy administration, international energy outlook 2021 (IEO2021 Reference case)

technologies that promote energy conservation and lower carbon emissions. Meeting future energy needs, while ensuring a balance between economic growth, social development, and environmental protection, will be a key focus in the coming decades. This can be achieved in short term with implementing the energy mix policies rather than energy transition. We will see the importance of this approach as we explore the capacity and current and near-future forecast of renewable energies in comparison to fossil fuels and world energy demand.

It's important to note that the expected 50% rise in energy use by 2050 depends on several factors like population growth, economic development, technological progress, policy decisions, and changes in energy consumption practices. Therefore, taking active steps and having a thorough energy plan are essential to guarantee a cost-effective and sustainable energy source that can sustain worldwide development.

Figure 1.11 presents the relationship between different sectors and fuel sources that contribute to the overall energy landscape. Understanding how energy is distributed among different sectors and the types of fuels powering these sectors is essential in developing sustainable energy policies. In Fig. 1.11, we are looking at end-use energy consumption by sector in the world, measured again in quadrillion British thermal units (QBTUs) over the period of 2010–2050. It provides a breakdown of energy consumption across different sectors including industrial, transportation, residential, and commercial (Patterson et al. 2022). Starting with the industrial sector, the graph demonstrates a projected increase of 52% in energy consumption from around 240 in 2020 to 365 QBTUs in 2050. This significant growth reflects the expanding industrial activities, and the rising energy demand associated with industrial processes and manufacturing. Moving on to the transportation sector, it is anticipated to experience a substantial increase of 46% in energy consumption from 110 QBTUs in 2020 to an estimated 160 QBTU by 2050. This significant increase can be attributed to the growing number of vehicles and the energy requirements for transportation services worldwide. The residential sector shows a forecasted rise of 38% in energy consumption, increasing from 60 QBTUs in 2020 to approximately 83 QBTUs in 2050. This elevation is driven by population growth, and increasing demand for residential energy services, including heating, cooling, and electricity. Similarly, the commercial sector is expected to see a 40% increase in energy consumption, rising from 35 in 2020 to around 49 QBTU by 2050. This rise can be attributed to the expansion of commercial activities and increasing energy needs for various commercial purposes. Again, remember, all these forecasts include all the policymakers, all the regulations, and everything that has been put toward the energy consumption forecast.

On the right-hand side of Fig. 1.11, we observe the breakdown of energy consumption by fuel type from 2010 to 2050. The graph illustrates the forecasted changes in energy consumption for petroleum and other liquids, natural gas, coal, and renewables. Petroleum and other liquids are shown to have the highest energy consumption, exceeding 180 QBTU by 2020. The forecast indicates a 40% increase in energy consumption from this source, reaching over 250 QBTU by 2050. This is primarily due to the widespread use of petroleum in transportation and various industrial applications. Natural gas with an energy consumption of approximately 100 QBTU in

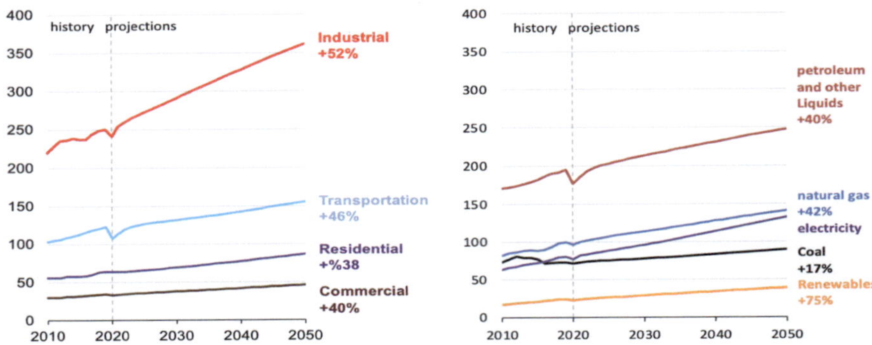

Fig. 1.11 World energy consumption by sector and fuel in quadrillion BTU. *Source* U.S. Energy Administration, International Energy Outlook 2021 (IEO2021 Reference case)

2020 is projected to experience a significant growth of 42% by 2050. This increase is related to the expanding use of natural gas in power generation, heating, and industrial processes. Coal, another significant source of fuel, had an energy consumption of around 70 QBTU in 2020. The graph suggests a relatively modest growth of 17% in energy consumption from coal by 2050, driven by factors such as industrial demand and power generation in certain regions. Renewables, represented by a relatively lower energy consumption of 22 QBTU in 2020, are expected to undergo a significant increase of 75% by 2050. This reflects the global shift toward cleaner and more sustainable energy sources, such as solar, wind, hydropower, and bioenergy.

These projected trends in energy consumption by sector and fuel type show the evolving energy landscape and the need to pay attention to meeting the growing energy demand. They highlight the importance of energy supply planning and technological advancement to ensure a sustainable energy supply for the future. Note that, even with the 75% increase in renewables with the planned investments, still we see the predictions and the forecast of the energy provided by renewables, much less than petroleum, natural gas, and coal. The insights from the details of World Energy Consumption by Sector and Fuel as presented in Fig. 1.11 help to extend our attention beyond sectors and analyze the primary energy sources that fuel our global energy demand. Understanding the mix of energy sources is fundamental in interpreting the broader trends and implications for sustainability.

1.3.2 Primary Energy Consumption

In Fig. 1.12, on the left, we are looking at the share of primary energy consumption by source in the world by the year 2020. It represents the distribution of energy consumption among renewables, coal, petroleum, natural gas, and nuclear energy sources. In 2020, petroleum had the highest share in primary energy consumption, accounting for 31% of the total energy consumption. This can be attributed to the extensive use of petroleum in transportation, industrial processes, and other applications. Following petroleum, coal held the second largest share of 27%, driving with use in electricity generation and the industrial sector. Natural gas accounted for 25% of the primary energy consumption share in 2020. Its popularity is due to its versatility and cleaner burning properties compared to coal and petroleum. The contribution of renewables, including sources such as solar, wind, hydropower, and bioenergy, is estimated to be 13%. This reflects the increasing global focus on transitioning toward cleaner and more sustainable energy sources. Nuclear energy, which relies on the process of nuclear fusion, is only 4% of the primary energy consumption share in 2020. Nuclear power plants generate electricity through controlled nuclear reactions. The society perception of nuclear energy and nuclear waste management seems to be the largest obstacle in new investments in nuclear power as this source of energy is the one that could respond to ever-increasing demand of energy.

The right graph represents a forecast for the year 2050, indicating the projected changes in the share of primary energy consumption by source. Despite policy changes and ongoing energy transition efforts, fossil fuels are still expected to dominate the energy landscape, contributing to over 70% of the total energy demand in 2050. Petroleum share is projected to decrease slightly to 28% by 2050. This reduction may be driven by efforts to reduce dependency on petroleum transportation to the adoption of electrical vehicles and implementation of alternative fuels.

Renewable energy sources, including solar, wind, hydropower, and bioenergy, are expected to experience significant growth, reaching a share of 27% in primary energy

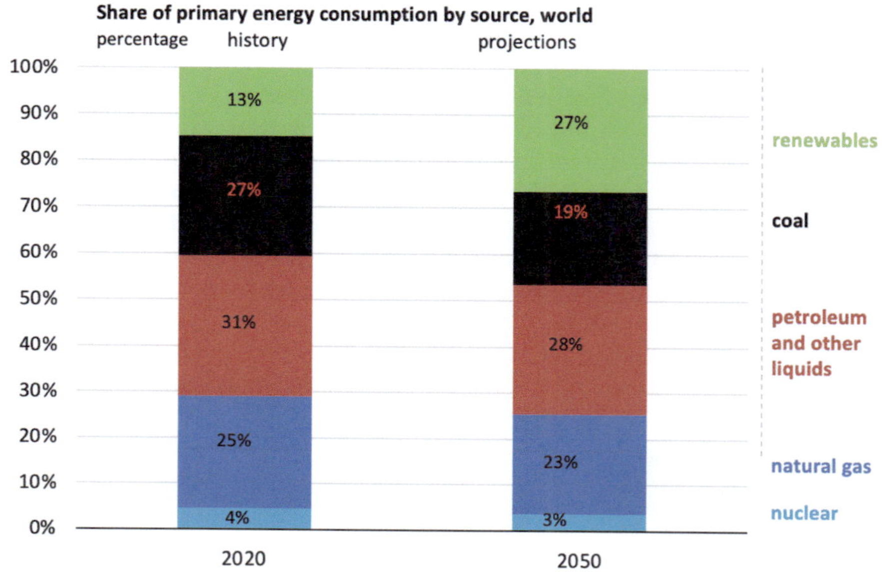

Fig. 1.12 Primary energy consumption by energy. *Source* U.S. Energy Administration, International Energy Outlook 2021 (IEO2021 Reference case)

consumption by 2050. This increase reflects the global push for decarbonization and increased investment in renewable energy infrastructure. Natural gas is projected to hold a share of 23% in primary energy consumption by 2050, maintaining its relevance due to its relatively lower carbon emissions compared to coal and petroleum. Coal share is expected to decline to 19% by 2050 from 27% because of efforts to reduce greenhouse gas emissions and transition to clean energy sources.

Nuclear energy is forecasted to represent a smaller share of 3% of primary energy consumption by 2050. This may be due to concerns regarding nuclear waste disposal, safety, society acceptance, and development of alternative energy options.

While there are efforts to shift toward cleaner and more sustainable energy sources, fossil fuels are forecasted to continue significantly contributing to meet global energy demands in 2050. The growing share of renewables indicates progress in the energy transition, but further advancements and policy supports are necessary to accelerate the shift toward a more sustainable and low-carbon energy future. This evidence supports the idea of energy mix rather than energy transition as in short term it does not seem to be economically possible for transitioning away from fossil fuels as assuming all the promises are held regarding investments in renewable energies still 70% of energy demand will be supported by fossil fuels in 2050 based on IEO 2021, Fig. 1.12.

1.3.3 Evaluating Energy Sources

In meeting the world's growing energy demand, it is evident that oil, natural gas, and coal will continue to serve as the primary energy sources. However, it is important to recognize that a diverse range of resources will be necessary to meet increasing energy demand. Figure 1.13 lists oil, coal, natural gas, nuclear, wind, solar, biofuel, hydro, and geothermal use in energy mix to respond increasing energy demand. Each energy source has its own set of benefits and challenges that must be addressed. For instance, oil provides high energy density and easy transportability, making it essential for transportation and various industrial uses. Nonetheless, the processes of extracting, transporting, and burning oil come with environmental consequences, such as greenhouse gas emissions.

Natural gas, recognized for its comparatively lower carbon emissions compared to coal and oil, serves as a multipurpose energy source for electricity generation, heating,

Fig. 1.13 Oil, coal, natural gas, nuclear, wind, solar, biofuel, hydro, and geothermal use in energy mix to respond increasing energy demand (Designed by Freepik www.freepik.com)

and industrial applications. However, concerns arise from extraction methods, like hydraulic fracturing required producing the natural gas from shale reservoirs, and still CO_2 emissions from natural gas. Coal has historically been a major energy source for generating electricity (Jorge Morales Pedraza 2019). However, burning coal releases a substantial amount of carbon dioxide, contributing to climate change and air pollution. Implementing technologies like carbon capture and storage can help in mitigating environmental impacts of burning coal. While oil, natural gas, and coal will remain significant in meeting energy demands, it is necessary to diversify the energy mix and incorporate alternative resources. Renewable energies, such as geothermal, solar, wind, hydropower, and bioenergy, offer the advantage of being abundant, and clean. They help reduce greenhouse gas emissions and provide opportunities for decarbonization and energy independence. However, renewable energy sources face challenges related to storage, infrastructure development, initial cost, and the fact that renewable energy sources may not generate a consistent supply of energy throughout the entire day, everywhere, and at all conditions.

Nuclear energy provides a reliable and low-carbon energy source. It generates electricity from controlled nuclear reactions. Nuclear power plants can produce large amounts of electricity without greenhouse gas emissions. However, concerns surrounding nuclear waste management, safety (potential for accidents of nuclear materials), and society acceptance need to be addressed. To meet the world's energy demand, a balanced and diversified approach is required. This involves maximizing the benefits of each energy source, while minimizing their environmental impacts. Continued research and development, technological advancement, and supportive policies are essential for transitioning to cleaner energy sources and improving the efficiency of existing ones. However, since then the best approach seems to adhere to energy mix concept.

1.3.3.1 Criteria for Selecting Optimal Energy Sources

When considering different sources of energy, several factors need to be considered to make informed decisions and develop sustainable energy systems, here are some key points to consider, cost, storage, scale, and environment.

Cost Considerations

An important aspect to consider when deciding on using a source of energy as backbone of supply is the cost of setting up and operating the energy source. This includes upfront capital expenditure (CAPEX), ongoing operational expenditure (OPEX), maintenance costs, and fuel prices. It is important to assess the financial feasibility and long-term economic viability of the energy source. In evaluating the costs associated with different energy sources, it is essential to consider the entire lifecycle costs and the evolving landscape of energy markets and policies. The decision-making process should align with broader sustainability goals and consider the social and

Fig. 1.14 Business ethics concept image. Designed by Freepik www.freepik.com

environmental impacts associated with each energy option. In the energy industry, business ethics defines the decision-making process that protects the environment while still providing the revenue. Ethical companies work to find a balance between making a profit and taking care of the planet as conceptualized in Fig. 1.14.

The Levelized Cost of Energy (LCOE) is an important metric for assessing the economic viability of various energy technologies. It provides a comprehensive view by considering the total lifetime costs divided by the energy production of a power plant. This calculation involves determining the present value of all costs associated with building and operating a power plant throughout its lifespan and total energy production. LCOE allows for comparisons between different technologies, such as wind, solar, geothermal, and natural gas, even when they have unequal lifespans, different project sizes, distinct capital costs, varying levels of risk and return, and diverse capacities. This comparative aspect assists policymakers, investors, and energy professionals in making informed decisions in the complex landscape of energy infrastructure.

In addition to LCOE, another important metric in the energy landscape is the Levelized Cost of Storage (LCOS). LCOS evaluates the cost of storing energy over its lifetime and is critical for understanding the economic feasibility of energy storage technologies. Like LCOE, LCOS considers factors such as initial investment, operational and maintenance costs, project lifespan, discount rate, and the impact of taxes and tax credits. When calculating LCOS, it is essential to assess the total cost of setting up and operating an energy storage system in relation to the amount of energy it can store and release. This metric is particularly valuable as energy storage is an essential component in enhancing the reliability and efficiency of renewable energy sources.

Investigating the future of energy resources, Table 1.2 by U.S. Energy Information Administration, Annual Energy Outlook 2022 provides a comprehensive overview of the Estimated Capacity-Weighted Levelized Cost of Electricity (LCOE) and Levelized Cost of Storage (LCOS) for upcoming energy sources expected to be operational in 2027. Capacity-weighted average represents the mean levelized cost per technology, considering the new capacity being introduced in each region. The calculation relies on capacity additions projected for each region during the period from 2025 to 2027. Technologies without anticipated capacity additions do not contribute to the capacity-weighted average and are labeled as NB, indicating they are not built. This table, expressed in 2021 dollars per megawatt-hour considers both the cost of electricity generation and the associated storage capabilities and provides valuable insights into the anticipated economic dynamics of these new resources. In Table 1.2, the tax credit is determined by specific federal tax credits like the Production Tax Credit (PTC) or Investment Tax Credit (ITC), applicable to certain technologies. This factor considers tax credits applicable exclusively to facilities commissioned in 2027 and incorporates the significant reduction in both the PTC and ITC as outlined in existing legislation. Technologies ineligible for PTC or ITC is denoted as NA. The outcomes are derived from a regional model, with LCOE and LCOS calculations excluding any state or local incentives (*Source* Levelized Costs of New Generation Resources in the Annual Energy Outlook 2022). Based on the capacity factor, which generally correspond to the high end of the technology likely utilization range and total LCOE geothermal energy and natural gas-fired combined cycle seems to be more volatile source of energy in years to come. Solar energy has also low LCOE of $33.83, however requires to be coupled with other sources as it has low-capacity factor of 29%.

Source Levelized Costs of New Generation Resources in the Annual Energy Outlook 2022

Energy Storage Solutions

The importance of energy storage becomes evident, especially for intermittent renewable sources like solar and wind. The storage systems enable the capture of excess energy for use during periods when these renewable sources are not actively generating electricity. The cost and efficiency of the storage technologies significantly impact the overall cost and reliability of the energy source. To assess the economic viability of energy storage technologies, the Levelized Cost of Storage (LCOS) discussed in Table 1.2 is an important metric. It evaluates the cost of storing energy throughout the project's lifespan. Particularly, geothermal energy emerges as advantageous when comparing capacity and LCOS with other renewable sources. Geothermal systems efficiently utilize underground formations, storing thermal energy in depleted oil and gas reservoirs and shallow sandstones. This capability ensures continuous power generation even during periods of low demand. Despite low LCOE for solar energy and its low-capacity factor, large-scale battery installations may pose challenges due to substantial land use requirements, raising

Table 1.2 2027 Estimates: LCOE and LCOS for new resources (2021 $/MWh)

Plant type	Capacity factor (percent)	Levelized capital cost	Levelized fixed O&M	Levelized variable cost	Levelized transmission cost	Total system LCOE or LCOS	Levelized tax credit	Total LCOE or LCOS including tax credit
Dispatchable technologies								
Ultra-supercritical coal	85%	$52.11	$5.71	$23.67	$1.12	$82.61	NA	$82.61
Natural gas-fired combined cycle	87%	$9.36	$1.68	$27.77	$1.14	$39.94	NA	$39.94
Advanced nuclear	90%	$60.71	$16.15	$10.30	$1.08	$88.24	($6.52)	$81.71
Geothermal	90%	$22.04	$15.18	$1.21	$1.40	$39.82	($2.20)	$37.62
Resource-constrained technologies								
Biomass	83%	$40.80	$18.10	$30.07	$1.19	$90.17	NA	$90.17
Wind, onshore	41%	$29.90	$7.70	$0.00	$2.63	$40.23	NA	$40.23
Wind, offshore	44%	$103.77	$30.17	$0.00	$2.57	$136.51	($31.13)	$105.38
Solar, standalone	29%	$26.60	$6.38	$0.00	$3.52	$36.49	($2.66)	$33.83
Solar, hybrid	28%	$34.98	$13.92	$0.00	$3.63	$52.53	($3.50)	$49.03
Hydroelectric	54%	$46.58	$11.48	$4.13	$2.08	$64.27	NA	$64.27
Capacity resource technologies								
Combustion turbine	10%	$53.78	$8.37	$45.83	$9.89	$117.86	NA	$117.86
Battery storage	10%	$64.03	$29.64	$24.83	$10.05	$128.55	NA	$128.55

Fig. 1.15 Cartoon illustarting the need for intermittent renewable. *Source* Designed by Freepik www.freepik.com

environmental concerns and potential land use conflicts, see cartoon presentation in Fig. 1.15.

From the conventional sources of energy, natural gas storage offers a dependable and dispatchable energy source, allowing for effective management of supply and demand fluctuations. However, it is not without environmental challenges. The extraction, storage, and combustion of natural gas contribute to greenhouse gas emissions, primarily in the form of carbon dioxide (CO_2). Additionally, concerns arise over methane leakage during the extraction and storage processes.

Scalability of Technologies

Ensuring that an energy source can meet the demands of a growing society requires an assessment of its scalability. This involves considering resource availability, infrastructure requirements, and technological capabilities to guarantee that scalability doesn't compromise efficiency or environmental sustainability.

Conventional oil and gas hold significant potential for scalability as it can be quickly adjusted to respond to changes in electricity demand. This flexibility enhances its scalability, and the construction of infrastructure, such as power plants, is generally faster than large-scale renewable projects. These conventional sources of energy have well-stablished supply chains and leave small footprint. While oil and natural gas provide dispatchability and rapid deployment, its scalability may be constrained by environmental concerns where scaling translates to more emissions.

On the other hand, alternating renewables claimed to be abundant resources with adaptable applications face with challenges related to intermittency and land use.

Technologies like batteries for storing wind and solar energy showcase excellent scalability. They can be easily adapted to various system sizes, making them suitable for both residential-and commercial-scale applications. This characteristic extends to diverse locations, providing flexibility in energy generation and storage placement to meet specific regional or project requirements. In contrast, geothermal energy faces limitations due to its dependence on specific locations. Practical geothermal resources may not be universally available, restricting the geographic areas where geothermal storage can be effectively implemented. The future energy landscape is expected to comprise a combination of both renewable and conventional energies, with ongoing advancements in technology and infrastructure addressing scalability challenges for each. However, it's critical to note that the scalability of both natural gas and intermittent renewable energies is influenced by various factors, including technology, infrastructure, regulations, and environmental considerations.

Environmental Impact Assessment

Assessing the environmental impact of energy production is also one of the critical aspects in choosing the proper energy source. This includes evaluating greenhouse gas emissions, air, noise and water pollution, land use, potential for habitat destruction or biodiversity loss as shown in Fig. 1.16. Sustainable energy sources should minimize negative environmental impacts and try to reduce carbon emissions and mitigate pollution to alleviate the environmental concerns. The entire life cycle of energy sources should be considered including exploration, resource extraction, manufacturing, operations, and abandonment.

Evaluating the environmental impact throughout the life cycle helps identify the overall sustainability of the energy source. It is essential to consider social acceptance, policy framework, regulatory requirements, and technological advancements associated with each energy source. A stakeholder agreement, public perception, and community involvement are critical for successful and sustainable energy mix policies. As discussed earlier, the society perception against the nuclear power or fracking concerns that limit the development of geothermal systems has been derived due to lack of community engagement and stakeholder awareness.

1.4 Environmental Concerns of Renewable Energy Technologies

While we have explored the environmental concerns associated with the use of conventional energy sources such as oil, gas, and coal, it is important to recognize that renewable energy resources are not free of environmental impacts. Although

Fig. 1.16 Cartoon showing different environmental impacts of different energy sources. Image: Designed by Freepik www.freepik.com

renewables generally offer a cleaner and more sustainable alternative to fossil fuels, their development and deployment can still pose environmental challenges. In this section, we will examine several key concerns related to the expansion of renewable energy technologies. This section serves to remind that environmental impacts are not exclusive to conventional energy sources, and that renewables are not without consequences.

1.4.1 Overview of Renewable Energy Options

1.4.1.1 Solar Energy Technologies

Even though new generations of solar panels are designed with higher standards and safety requirements still they contain different concentrations of toxic materials and heavy metals. Their mining, processing, and disposal also raise environmental and health concerns. When heavy metals and toxic materials in solar panels leach into the water table, it can have detrimental effects on water supplies, the environment,

Fig. 1.17 Solar energy farm (Designed by Freepik www.freepik.com)

and human health. The disposal of solar panels in landfills contributes to the accumulation of toxic sludge, posing further environmental risks. Large-scale deployment of renewable projects, especially solar and wind farms, may require significant land area as shown in Fig. 1.17. Spatial constraints and potential conflicts over land use can affect scalability.

Environmental Impacts of Solar Panels

In Puerto Rico, a solar farm was severely damaged by the impact of Hurricane Maria and left panels scattered and destroyed across the ground. The supporting structures were ruined, and debris was spread throughout the surrounding area.

This incident, illustrated in Fig. 1.18, highlights the weakness of solar infrastructure to extreme weather events and underlines the challenges faced in maintaining and protecting renewable energy installations in disaster-prone regions. The destruction not only disrupted power generation but also raised additional environmental concerns. For instance, damaged solar panels may leach toxic materials into the soil and potentially into groundwater supplies. The widespread scattering of panel fragments also complicates waste management and poses long-term contamination risks to nearby ecosystems.

This case presents the need to carefully consider the environmental impact of renewable energy systems as part of the broader energy mix process. While renewable energy technologies reduce carbon emissions, they are not without their own environmental and logistical challenges.

Fig. 1.18 In Puerto Rico, a solar farm was severely damaged by the impact of Hurricane Maria and left panels scattered and shattered across the ground. *Source* National Weather Service San Juan, PR; NOAA Climate.gov

1.4.2 Wind Energy Technologies

1.4.2.1 Environmental Impacts of Wind Turbines

Wind farms, with their gigantic turbines and extensive layouts, can significantly transform the visual and environmental character of a landscape. The San Gorgonio Pass Wind Farm, located north of Palm Springs, California, is one of the many examples (Fig. 1.19). The landscape with its natural beauty and open views now hosts hundreds of wind turbines that dominate the horizon, reshaping the visual identity of the region. The construction of the wind farms and the presence of the turbines have also impacted the local wildlife. The spinning blades on the turbines pose a potential risk to birds and bats which can collide with them during flight (Jeffrey Lovich 2015). The wind farm's location along the migratory bird route further amplifies this issue.

One of the less visible but increasingly important concerns is the environmental footprint associated with the manufacturing, installation, and disposal of wind turbine components such as the blades. These blades are typically made from composite materials such as fiberglass reinforced epoxy, which are difficult to recycle. Their massive size and structural complexity make both transportation and disposal challenging.

At the end of their operational life, wind turbine blades cannot be conventionally recycled or destroyed. Specialized methods are required for decomposition such as diamond saws to cut the blades into manageable sections for transportation. These sections are then typically sent to landfills such as the landfill in Casper, Wyoming, which now holds over 800 wrecked wind turbine blades (Fig. 1.20). Once these blades

Fig. 1.19 Wind farm, Palm Springs images. Free to use under the Unsplash License

buried, they do not break down over time that means they will remain in landfills indefinitely and contribute to long-term solid waste accumulation. Due to extremely large sizes cutting and burying them takes up significant volume of landfill space. While turbine blades are generally considered chemically stable, some concerns remain about leaching over time (Mark Mallet 2024). Potential leaching of additives, resins, or degradation byproducts could pose risks to soil and groundwater quality.

1.4.3 Hydropower Systems

Hydropower is one of the oldest and most widely used renewable energy sources. It harnesses the energy of flowing or falling water to generate electricity. Figure 1.21 shows the hydroelectric power plant in Shasta County, California. While it is a clean energy source in terms of carbon emissions, it comes with significant environmental and ecological impacts, especially in relation to aquatic ecosystems and water flow. Hydropower projects, especially large dams, can significantly alter local aquatic ecosystems. These reservoirs can change the temperature, sedimentation, and water chemistry, which can impact aquatic species that rely on specific conditions for survival. Hydropower systems also regulate water flow to optimize power generation, which can lead to significant changes in natural river systems. The altered flow can result in reduced water quality, lower dissolved oxygen levels, and altered sediment transport (Abbott et al. 2022). Moreover, changes in river flow can lead to erosion

Fig. 1.20 Landfill in Wyoming the destination of wind turbine blades. *Source* (© Google Earth)

or sediment buildup downstream, which can degrade ecosystems and affect human activities like agriculture. In the case of deep reservoirs, stratification can occur, where warmer, oxygen-poor water sits on top of cooler, oxygen-rich water. While hydropower has significant ecological drawbacks, there are efforts to mitigate these effects. Modern hydropower systems include environmental flow management which is being integrated into hydropower systems to ensure that rivers maintain a minimum flow even when water is being diverted for power generation (Widen et al. 2022).

1.4.4 Geothermal Energy Systems

Geothermal energy extracts the Earth's natural heat to generate electricity or provide direct heating. It is considered a sustainable and renewable energy source, but it also presents some environmental challenges, primarily related to land and water use. Geothermal power plants require a significant amount of land to develop, especially for large-scale systems. In many cases, the land surrounding geothermal plants is used for farming or is ecologically significant. Figure 1.22 shows a schematic of geopressured reservoir geothermal that produces hot water to be used to generate electricity. These plants typically use large amounts of water in the range of more than 100,000 bbls/day of injection to sustain production (Iyegbekedo et al. 2025). This can lead to

Fig. 1.21 Hydroelectric power plant in Shasta County, California (Designed by Freepik www.freepik.com)

groundwater depletion, especially in regions where water sources are already limited. In closed-loop systems, water use is minimized, but has not been proven economically viable, open-loop systems (which involve withdrawing water from underground reservoirs) can lead to a reduction in groundwater supplies. Additionally, the water used in geothermal power plants can sometimes be contaminated with minerals, and the disposal of this water can create pollution if not properly managed. Geothermal power plants can also emit gases such as sulfur dioxide, carbon dioxide, and methane (Holm and Blodgett 2012). The emissions arise from the release of gases trapped deep in the Earth's crust. Geothermal energy extraction can also cause seismic events, known as induced seismicity, although they are generally of low magnitude during the hydraulic fracturing in dry hot rock areas or pressure maintenance in geopressured reservoirs (Rathnaweera et al. 2020). While these events are generally not destructive, they can be a concern for communities living near geothermal power plants. The environmental impacts of geothermal energy can be minimized through careful site selection and technology improvements.

1.4.5 Biomass Energy

Biomass energy is derived from organic materials such as wood, agricultural residues, and animal waste, which are burned to generate heat or electricity. Figure 1.23 shows the schematic of biomass fuel-electricity generation. This renewable source of

Fig. 1.22 Schematic of geopressured reservoir and geothermal power plant. Freepik www.freepik.com

energy is considered as near-zero or carbon-neutral energy source; however, biomass energy has significant sustainability and emissions concerns that must be addressed. If biomass energy is produced using agricultural residues and forestry waste and integrated with carbon capture and sequestration it can be also considered as carbon negative; however, many real-word applications do not meet this criterion. When biomass is obtained from malpractices, such as clear-cutting forests or overharvesting crops, it can result in deforestation and soil degradation (Elliot and Rhee 2022). In some cases, large-scale biomass production can compete with food production for land, which may lead to disruption in food chain production leading to higher food prices or food insecurity (Muscat et al. 2020). For biomass to be truly sustainable, it must be sourced from waste products, such as agricultural residues or forestry byproducts, or from dedicated energy crops grown using sustainable practices. While biomass is often considered carbon neutral because the CO_2 released during combustion is theoretically offset by the CO_2 absorbed during the growth of the biomass, this balance is not always achieved. The efficiency of biomass combustion, transportation, and processing all contribute to its overall carbon emissions. If biomass is not combusted efficiently or if it is transported over long distances, the carbon emissions can be higher than those from coal or natural gas, which biomass is intended to replace. Additionally, the combustion of biomass can release other pollutants, including particulate matter, nitrogen oxides, and volatile organic compounds (VOCs), which can degrade air quality and pose health risks (Desservettaz et al. 2023). Sustainable biomass

Fig. 1.23 Biomass fuel-electricity generation. Freepik www.freepik.com

farming practices, such as agroforestry or polyculture, can help improve biodiversity and soil health (Fahad et al. 2022). Additionally, technologies like combined heat and power (CHP) systems can improve the efficiency of biomass combustion, reducing emissions per unit of energy produced.

1.5 Mitigation Strategies

1.5.1 Global Carbon Capture and Storage

While Carbon Capture and Storage (CCS) is commercially applied in different countries, it hasn't reached the necessary scale to significantly reduce global CO_2 emissions. The development of different CCS projects globally can be monitored using the interactive Global Map for CCS from Scottish Carbon Capture & Storage.

Important information about the status and progress of CCS implementation can be obtained globally exploring the Global CCS Map, Fig. 1.24. Some key details about CCS projects that can be obtained include the project location, types of carbon storage sites, volume of the CO_2 injected, storage, and infrastructure and networks.

The CCS map highlights the geographic location of CCS projects, where significant CCS activities are taking place. This information shows the distribution and concentration of CCS efforts globally. The CCS map also highlights the different types of carbon storage sites being utilized, such as depleted oil and gas reservoirs, saline aquifers, and other geological formations. Understanding the availability and suitability of these storage sites is significant for successful CCS development. Saline formations are compacted rocks deep underground that have a high capacity

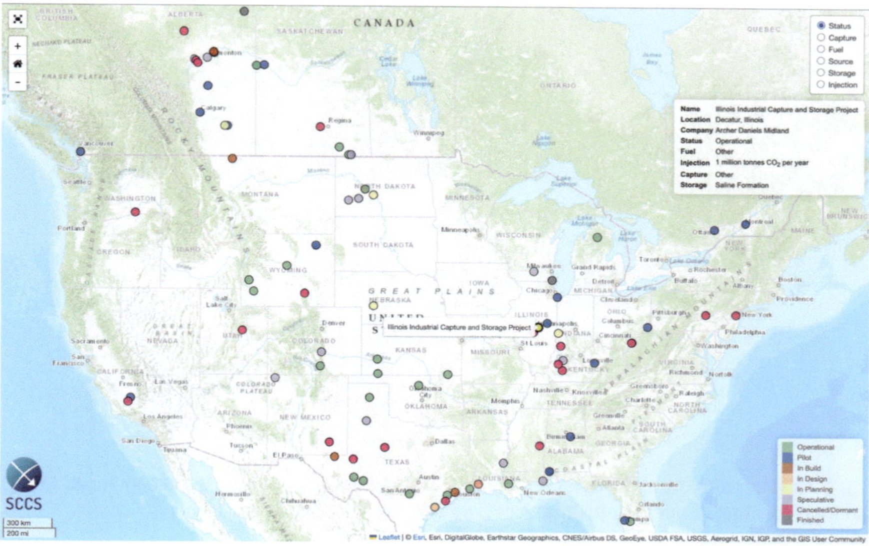

Fig. 1.24 Status map of CCS projects in North America. Scottish Carbon Capture & Storage http://sccs.org.uk/resources/global-ccs-map

to permanently trap CO_2. Identifying these sites on the map helps users to visualize the widespread adoption of saline formations as a key storage method in CCS projects globally. Many CCS projects repurpose depleted oil and gas reservoirs for CO_2 storage (Callas et al. 2022). The map can provide information on sites where such reservoirs are used, quantifying their potential for storing large quantities of CO_2. This demonstrates the utilization of existing infrastructure and the possibility of repurposing depleted hydrocarbon reservoirs for long-term CO_2 storage (Eigbe et al. 2023). The map can also indicate storage sites where CO_2 is injected into oil reservoirs to enhance oil recovery. EOR, or enhanced oil recovery, involves injecting CO_2 into oil-bearing formations, enabling the extraction of additional oil while simultaneously storing the CO_2 underground this shows the dual benefit of CO_2 storage and increased oil production through this application.

Clicking on specific project spots on the map provides additional information and details about the individual CCS projects. This includes the project name, participation or organization, project stages, for example, if it's in the planning, construction, or operation, capture and storage capacities, and other relevant details such as information that enable a deeper understanding of scale and the scope of each project. The map also provides data on the volume of CO_2 that has been injected at various storage sites that can give us information regarding the scale of carbon capture and storage activities at each location. Visualizing the distribution of CO_2 injection volumes can also help to identify regions where CCS significance is recognized and use the storage capacity to gauge the overall progress of CCS implementation.

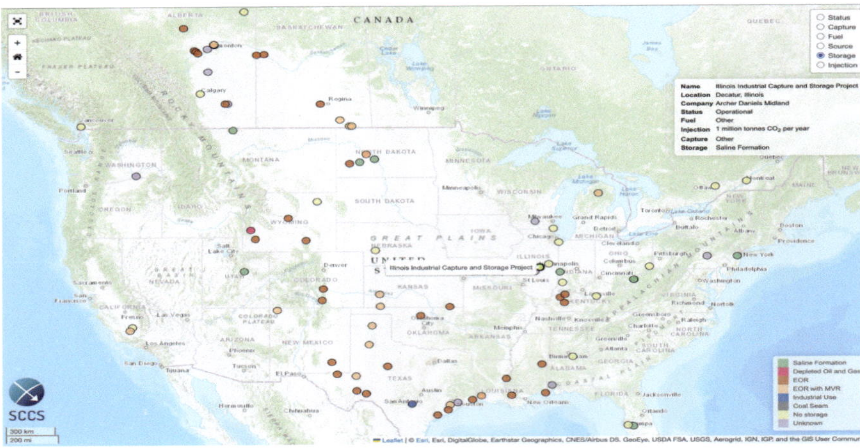

Fig. 1.25 Map of North America CCS projects storage types. Scottish carbon capture & storage and include a link to our website http://sccs.org.uk/resources/global-ccs-map

CCS map highlights the projects focused on capturing and storing CO_2 emissions from what so called "hard-to-decarbonize" industries such as cement, steel, or chemical processing like fertilization. This can also be used to gauge how much progress has been made to decrease the CO_2 emissions from these industries. The map can also provide information on storage sites that utilize coal seams for CO_2 storage. In this method, CO_2 is injected into unmineable coal seams where it is absorbed and stored within the coal. Identifying these sites on the map (Fig. 1.25) illustrates the potential for utilizing coal seams as a storage portion, as a storage option, particularly in the regions with significant coal reserves such as those in China including Shenhua Ningxia CTL Project and Shanxi International Energy Oxyfuel Project with each two million tons of CO_2 injection per year (Dziejarski et al. 2023). This can provide a comprehensive understanding of the diverse approaches being employed in carbon capture and storage projects globally. This information can guide decision-making, facilitate knowledge sharing, and promote the adoption of appropriate storage methods based on regional geology and project objectives. The map also shows the existing or planned infrastructure associated with CCS, such as transporting captured CO_2 from emission sources to storage sites. This approach helps us in assessing the logistical considerations and connectivity of CCS system.

1.5.2 Strategies for Achieving Carbon Neutrality

Reaching a neutral carbon balance is a critical step in reaching the objectives of limiting the global warming and avoiding the most severe consequences of climate change, including extreme weather events, rising sea levels, and ecosystem destruction. A neutral carbon balance also requires reducing the fossil fuel dependencies toward renewables and low-carbon energy sources. This transition promotes the development and adoption of clean technologies, energy efficiency measures, carbon capture technologies, and sustainable practices. We can take responsibility for the environmental impact of our actions and attempt to minimize our carbon footprint. Involving in neutral carbon balance practices in multiple scales from individual to family, city and country will help in protection and restoration of ecosystems, sustainable land management, and preservation of biodiversity, promoting a healthier planet for current and future generations. So, what is the solution and how we can achieve that?

What Is the Solution?

1. We have seen the correlation between global warming and carbon dioxide concentration in atmosphere.
2. We have discussed the events and the consequence of increasing temperature.
3. We have looked at the energy demand, and we have also looked at the energy demand and production forecasts.
4. We see that even considering all these regulations, still fossil fuels are needed, and they will be a major part of the energy portfolio.

Therefore, in near future fossil fuels are major source of energy production and carbon dioxide emission is inevitable thus the conclusion is that carbon capture and storage is a key solution for achieving net-zero emissions. Carbon capture and storage or CCS technologies have emerged as a critical solution in addressing the challenges of climate change and achieving net-zero emissions, Fig. 1.26. The technologies work by capturing carbon dioxide emissions from various industry processes, such as power generation and industrial manufacturing, and then securely storing them underground or utilizing them for other purposes if CO_2 permanently stays out of atmosphere. The atmosphere's carbon budget is the highest amount of carbon dioxide emissions that can be released while keeping global warming within a specific target, like the 1.5 or 2 degrees Celsius limit outlined in the Paris Agreement. After the Carbon budget is reached, the net-zero emission is inevitable to avoid temperature rise more than 1.5 or 2 degrees Celsius. In this journey, the CCS can seriously participate in several ways to help us attain a net-zero carbon feature by reducing emissions, transitioning existing infrastructure, enabling carbon removal, and supporting a smooth transition. In conclusion, the CCS technology holds great promises in our journey toward achieving net-zero carbon emissions. Capturing and securely storing CO_2 emissions, CCS can significantly contribute to mitigating climate change, facilitating the transition to a low-carbon economy, and ensuring a sustainable future for generations to come.

Fig. 1.26 Net-zero carbon footprint image (Designed by Freepik www.freepik.com)

1.5.3 Climate Action Tracker

One of the ways to quantify the progress of countries toward net-zero carbon emission goal is to monitor the progress of each country according to climate policies. The Climate Action Tracker, or CAT, provides an independent scientific analysis of countries' climate policies, tracking their progress since 2009. It offers insights into whether a country's climate actions are insufficient, on track for limiting global warming to well below 2 degrees C, or compatible with a more ambitious goal of 1.5 degrees C. Monitoring and evaluating countries' climate policies and actions are essential in assessing their alignment with the Paris Agreement's goals. Tools like CAT help in holding nations accountable and promoting transparency in the global efforts toward mitigating the climate change. The CAT provides information on several key aspects, including country, climate policies, and the emission pathways. It assesses the emissions trajectories of countries and compares them to what is required to meet the temperature goals of the Paris Agreement. CAT also provides insights into the progress made by countries in the reduction of greenhouse gas emissions and transitioning to cleaner and more sustainable energies. It evaluates the performance of countries in terms of their climate action, considering factors such as emission reduction efforts, renewable energy deployment, energy efficiency, and policy implementation. CAT offers an assessment of how well countries are meeting their climate targets and commitments. It also conducts in-depth policy analysis, examining the effectiveness and impact of specific climate policies and measures implemented by countries.

You can check if your own country's climate policies are insufficient, on track for "well below 2°C," or compatible with 1.5 °C, using the Climate Action Tracker (CAT) at the following website: https://climateactiontracker.org/.

By keeping track of countries' climate policies and actions, we can gauge the collective progress toward the objective of the Paris Agreement and identify areas where more ambitious measures are needed to address the urgent challenge of climate change.

Changes in CO_2 emissions from fossil fuels by regions during the periods between 2019–2020, 2020–2021, and 2019 all the way to 2021 studied by CarbonBrief Global emissions in November 2021, highlighting the impacts of the COVID-19 pandemic and other factors on global emissions. In the first period of 2019–2020, the world experienced a significant decrease in CO_2 emissions by 5.4%. This reduction can be attributed to various factors including the COVID-19 pandemic, which basically led to lockdowns, reduced economic activities, and restricted travel. Some highlights are as follows:

- The United States saw a substantial drop in emissions by 10.6%, while the European Union witnessed a similar decline of 10.9%.
- India and the rest of the world also experienced notable reductions, with emissions decreasing by 7.3% and 7%, respectively.
- Surprisingly, China was the only region that saw a slight increase in emissions by 1.4%, even during the COVID-19 pandemic.

However, the second period of 2020–2021 presented a different trend, as all regions showed an increase in CO_2 emissions.

- Global emissions rose by 4.9%. The United States and the European Union both experienced a 7.6% increase in emissions.
- China, despite its initial increase in the first period, still showed a relatively lower increase of 4%.
- India, on the other hand, observed a significant rise in emissions by 12.6%.
- The rest of the world experienced a comparatively moderate increase of 2.9%.

These changes in emissions reflect the complex relationship of various factors. The initial decline in emissions during the first period can largely be attributed to the COVID-19 pandemic and associated restrictions. However, as economies began to recover and restrictions eased in the second period, emissions started to rise again. The increase in emissions across regions highlights the ongoing challenges in achieving long-term sustainable reductions and the need for a concrete global effort to address climate change.

1.5.4 The Carbon Budget and its Ethical Implications

The carbon budget of the atmosphere is the maximum amount of carbon dioxide emissions that can be released into the atmosphere while still limiting global warming to a specific target. It represents the total amount of CO_2 that can be emitted before exceeding the desired temperature threshold. The exact value of the carbon budget depends on several factors, including the target temperature increase and the level of confidence in climate models. However, it is generally expressed as a total cumulative emission of CO_2 for a specific starting year until the target is reached.

The carbon budget indicates the overall quantity of carbon dioxide emissions allowed to be released into the atmosphere while maintaining global warming within a specific target (Rogelj et al. 2019). In the case of 2 degrees C, global warming, the estimated carbon budget is around 888 gigatons or billion tons (10^9 tons) of carbon dioxide. This means that cumulative emissions should not exceed this amount if the goal is to stay within the 2 degrees C target. To understand the scale of emission, it is important to consider the annual emission of CO_2. The emissions come from various sources, including burning fossil fuels for energy, industrial processes, and land use changes such as deforestation. The estimated annual emissions of CO_2 are around 42.2 gigatons per year that translates to an average of 1337 tons of CO_2 being released into the atmosphere every second. This stunning rate highlights the continuous and substantial release of CO_2 into the atmosphere, contributing to the accumulation of greenhouse gases and the warming of the planet while with 1270 gigatons or billion tons of carbon dioxide is as the total carbon budget.

The significance of these figures lies in the urgency to reduce emissions and manage carbon storage, the Carbon budget. The push for a future with 100% renewable energy and elimination of the fossil fuel is a posture advocated by environmental organizations in the Western world. However, the practical feasibility and ethical implication of such a transition are subjective as debated in COP28 Climate Summit.

Is It Ethical to Deny Access to the Wealth Associated with the Energy Provided by Fossil Fuels to Billions of People?

Currently, fossil fuels, including coal, oil, and natural gas, account for over 83% of the world's primary energy consumption. They are essential in driving economic growth, industrial development, and improving living standards in many parts of the world. One argument against the complete abandonment of fossil fuels is the issue of energy poverty. Approximately, 1.2 billion people counting for 16% of the global population lack access to electricity by 2010 (Nick Routley 2019), this number is approximately 675 million people (about 8% of the global population) reported by 2025. Additionally, around 2.3 billion people, constituting 30% of the world's population, lack access to clean cooking facilities (IEA 2023). Denying these individuals access to the energy provided by fossil fuels can lead to inequality and delay their ability to escape poverty.

Ethical considerations arise when determining whether it is justifiable to deny billions of people the opportunity and benefit associated with energy access that fossil fuels provide. Access to reliable and affordable energy is essential for powering essential services, improving health care, facilitating education, and fostering economic growth in developing regions. So, the question is if it is ethical to deny access of billions of people the wealth associated with the energy provided by these fuels.

The idea of leaving fossil fuels underground has gained momentum in recent years, leading to a global movement advocating for the divestment of fossil fuels. This approach recognizes the finite nature of the carbon budget and emphasizes the need to transition away from fossil fuels once the budget has been depleted. On the other hand, an alternative response to the carbon budget constraint is to **continue accessing the carbon reserves but implement carbon capture and storage technology.**

1.6 International Climate Agreements and Summits

1.6.1 Paris Agreement on Climate Change

The Paris Agreement on Climate Change, established in 2015, is an international agreement approved by almost all nations under the United Nations Framework Convention on Climate Change (UNFCCC). The primary goal of the Paris agreement is to restrict global warming to under 2 degrees Celsius above pre-industrial levels. Some countries are taking more aggressive efforts to cap the temperature increase at 1.5 degrees Celsius (UNFCCC, 2015).

1.6.1.1 Key Developments Post Paris Agreement

The Paris Agreement was critical international commitment with key objectives to address climate change comprehensively. These objectives include increasing the efforts to significantly reduce greenhouse gas emissions and providing financial and technological supports to developing nations to gain capacity to reduce their emissions. The primary temperature goal in the Paris Agreement is set to limit the global temperature increase to less than 2 degrees Celsius compared to pre-industrial levels. However, there is an even more ambitious target of limiting the rise to 1.5 degrees Celsius that emphasizes the fact that even 2 degree Celsius increase in temperature could have significant impact on climate change.

Each participating country in the Paris Agreement pledged to voluntary contribution by submitting its climate action plan known as Nationally Determined Contributions (NDCs). NDCs outline a country's strategies for reducing emissions, adapting to climate impacts, and supporting other nations financially and technologically. A significant aspect of the agreement is the Global Stock take that will be performed every 5 years. This process evaluates overall progress to mitigate climate change, enabling countries to enhance their efforts based on the latest scientific knowledge and technological advancements.

Developed countries obligated to provide financial resources, with a target of $100 billion annually by 2020, and an intend to increase that pledge annually after that (UNCC 2024). For this purpose, a robust system for tracking countries' progress toward net-zero emission is placed for transparency and accountability of countries receiving the fundings and promote the trust among nations. Nearly all countries have approved the Paris Agreement, marking it as a global collaboration to mitigate climate change.

The Paris Agreement officially started on November 4, 2016, when 55 countries, representing at least 55% of global emissions, joined the agreement. A significant moment in the history of Paris Agreement occurred when on July 1, 2017, the United States announced its intention to withdraw. Despite this, emissions from the remaining countries, constituting 73% of global emissions, surpassed the required threshold. Later, on February 19, 2021, the United States rejoined the Paris Agreement, reinforcing its commitment to the Global Climate Solution. This move added substantial momentum to the global climate effort. However, USA officially notified United Nations regarding it is intention to withdraw from agreement which will be in effect on January 26, 2026 (Library of Congress (2025). This decision is expected to have a significant impact on the effort in fighting against climate change.

1.6.2 Overview of COP28 Climate Summit

In November 2023, the 28th United Nations climate conference, known as COP28, took place in Dubai, UAE. A significant agreement was reached where almost every country pledged to "move away from fossil fuels." The commitment is part of a big review called the "global stock take," which checks how countries are doing in achieving the goals set by the Paris Agreement. The "global stock take" is like a big evaluation of worldwide efforts to fight climate change. Over the past 2 years, lots of information and ideas are discussed and findings show: 1—Not enough reduction in emissions, 2—Not being ready for climate problems, and 3—Developed countries not helping developing ones enough.

Countries suggested different plans in response to the stock take, like stopping the use of fossil fuels, increasing renewable energy a lot, and providing trillions of dollars for developing nations. Negotiations at COP28 decided which ideas would be part of the final plan, affecting how fast global changes would happen. Despite some progress, there's frustration because there isn't a clear agreement to stop using fossil fuels in this decade. People argue about whether to reduce or completely stop using them. Some want stricter rules on fossil fuels, while others talk about using more renewable energy instead. Words like "unabated" and "abated" are causing debates because they have different meanings. "Unabated" means releasing greenhouse gases directly into the air from burning fossil fuels, making global warming worse. "Abated" means burning fossil fuels but also capturing and storing some greenhouse gases. The lack of clear definitions makes discussions tricky. Even though there are challenges and arguments about terms like "phase down" and "phase out," **COP28 aims to tackle how we produce and use fossil fuels to meet global climate goals.**

A positive step at COP28 is the launch of a fund to help countries dealing with "loss and damage" from climate change. However, developing countries are disappointed because there aren't new money commitments to help them move away from fossil fuels. Talks about where the fund should be and where the money should come from, like the World Bank, show there are still challenges ahead (www.carbonbrief.org/cop28).

1.6.2.1 Analysis of Agreements and Goals

Analyzing the outcomes of both the Paris Agreement and COP28, separated by a 7-year span, offers valuable insights into the movement toward achieving net-zero emissions. The key observation is that while there's been good progress and increased global dedication to mitigate climate change, substantial problems persist. The results of COP28 emphasize a sustained global commitment to shift away from fossil fuels. Numerous nations have pledged to move toward net-zero emissions, showcasing an enhanced recognition of the urgency to address climate change. Moreover, there's a notable improvement in the emphasis on renewable energy sources, such as proposals to triple renewable energy capacity. The introduction of the "global stock take" mechanism shows the commitment to routinely assess and evaluate global efforts. However, challenges remain in achieving meaningful emission reductions, as revealed by the stock take's findings, suggesting that the existing measures may not be sufficient to address the challenges at hand, and a higher level of commitment, innovation, or intensity is required to effectively tackle climate change. The gaps in funding and disagreements on where the money should come from are big problems. Worries about unclear language and loopholes in agreements make people question how well the promises will work. The primary takeaway is that while there's evident progress in global climate commitments and an increased focus on renewable energy, persistent challenges exist in achieving emission reductions and securing essential financial support. Transitioning away from fossil fuels remains a contentious issue that requires enhanced international cooperation and accelerated actions to bridge the gap between commitments and effective implementation (IEA 2024, COP28 Tripling Renewable Capacity Pledge, Executive summary).

In moving forward and striving to reach the goals of the Paris Agreement and achieve net-zero emissions, a practical and applied approach is required. Focusing on carbon capture and sequestration while simultaneously investing in renewables appears to be a more feasible option, considering global energy consumption projections and the reliability associated with renewable energies. This strategic approach aligns with the interests of most countries, providing a realistic pathway toward a sustainable and low-carbon future and it is in line with our previously discussed "*energy mix*" policy.

1.6.3 Overview of COP29 Climate Summit

In November 2024, COP29 is held in Baku, Azerbaijan where Delegates from around the world agreed on new 300 billion annual climate finance target by 2035 replacing the outdated 100 billion goal. One other major accomplishment in COP 29 was the launch of Loss and Damage Fund to assist affected nations with climate change disasters to be started in 2025. IN COP 29 the global carbon market discussed in (Article 6.4) were finalized, that allows countries and private sectors to trade carbon credits to help them achieve their emission reduction targets. In compared to Paris Agreement and COP28, COP29 stands out with major increase in support for under-development countries; however, no progress has been made regarding restriction or limitation on fossil fuels. Despite these achievements major concerns remained regarding insufficient finances for developing countries to move away from fossil fuels. Even with finalizing the global carbon market still major concerns remain in terms of claims, monitoring, and validation of the emission cuts (UNCC 2024, COP 28: What Was Achieved and What Happens Next?).

1.7 Case Study

1.7.1 U.S. Methane Emission Reduction Action Plan

The U.S. Methane Emission Reduction Action Plan, released in November 2021, summarizes a strategy planned by USA government to cut methane emissions by 30% below 2020 emission levels by 2030. The plan targets key industries including oil and gas, landfills, agriculture, and coal mining. In oil and gas industry, the government plans more restricted EPA rules to reduce the venting and flaring of natural gas and to cut emissions by 75%. This includes $4.7 billion investment in plugging the abandoned wells and enhancing pipeline safety to minimize the possibility of leakage (The White House Office of Domestic Climate Policy 2021).

For landfills and agriculture industry, the government planned to reduce the emission by 70% toward implementing carbon capture and sequestration, restricting the food waste, investing in initiatives such as smart farming, manure management upgrades, and biogas production. The United States Department of Agriculture (USDA) also planned to promote programs to incentivize farmers to adopt sustainable practices through methane-reducing technologies.

1.7.2 Analyzing a Country's Energy Transition

The U.S. is advancing toward more secure and affordable energy solution for future by adjusting its goal from more aggressive move toward net-zero emission energy to more energy independence policies. The move toward net-zero emission is planned earlier by Bipartisan Infrastructure Law (BIL) in 2021 and application of Inflation reduction act law in 2022 under President Biden where the focus was on rapid renewable expansion and balancing short-term oil supply needs. While with President Trump's policy on energy dominance, maximizing oil and gas production, by accelerating the oil and gas permits on public lands and promoting offshore drilling the country is moving toward more energy security policies.

The International Energy Administration (IEA) 2024 energy policy review indicated that the United States effort in mitigating the climate change resulted in 18% reduction in net greenhouse gas emissions from 2005 to 2022. The plans were also set to achieve the 100% carbon-free power generation by 2035 and net-zero emission by 2050 as shown in Fig. 1.27. This has been achieved by 58% reduction in electricity generation using coal between 2005 and 2022, by partially replacing the coal with natural gas and increasing the share of renewables in proving electricity as shown in Fig. 1.28.

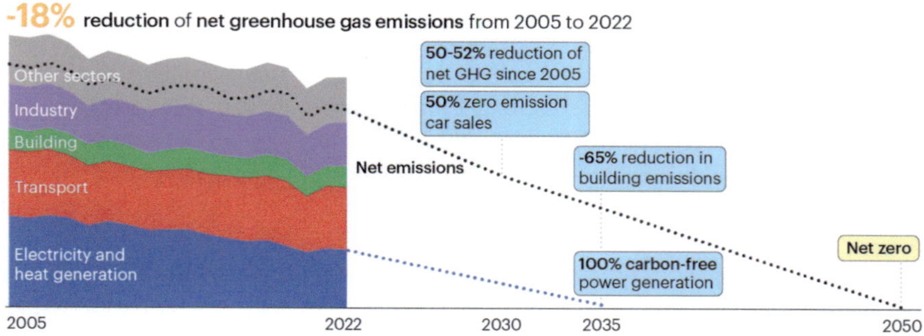

Fig. 1.27 The U.S.A effort in mitigating greenhouse gas emissions in different sectors from 2005 to 2022 and forecasts for net-zero emission by 2050. IEA 2024 energy policy review

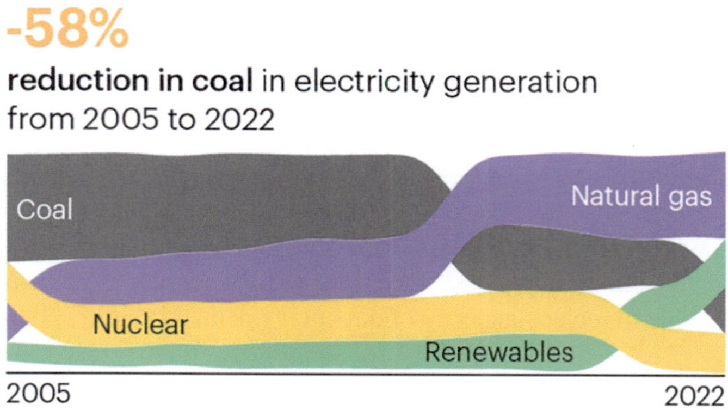

Fig. 1.28 Share of different resources in electricity generation with major emphasize on reducing coal and increasing natural gas and renewables. IEA 2024 energy policy review

1.8 Conclusion

This chapter has established the critical relationship between climate change, energy mix, and CO_2 sequestration. We emphasize the urgent need for informed decision in global energy systems based on energy demand, current state of energy production, and climate change. The discussion began by outlining the scientific basis of climate change, particularly the role of CO_2 emissions from fossil fuels in driving global warming. As energy demand continues to rise, the implementation of sustainable energy sources becomes inevitable. However, while renewable energy technologies such as solar, wind, hydropower, geothermal, and biomass offer cleaner alternatives to fossil fuels, their deployment must be carefully managed to mitigate environmental compromises, including land use, resource depletion, and ecosystem disruption.

A key takeaway is that no single energy source can fully replace fossil fuels; instead, a diversified energy mix, combined with carbon capture and storage (CCS), is essential for achieving deep decarbonization (Professor Ben Ebenhack 2024). The chapter highlighted the importance of technological advancements and policy frameworks in scaling CCS. International climate agreements, particularly the Paris Agreement and subsequent COPs, are significant in aligning global efforts toward carbon neutrality. However, as seen in COP28 and COP29 analyses, gaps remain in enforcement mechanisms and equitable climate financing, underscoring the need for stronger accountability measures.

The ethical dimensions of the carbon budget further complicate the discussion, as developed and developing nations face differing responsibilities in emission reductions. Case studies, such as the U.S. Methane Emission Reduction Action Plan, demonstrate practical approaches to policy-driven decarbonization. Meanwhile, forecasting energy demand reveals the challenges of balancing economic growth with sustainability, necessitating innovations in energy efficiency and storage.

Ultimately, this chapter highlights that mitigating climate change requires an integrated strategy combining CCS deployment, international cooperation, and renewable energy expansion. As we approach the limit of the carbon budget, the application of CCS must increase to eventually reach 100% carbon capture. CCS allows the utilization of additional fossil fuel resources that would otherwise remain unexploited, therefore providing the potential economic benefits associated with fossil fuels. CCS serves as a mechanism to extend the lifespan of fossil fuels within the carbon budget constraint, by preventing a significant portion of carbon dioxide from entering the atmosphere, while other renewable resources are developing in all aspects to be able to replace majority of our energy demand (Fig. 1.29).

Fig. 1.29 Images (Designed by Freepik)

References

K.M. Abbott, P.A. Zaidel, A.H. Roy, K.M. Houle, K.H. Nislow, Investigating impacts of small dams and dam removal on dissolved oxygen in streams. PLoS One **17**(11), e0277647 (2022). https://doi.org/10.1371/journal.pone.0277647

C. Callas, S.D. Saltzer, J.S. Davis, S.S. Hashemi, A.R. Kovscek, E.R. Okoroafor, G. Wen, M.D. Zoback, S.M. Benson, Criteria and workflow for selecting depleted hydrocarbon reservoirs for carbon storage. Appl. Energy **324**, 119668 (2022). https://doi.org/10.1016/j.apenergy.2022.119668

M. Desservettaz, M. Pikridas, I. Stavroulas, A. Bougiatioti, E. Liakakou, N. Hatzianastassiou, J. Sciare, N. Mihalopoulos, E. Bourtsoukidis, Emission of volatile organic compounds from residential biomass burning and their rapid chemical transformations. Sci. Total Environ. **903**, 166592 (2023). https://doi.org/10.1016/j.scitotenv.2023.166592

B. Dziejarski, R. Krzyżyńska, K. Andersson, Current status of carbon capture, utilization, and storage technologies in the global economy: a survey of technical assessment. Fuel **342**, 127814 (2023). https://doi.org/10.1016/j.fuel.2023.127814

P.A. Eigbe, O.O. Ajayi, O.T. Olakoyejo, O.L. Fadipe, S. Efe, A.O. Adelaja, A general review of CO_2 sequestration in underground geological formations and assessment of depleted hydrocarbon reservoirs in the Niger Delta. Appl. Energy **350**, 121723 (2023). https://doi.org/10.1016/j.apenergy.2023.121723

W.J. Elliot, H. Rhee, Impacts of forest biomass operations on forest hydrologic and soil erosion processes. Trees, For. People **7**, 100186 (2022)

D. Erik, *Craft, An Economic History of Weather Forecasting* (University of Richmond ECONOMIC HISTORY ASSOCIATION, 2023)

S. Fahad, S.B. Chavan, A.R. Chichaghare, A.R. Uthappa, M. Kumar, V. Kakade, A. Pradhan, D. Jinger, G. Rawale, D.K. Yadav, et al., Agroforestry systems for soil health improvement and maintenance. Sustainability **14**(22), 14877 (2022). https://doi.org/10.3390/su142214877

A. Holm, D. Jennejohn, L. Blodgett, *Geothermal Energy and Greenhouse Gas Emissions* (Geothermal Energy Association (GEA), 2012)

IEA, in *A Vision for Clean Cooking Access for all*, ed. by Licence: CC BY 4.0, (IEA, Paris. https://www.iea.org/reports/a-vision-for-clean-cooking-access-for-all, 2023)
I.B. Iyegbekedo, S.K. Yerravally, E. Fathi, N. Garapati, Sub-surface analysis for geothermal hot water system for Morgantown, West Virginia, in *Proceedings of the 50th Workshop on Geothermal Reservoir Engineering*, (Stanford University, 2025)
M.O. Jeffries, J.E. Overland, D.K. Perovich, The Arctic shifts to a new normal. Phys. Today **66**, 35–40 (2013)
M.W. Jones, G.P. Peters, T. Gasser, R.M. Andrew, C. Schwingshackl, J. Gütschow, R.A. Houghton, P. Friedlingstein, J. Pongratz, C. Le Quéré, National contributions to climate change due to historical emissions of carbon dioxide, methane and nitrous oxide. Scientific Data. Zenodo, November 13, (2025). https://doi.org/10.5281/zenodo.16640595
Library of Congress, U.S. withdrawal from the Paris Agreement: Process and potential effects (Report No. R48504), (2025). https://www.congress.gov/crs-product/R48504
J.E. Lovich, Golden Eagle mortality at a wind-energy facility near Palm Springs, California. U.S. Geological Survey, Southwest Biological Science Center, Western Birds **46**, 76–80 (2015)
M. Mallet, home, environmental, human Healthon posted on November 28, 2024, in *Toxic blade time bomb*, (2024)
W.N. Meier et al., Arctic Sea ice in transformation: A review of recent observed changes and impacts on biology and human activity. Rev. Geophys. **52**, 185–217 (2014). https://doi.org/10.1002/2013RG000431
M. Mizanuzzaman, M.H. Kabir, Challenges and threats of rising sea level: a scenario assessment on the coastal areas in Bangladesh. Open Access Libr. J. **12**, 1–1 (2025). https://doi.org/10.4236/oalib.1112814
J. Morales Pedraza, *Status and Perspective in the Use of Coal for Electricity Generation in the North America Region* (Elsevier, 2019)
A. Muscat, E.M. de Olde, I.J.M. de Boer, R. Ripoll-Bosch, The battle for biomass: a systematic review of food-feed-fuel competition. Glob. Food Sec. **25**, 100330 (2020). https://doi.org/10.1016/j.gfs.2019.100330
NOAA National Centers for Environmental Information, 2021 was world's 6th-warmest year on record (2025b). https://www.noaa.gov/news/2021-was-worlds-6th-warmest-year-on-record
NOAA National Centers for Environmental Information, Climate at a glance: Global rankings (2025a). https://www.ncei.noaa.gov/access/monitoring/climate-at-a-glance/global/rankings
C.L. Parkinson, A 40-y record reveals gradual Antarctic Sea ice increases followed by decreases at rates far exceeding the rates seen in the Arctic. Proc. Natl. Acad. Sci. USA **116**(29), 14414–14423 (2019). https://doi.org/10.1073/pnas.1906556116
M. Patterson, P. Singh, H. Cho, The current state of the industrial energy assessment and its impacts on the manufacturing industry. Energy Rep. **8**, 7297–7311 (2022). https://doi.org/10.1016/j.egyr.2022.05.242
T.D. Rathnaweera, W. Wu, Y. Ji, R.P. Gamage, Understanding injection-induced seismicity in enhanced geothermal systems: From the coupled thermo-hydro-mechanical-chemical process to anthropogenic earthquake prediction. Earth Sci. Rev. **205**, 103182 (2020). https://doi.org/10.1016/j.earscirev.2020.103182
J. Rogelj, P.M. Forster, E. Kriegler, et al., Estimating and tracking the remaining carbon budget for stringent climate targets. Nature **571**, 335–342 (2019). https://doi.org/10.1038/s41586-019-1368-z
V. Smil, Energy transitions: global and national perspectives, 2nd ed. Santa Barbara, CA, USA: Praeger (2017)
N. Routley, Mapped: The 1.2 billion people without access to electricity. Visual Capitalist (2019). https://www.visualcapitalist.com/1-2-billion-people-without-access-to-electricity/
The White House Office of Domestic Climate Policy, Methane emissions reduction action plan: Critical and commonsense steps to cut pollution and consumer costs, while boosting good-paying jobs and American competitiveness (2021)

K.W. Thoning, P.P. Tans, W.D. Komhyr, Atmospheric carbon dioxide at Mauna Loa observatory: 2. Analysis of the NOAA GMCC data, 1974–1985. J. Geophys. Res. **94**(D6), 8549–8565 (1989). https://doi.org/10.1029/JD094iD06p08549

United Nations Climate Change. (2024). From Billions to Trillions: Setting a New Goal on Climate Finance

D.P. Van Vuuren et al., The representative concentration pathways: An overview. Clim. **109**(1–2), 5–31 (2011)

J.E. Walsh, Melting ice: What is happening to Arctic Sea ice, and what does it mean for us. Oceanography **26**, 171–181 (2013)

F.J. Wentz, M. Schabel, Effects of satellite orbital decay on MSU lower tropospheric temperature trends. J. Clim. **11**(1), 1–9 (1998). https://doi.org/10.1175/1520-0442(1998)011<0001:EOSODO>2.0.CO;2

Å. Widén, B. Malm Renöfält, E. Degerman, D. Wisaeus, R. Jansson, Environmental flow scenarios for a regulated river system: Projecting catchment-wide ecosystem benefits and consequences for hydroelectric production. Water Resour. Res. **58**, e2021WR030297 (2022). https://doi.org/10.1029/2021WR030297

D.J. Wuebbles, D.W. Fahey, K.A. Hibbard, D.J. Dokken, B.C. Stewart, Maycock, T. K. (Eds.), *Climate Science Special Report: Fourth National Climate Assessment* (Volume I. U.S, Global Change Research Program, 2017). https://doi.org/10.7930/J0J964J6(USCGRP)

Chapter 2
Carbon Capture and Storage (CCS) Technologies, Challenges, and Future Outlook

2.1 Chapter Overview and Objectives

2.1.1 Overview

In this chapter, we will discuss the classification of CCS facilities into carbon negative, carbon neutral (or near carbon neutral), and carbon positive categories based on their overall carbon balance. This classification helps in understanding the strategies and the challenges in scaling CCS technologies for different applications and is significantly important in quantifying the environmental impact and effectiveness of CCS projects. Throughout the chapter, we will explore different CCS facilities, their applications, successful case studies from around the world, technological advancements, environmental benefits, and the challenges and opportunities they present. We then examine the CCS landscape's evolution and will illustrate how advancements in technology are driving more efficient carbon capture and storage processes. Next, we highlight CCS applications across hard-to-decarbonize industries, such as steel, cement, and chemical manufacturing. This chapter provides a fundamental for understanding the various classifications in CCS, which are essential for promoting the adoption of CCS technologies.

2.1.2 Learning Objectives

Upon completion of this chapter, you will be able to:

1 Examine the classification of CCS facilities and understand their implications for carbon balance.
2 Analyze real-world CCS applications and case studies.
3 Assess the economic and technical challenges of CCS implementation.

E. Fathi and F. Belyadi, *Subsurface CO_2 Sequestration*,
https://doi.org/10.1007/978-3-032-28296-5_2

4 Explore future trends and policy implications in CCS development.

2.2 Classifications of CCS Facilities

Classifications of CCS facilities include Carbon-Negative CCS facilities, Near Carbon Neutral CCS facilities, and Carbon Positive CCS facilities.

Carbon-Negative CCS facilities employ advanced technologies like Bioenergy with Carbon Capture and Storage (BECCS) or Direct Air Carbon Capture and Storage (DACCS). Their primary goal is to achieve a carbon-negative balance by removing more carbon dioxide from the atmosphere than they emit. Near Carbon Neutral CCS facilities focus on producing carbon-free energy, such as electricity, hydrogen, or heat, derived from fossil fuels like natural gas, oil, and coal. While they attempt to be carbon neutral, they are considered "near" carbon neutral due to emissions released during the extraction and conversion of fossil fuels to carbon-free energy.

Carbon Positive CCS facilities are involved in producing processed fuels, including liquefied natural gas, coal-to-liquid fuels, or oil sands. In these facilities, CCS technology is applied to reduce the carbon footprint of the energy generated. While emissions are significantly reduced, some carbon remains in the fuel, leading to their classification as carbon positive. CCS facilities typically aim to capture and store 80–95% of carbon dioxide emissions. Despite their efforts, a portion (5–20%) of carbon dioxide may still be released into the atmosphere from these facilities.

2.2.1 Carbon-Negative CCS Facilities

The process of achieving carbon-negative balance involves a series of interconnected steps that result in reducing the overall carbon content in the atmosphere. This process starts with the carbon sequestration through which the process of photosynthesis by plants absorbs atmospheric CO_2 and converts it into biomass. A portion of the carbon stored in the biomass, such as forests, is then harnessed to produce ethanol, which is a renewable energy source. This conversion is through the process called Biomass Conversion. This biomass-derived ethanol enters the global energy system as an additive in co-fired with coal, fuel cells, or gasoline blending and eventually releases CO_2 into the atmosphere when used. During the conversion of biomass into ethanol in a biorefinery, the remaining fraction of carbon is captured as a pure stream of CO_2. This carbon capture process prevents it from being released into the atmosphere. The captured CO_2 is then transported and permanently stored in geological formations deep underground. This geological carbon storage ensures that the captured carbon remains indefinitely underground and does not contribute to atmospheric CO_2 levels. This removal of carbon contributes to achieving a negative carbon balance, where the amount of carbon captured and stored exceeds the carbon released from biomass-derived ethanol (Fig. 2.1).

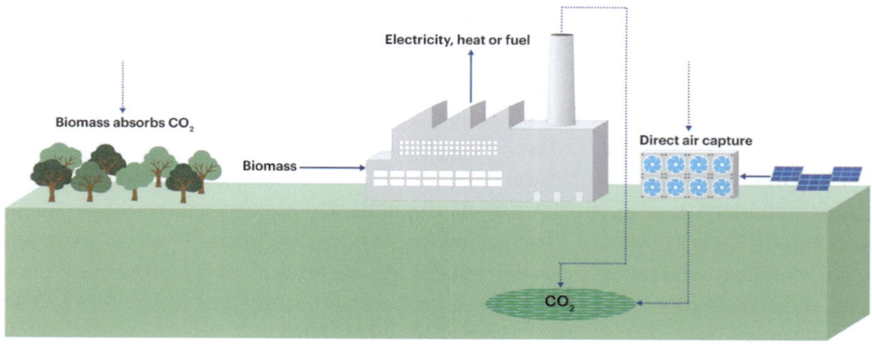

Fig. 2.1 Schematic of bioenergy with carbon capture and storage. *Source* IEA (2020)

Among different carbon-negative projects, there are three important projects that stand out, including the Illinois Industrial Carbon Capture and Storage (CCS) project, Drax, a special power plant using biomass, and Orca, known as the world's biggest direct air capture carbon removal plant. Following, we briefly explore each one of these projects.

2.2.1.1 Illinois Basin—Decatur Project

In Decatur, Illinois, there are two CCUS projects. The first one is the Illinois Basin Decatur Project (IBDP), led by the Midwest Geological Sequestration Consortium (MGSC) and funded by the U.S. Department of Energy, demonstrated the safe and effective storage of carbon dioxide (CO_2) deep underground. The project effectively captured, transported, stored, and monitored one million tons of CO_2 into the Mt. Simon Sandstone (an onshore Saline Reservoir) from the Archer Daniels Midland Company (ADM) (Iyegbekedo et al. 2024).

The Illinois Basin, spanning 155,000 square kilometers, contains the Mt. Simon Sandstone, a massive saline reservoir with no major faults that could risk leakage, and can store 11–150 billion tons of CO_2. The Mt. Simon Sandstone is over 457 meters thick, a widespread rock layer from the Cambrian period, found across the Midwestern U.S. Above the sandstone, the Eau Claire Formation, a 212-meter (695-foot) thick layer of shale and fine sandstone, exists, which acts as perfect cap rock confining the movement of injected CO_2. The extremely low permeability of the Eau Claire Formation confirms long-term storage of carbon dioxide. CO_2 injection in IBDP began in November 2011 after drilling test wells and building a compression plant. To ensure safety, the project performed both pressure and seismic monitoring to confirm permanent storage of CO_2, proving the feasibility of large-scale carbon capture and storage (CCS). Microseismic monitoring is performed in both the injection well CCS#1 and the nearby monitoring well VW#1 from November 2011 to July 2018, Fig. 2. CCS1 in Fig. 2 shows the CO_2 injection well, and VW1 indicates the verification monitoring well. The CO_2 pipeline (orange line in Fig. 2) connects to compression facilities (orange rectangles) in the southeast sector. The demonstration site boundary spans 800 m per side.The IBDP successfully showed that deep geological formations can safely hold CO_2, helping reduce greenhouse gas emissions (Carr et al. 2024, b). This extensive three-year demonstration, featuring a daily injection rate of 1000 tons and compression capacity of 1100 tons, overcame various challenges, obtained Class VI permitting, and implemented monitoring, verification, and accounting (MVA) programs. The project contributed significantly to geologic and social site characterization, reservoir modeling, uncertainty and contingency planning, and stakeholder engagement. Post-injection monitoring concluded in July 2018 (Hnottavange-Telleen et al. 2011; Iyegbekedo et al. 2024).

The second project, the Illinois Industrial CCS Project, is an industrial-scale demonstration with a potential volume of up to five million tones. In this project, CO_2 is injected for three years, starting in April 2017 with an injection rate of 3000 tons per day and a compression capacity of 2200 tons per day. Continuous optimization of the capture process has effectively managed approximately 2,400,000 tons of CO_2 as of June 2021 (McDonald 2017; Sallie 2021).

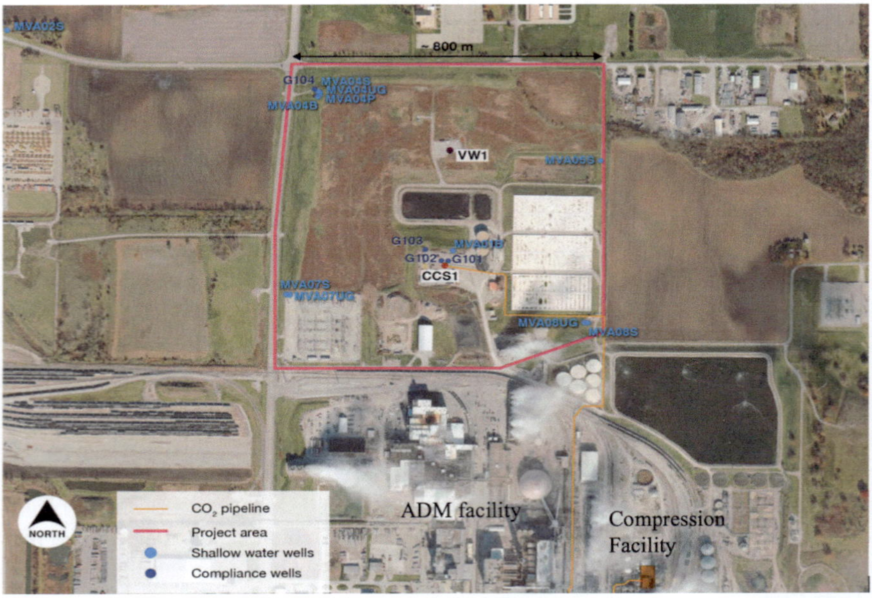

Fig. 2.2 The Illinois Basin Decatur Project (IBDP) carbon capture and storage (CCS) project, CCS1 is the injection well, and VW1 is the monitoring well, (Greenberg, 2017, NETL, DE-FC26-05NT42588)

2.2.1.2 Drax BECCS

In the United Kingdom, Drax, which is a biomass-fueled power plant, launched a pilot project in 2019 focused on a small-scale trial to test capturing CO_2 emissions. Drax pilot project aims to capture 1 ton of CO_2 per day from biomass combustion as a proof of concept project. The pilot bioenergy with carbon capture and storage (BECCS) facilities, operated by Drax Power limited in North Yorkshire, are part of a broader plan for commercial-scale deployment by 2027, where the goal is to capture and store more than eight million tons of CO_2 per year (IEA, CCUS around the world 2021, Drax BECCS). As a significant contributor to the UK's electricity supply, Drax power station has already converted a substantial portion of its units from coal to renewable fuel. The pilot CO_2 capture facilities, utilizing technologies from C-Capture and Mitsubishi Heavy Industries, are currently testing their efficiency and releasing small amounts of captured CO_2 for various uses (www.drax.com). The planned commercial-scale BECCS project announced in 2022 aims to establish a net-zero industrial cluster by 2040, featuring a dedicated CO_2 and hydrogen pipeline network for emission capture and storage beneath the North Sea. If successful, Drax power station could achieve carbon negativity by 2030 (Drax Group 2019), which is a critical step in the UK's industrial sector decarbonization. However, since then, the proposed commercial-scale project has been facing challenges regarding the final government investment decision, the high cost of BECCS, and concerns related to sustainable biomass sources and pipeline for CO_2 transportation to the North Sea. Figure 3 shows the Drax biomass-fueled power plant.

Fig. 2.3 Drax, biomass-fueled power plant, by Paul Glazzard

2.2.1.3 Orca: The First Large-Scale Plant

The world's largest carbon removal plant, known as "Orca," has set a significant target of capturing 4000 metric tons of CO_2 directly from the air each year, initiated in Iceland (www.Climeworks.com). Orca was launched in September 2021, where direct air capture technology is used to extract carbon dioxide from the atmosphere. The CarbFix technology is then used to inject the captured atmospheric CO_2 into basalt rock formations for storage. This technology will lead to mineralization storage of CO_2 that permanently removes CO_2 from the atmosphere and solidifies under the ground. The unique characteristic of this project is that the geothermal energy is used as a source of energy to run the project, showing fully commitment to the path to net-zero energy in this project. It is important to note that despite its scale, the Orca direct air capture project is estimated to capture an amount of CO_2 that is equivalent to just three seconds of global emissions. This emphasizes the challenges associated with addressing the magnitude of global emissions. Therefore, in 2024, the same operator that initiated the Orca up-scaled the Orca project by ten-fold to a new project called Mammoth with a capacity of 40,000 tons of CO_2 per year when fully operational (https://www.carbfix.com/the-new-plant-called-mammoth-increases-the-direct). A mammoth project like Orca is using direct air capture and Carbfix CO_2 mineralization technologies.

Over 10 direct air capture plants are currently in operation across Europe, the United States, and Canada, with a focus on small-scale applications such as carbonation for beverages. The United States is pioneering a large-scale direct air capture plant in partnership between Carbon Engineering and Occidental Petroleum. Expected to be operational by 2026–2027, it aims to capture up to one million tons of CO_2 annually for an enhanced oil recovery project in Texas. Figure 2.2.4 shows the schematic of the Direct Air Capture Facility. Details of DAC operations are discussed in Chap. 9 (August 15, 2023, Occidental (NYSE: OXY)).

2.2.2 *Carbon Neutral or Near Carbon Neutral CCS Facilities*

Carbon neutral or near carbon neutral CCS facilities refer to facilities that aim to achieve a balance between the carbon emissions they generate and the carbon they remove from the atmosphere. These facilities produce energies such as electricity, hydrogen, or heat from sources like natural gas, oil, or coal (Chalmers and Gibbins 2010). While they are not completely carbon neutral because of CO_2 emissions during primary energy extraction and conversion, they use carbon capture and storage (CCS) technology to capture a significant portion (typically 80–95%) of the CO_2 emissions produced. This captured CO_2 is then stored permanently underground, leading to a reduction in the net carbon footprint of these facilities.

Fig. 2.4 Schematic of direct air capture (DAC) facility. (Los Alamos National Laboratory, nss-summer-2022)

2.2.2.1 The Boundary Dam CCS Facility

The Boundary Dam CCS facility in Saskatchewan, Canada (Fig. 2.5) is an example of a near-carbon-neutral CCS facility. It is the first power station with CCS and has been operational since 2014. The facility utilizes coal to generate electricity, which is a carbon-free energy vector.

Fig. 2.5 The Boundary Dam CCS facility in Saskatchewan, Canada image by Wtshymanski

The boundary dam carbon capture and storage project are enabling continuous use of coal for power generation when it's combined with commercial-scale post-combustion carbon capture and sequestration. This project converted the old coal-fired power plant to a new unit with a capacity of 120 MW, achieving a significant reduction of one million tons in emissions annually (www.saskpower.com/). The post-combustion CCS facility resulted in providing affordable electricity for over 100,000 homes and businesses, while being ten times cleaner than other coal units and four times cleaner than a comparable natural gas electricity generation unit. Before implication of CCS, the coal-fired power plant was generating 1.15 million tons of CO_2 annually; the CCS technology has successfully reduced emissions by 90%. This number has also improved, adding dry scrubbers to capture over 99% of fly ash and eliminating harmful pollutants before releasing the discharge gas to the atmosphere. The captured CO_2 is then permanently stored underground in a deep aquifer or sold for enhanced oil recovery purposes. The Boundary Dam power station stands as a model for a complete integration of large-scale carbon capture into coal-fired power stations globally. However, it's worth noting that the actual capture was less than 90% in cases of instrument failure, corrosion, or maintenance. Compared to DAC projects like Orca, the boundary dam project stored ¼ of Orca's capacity, however, at a four-fold lower price. So even though new technologies used in recent projects, like Orca, offer greater capture and storage capacity, they have come with much higher prices.

2.2.2.2 Blue Hydrogen Production

Hydrogen is a major component in the oil refining and chemical industries. It is extensively used to facilitate the conversion of heavy oils into lighter petroleum products. These lighter products, such as gasoline and diesel, are widely utilized across various sectors, including transportation and manufacturing.

The production of hydrogen typically involves using fuels as a feedstock, which can result in the release of a nearly pure stream of carbon dioxide (CO_2) as a byproduct. This CO_2 emission is a significant concern in terms of environmental impact and contributes to greenhouse gas emissions. Despite the CO_2 emissions associated with hydrogen production, it can still be considered a clean fuel option when utilized for energy generation. Hydrogen can be burned with oxygen (combustion) to release heat energy, and this process does not produce carbon dioxide emissions. It offers a potential solution for decarbonizing energy systems and reducing reliance on fossil fuels. Hydrogen technology is generally considered "Near Carbon Neutral."

There are three primary types of hydrogen production: gray hydrogen, blue hydrogen, and green hydrogen, distinguished by their sources and carbon emissions management (White hydrogen, which is a natural source of hydrogen, is excluded). Gray hydrogen, which is a traditional technology, is produced from natural gas through steam methane reforming (SMR), releasing CO_2 into the atmosphere without capture. Blue hydrogen follows the same production process as gray hydrogen from natural gas through steam methane reforming (SMR) or autothermal reforming

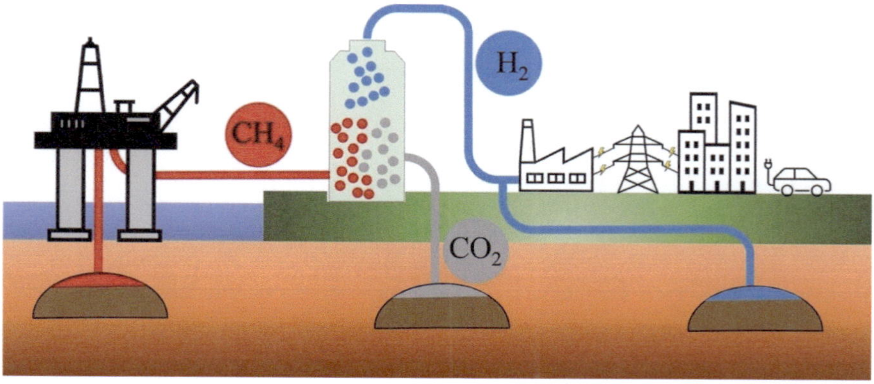

Fig. 2.6 Blue hydrogen production from natural gas. (Massarweh et al. 2023)

(ATR). However, blue hydrogen production is combined with carbon capture and storage (CCS) technology to sequester CO_2 underground and significantly reduce CO_2 emissions (Massarweh et al. 2023). Blue hydrogen is considered a transitional solution to decarbonize the hydrogen sector while enabling the development of a hydrogen economy. The term "Blue" is used to indicate that the hydrogen is derived from fossil fuels with carbon capture, distinguishing it from "Green" hydrogen produced through renewable energy sources. Figure2.6 shows the blue hydrogen production from natural gas, where the natural gas is produced from an offshore gas reservoir, and blue hydrogen is used to generate electricity. The CO_2 emissions are also captured and stored in subsurface reservoirs, ensuring the permanent storage of CO_2 (Massarweh et al. 2023).

Green hydrogen, in contrast, is generated through water electrolysis powered entirely by renewable energy, resulting in zero CO_2 emissions. While mass-scale production demands significant renewable electricity inputs, many nations are actively evaluating their economic potential. Numerous countries have already established strategic roadmaps incorporating defined targets for domestic green hydrogen capacity expansion (Bartels et al. 2025). Figure 2.7 shows the schematic of green hydrogen production using water and wind, and solar power for the electrolysis process, leaving zero CO_2 emissions.

In addition to combustion, hydrogen can also be used in fuel cells to generate electricity. Fuel cells convert the chemical energy of hydrogen and oxygen directly into electrical energy, with the only byproducts being water vapor and heat. This makes hydrogen fuel cells a promising technology for various applications, including transportation and stationary power generation (Manzo et al. 2025).

The Arthur CCS facility, located in Texas (Fig. 2.8), is one of the pioneering hydrogen facilities that incorporates carbon capture and storage (CCS) technology. It aims to capture and store carbon dioxide emissions generated during the blue hydrogen production process. This facility demonstrates the integration of hydrogen production and CCS for reducing carbon emissions and mitigating environmental

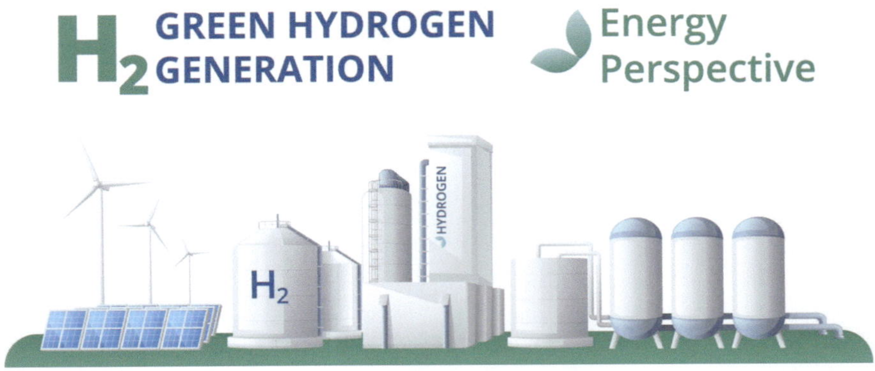

Fig. 2.7 Green hydrogen production green hydrogen designed by Freepik

Fig. 2.8 Air products and chemicals Inc.'s refinery in Port Arthur, Texas, (http://www.netl.doe.gov)

impact. The Port Arthur facility captures CO_2 from steam methane reforming using amine-based solvent capture technology, dries, and purifies one million metric tons of CO_2 every year, reaching an impressive 95–97% purity level (Preston 2018). Using a 12-mile connector pipeline, the captured CO_2 is efficiently sent to Denbury's Green Pipeline-Texas. After traveling 101–150 km through the pipeline, the CO_2 is injected for Enhanced Oil Recovery (EOR) in Denbury's onshore operations, helping recover an extra 1.6–3.1 million barrels of domestic oil annually.

2.2.2.3 Steam Methane Reforming (SMR) with CCS

Steam Methane Reforming (SMR) is a chemical process used to produce hydrogen from natural gas for industrial applications. The traditional application of the process

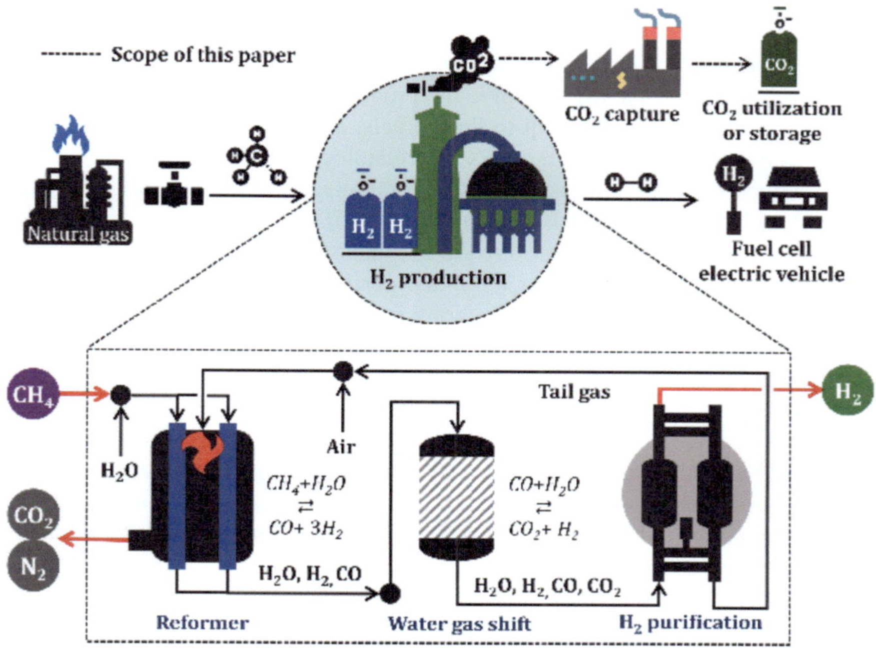

Fig. 2.9 Multi-step process for blue hydrogen production, carbon capture and utilization/storage (Lee et al. 2021)

will produce gray hydrogen and emit CO_2; however, when integrated with CCS facilities results in blue hydrogen production. The primary chemical reactions involved in SMR for blue hydrogen production and utilization with carbon capture, utilization, and storage are presented in Fig. 9 (Lee et al. 2021). In the first reactor, "Methane Reforming Reaction," steam "H_2O" is introduced from the bottom, while natural gas (likely methane "CH4") is introduced from the top. In this reaction, methane (CH_4) reacts with steam (H_2O) to produce carbon monoxide (CO) and hydrogen gas (H_2). This process is called the Methane Reforming Reaction:

CH4 (Methane) + H_2O (Steam) → CO (Carbon Monoxide) + 3 H_2 (Hydrogen).

In the SMR process, 1 mole of CH4 will result in 4 moles of H_2 and 1 mole of CO_2, which, if not captured, will be released in the atmosphere (Ferreira-Aparicio et al. 2005).

Next in the water-gas shift reaction unit, more steam "H_2O" is added to the carbon monoxide (CO), to produce additional hydrogen gas (H_2), and carbon dioxide (CO_2).

CO (Carbon Monoxide) + H_2O (Steam) → CO_2 (Carbon Dioxide) + H_2 (Hydrogen).

The carbon dioxide formed is removed, transported, and permanently stored or utilized in different industries. Finally, in the last chamber, the hydrogen purification unit, pure hydrogen (H_2) is collected and transported for different applications such as fuel cells (Lee et al. 2021).

2.2.3 *Carbon Positive CCS Facilities*

Carbon positive CCS facilities involve the production of processed fuels, such as liquefied natural gas (LNG), coal-to-liquid fuels, and oil sands. The implementation of carbon capture and storage (CCS) technologies in these facilities helps to reduce the carbon emissions associated with these energy sources, but it does not fully eliminate the emissions. By integrating CCS into carbon-positive facilities, a significant portion of the carbon dioxide emissions generated during the production of processed fuels can be captured and permanently stored. Carbon positive CCS facilities contribute to offsetting emissions by capturing and storing carbon dioxide, therefore, reducing the overall carbon footprint of the energy they produce (Schwander 2012). These facilities will help to extend the viability of fossil fuels proving cheaper energy for underdeveloped and developing countries while reducing the environmental concerns around the application of fossil fuels. This technology is much cheaper and easier to deploy compared to near-zero and carbon-negative technologies discussed earlier. The main reason is the use of existing infrastructures, including the industrial plant, pipelines, and storage sites, which significantly reduces the cost of the project. The implementation of this CCS facility is much easier compared to other technologies. For example, in LNG refinery facilities, the CO_2 is already separated in the gas sweetening process, and adding a compressor to convert that to supercritical and use the existing pipeline to transfer to the storage site is straightforward. There are also established processes and best practices in the oil and gas industry to handle CO_2 transport, injection, and monitoring. The oil and gas industry has implemented these practices in CO_2 enhanced oil recovery procedures for decades. In terms of the regulatory framework, this technology faces fewer difficulties and receives support, such as the 45Q Tax credit, which provides incentives for CCS projects.

2.2.3.1 The Quest CCS Facility

The Quest CCS facility in Alberta, Canada, operated by Shell Canada in partnership with Chevron Canada and Marathon Oil Sands, captures CO_2 emissions from oil sands refining. Bitumen's heavy composition requires coking or hydrocracking, using a process enabled by hydrogen produced via steam methane reformers (SMRs), where bitumen is upgraded into lighter synthetic crude oil (Stephens 2010). While SMRs generate the necessary hydrogen for this process, they also emit significant CO_2.

The Quest CCS facility was launched in 2015 and captures up to 1.2 million tons of CO_2 released per year due to the bitumen upgrading process. This accounts for up to 35% of emissions. The captured CO_2 is transported via a 64-km pipeline and stored in the Basal Cambrian Sandstone saline aquifer at depths of 2000–2500 meters, with potential for enhanced oil recovery (EOR).

Despite reducing direct emissions, the facility maintains a positive carbon balance because the final petroleum products still release CO_2 when combusted. The Quest facility was completed in 30 months from the start of construction at a cost of $1.35 billion CAD, which shows the CCS viability for industrial decarbonization (https://sequestration.mit.edu/).

2.2.3.2 Sleipner Project

The Sleipner CCS Project has been operating since 1996 in Norway's North Sea and is the first offshore CCS project in the world that continues capturing and storing CO_2 till now. Figure 2.10 shows the Sleipner, situated in the North Sea. The project aims to store natural CO_2 that comprises 4–9% of natural gas produced from the Sleipner West gas fields. This is unlike the CO_2, typically associated with burning fossil fuels; it's inherent in the natural gas mixture during gas production in the North Sea. The commercial pipeline limit for CO_2 content of natural gas is 2.5% that urges the CO_2 separation and capturing from the natural gas production stream (Cavanagh and Haszeldine 2014). The formerly Statoil (Equinor) is using the monoethanolamine (MEA) scrubbing to remove CO_2 from natural gas and then reinject that into saline aquifers 1000 meters below the North Sea. The facility stores one million tons of CO_2 annually (Energy.gov).

To ensure safe and permanent CO_2 storage in the Utsira sandstone formation, the Sleipner CCS Project implements monitoring techniques, including time-lapse seismic imaging, that reveal how injected CO_2 migrates and stabilizes in geological formations over time (Cavanagh and Haszeldine 2014). Figure 2.11 shows a seismic inline section of the reflective CO_2 plume and overburden at the Sleipner CO_2 Storage Site. The top shows the 1999 repeat survey, and the bottom shows the difference between the 1999 and 1994 baseline. Bright spots indicate natural gas in the overburden (yellow arrows) and CO_2 accumulation at the reservoir top

Fig. 2.10 Sleipner, situated in the North Sea, holds the title of the world's oldest CCS project. Attribution: Bair175

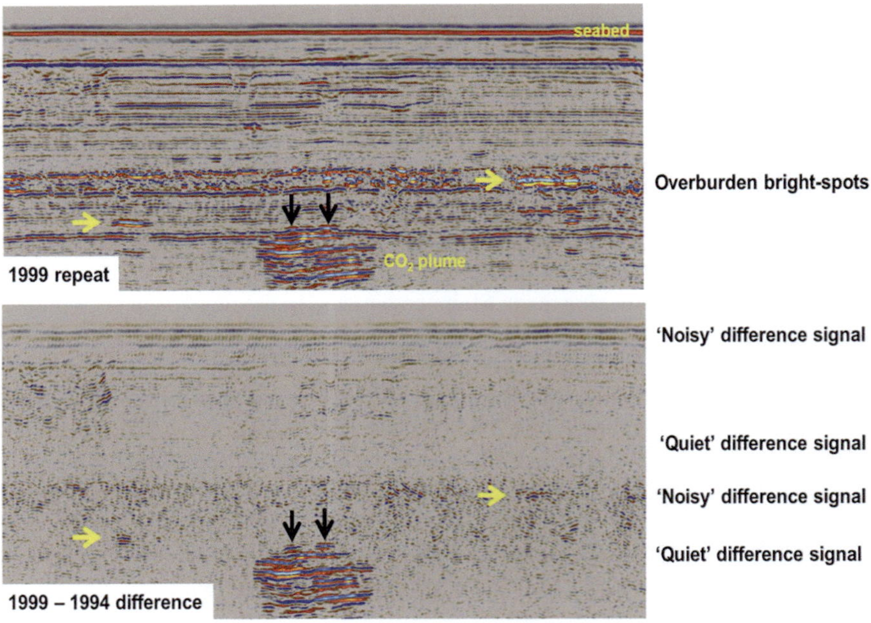

Fig. 2.11 Seismic inline section of the reflective CO_2 plume and overburden at the Sleipner CO_2 Storage Site. The top shows the 1999 repeat survey, and the bottom shows the difference between the 1999 and 1994 baseline (Chadwick et al. 2014)

(black arrows) (Chadwick et al. 2014). Figure 2.12 presents the calibrated reservoir simulation of the Sleipner CO_2 storage site. The CO_2 plume initially migrates upward due to buoyancy; however, its vertical movement is impeded by baffles. As a result, the plume spreads laterally beneath these barriers until sufficient pressure builds for the CO_2 to overcome or breach the restrictions, allowing continued upward migration. Light blue lines indicate migration pathways. Vertical profiles show reservoir temperature, hydrostatic pressure, and CO_2 density (Cavanagh and Haszeldine 2014). Similar observations are also made by Iyegbekedo et al. (2024), studying CO_2 plume dynamics at Illinois Basin Decatur Project Sites using microseismic data. The Sleipner's extensive dataset, which includes monitoring the dynamics of more than 16 million tons of CO_2 plume stored under the ground, has informed global CCS standards, particularly on geological storage of CO_2.

2.3 The Evolution of CCS Technologies

The evolution of carbon capture and storage (CCS) includes four distinct stages: rejection of technology, acceptance with cost concerns, recognition of the necessity of using the technology, and finally, routine adoption of the technology. Each stage of

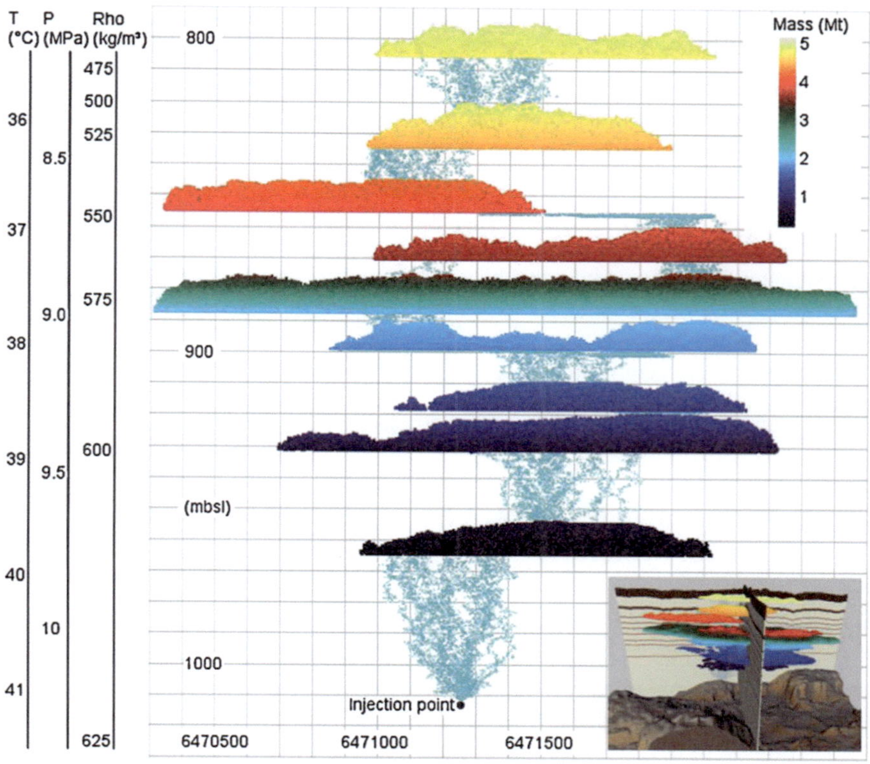

Fig. 2.12 Calibrated reservoir simulation of the Sleipner CO_2 Storage Site, showing the CO_2 plume within layered sandstones separated by thin baffles and overlain by thick caprock. Light blue lines indicate migration pathways. Vertical profiles show reservoir temperature, hydrostatic pressure, and CO_2 density (Cavanagh and Haszeldine 2014)

the technology evolution contains its own issues and difficulties that take significant time and effort to be established.

In the initial "Rejection" stage, CCS technology is seen with suspicion, due to the perception of impracticality and lack of a technological and regulatory framework. As the technology is shown to be feasible and pilot projects start gaining more attention, the critics shift to its economic feasibility in the "Cost Concerns" stage. Here, concerns arise about the perceived high costs and complexity of implementation, resulting in doubts about its practicality on an industrial scale.

As the evolution moves to Stage 3, already successful commercial examples are being developed and running, demonstrating the possibility of economically viable ways of implementing the technology. Moving into Stage 3, the "Recognition of Necessity," there is a growing acknowledgment that CCS represents a significant investment essential for addressing the consequences of CO_2 emissions and climate change. This shift in perception is driven by witnessing the consequences of climate change and the major environmental, social, and financial burdens of climate change for societies. Finally, in Stage 4, "Routine Adoption," CCS becomes an integral and routine component of CO_2 emission control measures. Examples such as catalytic converters for car exhaust emissions and sulfur removal to mitigate acid rain from coal-fired power plants illustrate technologies that have successfully navigated all four stages, becoming widely accepted and implemented over time (Moazzem et al. 2012).

The future of CCS is very much like any other pollution cleaning technology. It is anticipated that Carbon Capture and Storage (CCS) will follow this evolutionary trajectory, and with time, it is expected to become a standard component of fossil fuel power plants and industrial processes, serving as a routine measure for mitigating carbon emissions.

What is the current stage of CCS in your opinion, in terms of these four stages?

2.4 Economic and Technical Challenges in Decarbonizing Industrial Sectors

There is great hope and possibility in some industries, like power generation, to be replaced gradually with cleaner technologies as alternatives are advancing and becoming more economically viable. However, there are industries such as steel, cement, and chemicals, especially fertilizer production, that face significant barriers to transitioning to decarbonization. These sectors are major contributors to global greenhouse gas emissions due to their dependence on energy-intensive processes, high-temperature manufacturing, and chemical reactions that release carbon dioxide. The challenges within these industries lie in their production methods, making the implementation of decarbonization extremely difficult if not impossible, and that's why they have been referred to as "hard-to-decarbonize" industries.

The steel industry relies on traditional methods involving carbon-intensive blast furnaces, while cement production depends on the calcination of limestone, inherently releasing CO_2. Simultaneously, chemical processes in fertilizer production contribute to greenhouse gas emissions. Decarbonizing these industries demands innovative technologies, alternative production methods, and collaborative efforts across the supply chain to address the complexities and scale of emissions associated with their operations.

Figure 2.13 depicts the distribution of industrial sectors in the United States that are challenging to decarbonize, including iron and steel, cement, bioethanol, petrochemicals, ammonia, lime, and hydrogen production. It also shows the locations of CO_2 injection sites and their injection rates, measured in kilotons of CO_2 per year. This map clearly shows the widespread nature of these hard-to-decarbonize industries and their magnitude compared to CO_2 storage sites and amounts of current CO_2 volumes injected. The CO_2 pipelines are also shown as pink lines, which also

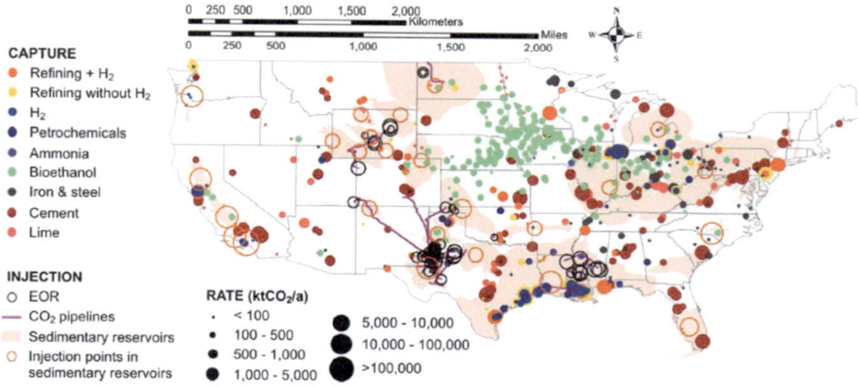

Fig. 2.13 Distribution of industrial sectors hard to decarbonize, USA Carbon capture and storage, US EPA, US DOE

indicate an urgent need for expanding the CO_2 transportation pipeline to be able to transport the CO_2 from the capture facilities to storage sites.

Figure 2.14 also showcases the air pollution sources in the United States. It highlights industrial greenhouse gas (GHG) emissions from various sectors, including ports of importing and exporting goods, power plants fueled by coal and natural gas, and industries. The map provides an overview of the geographical distribution of these emission sources and quantifies their magnitude in million metric tons of CO_2 per year ($MtCO_2$/year). The proximity of significant GHG emissions to areas where carbon management strategies, such as carbon capture and storage (CCS) or other mitigation measures, can be implemented can help address the requirements for proper mitigation strategies. The map is also valuable for targeted carbon management initiatives in regions where the concentration of air pollution sources is high.

The subsequent sections explore the specific challenges and potential solutions for each of these hard-to-decarbonize industries, providing an overview of the industry and its source of emissions, and successful stories and case studies to capture and store their emissions.

2.4.1 The Steel Industry

The steel industry is extremely important as it is the largest energy-consuming manufacturing industry worldwide (Guo and Fu 2010). The steel industry serves as a fundamental building block for various industries and infrastructure development. The amount of steel production is so important that once used as one of the indicators to distinguish between developed and developing countries. Historically, the U.S., Japan, and Germany had massive steel production during their peak industrialization,

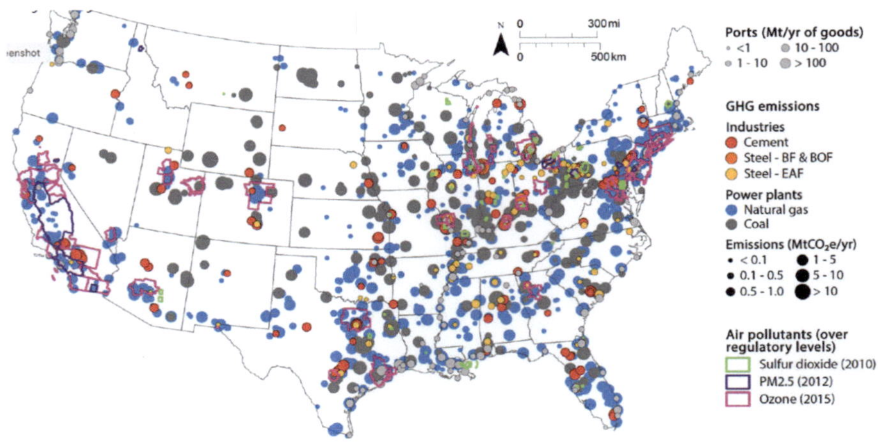

Fig. 2.14 Co-location of air pollution and carbon management, carbon capture and storage, US EPA, US DOE

which resulted in their advanced infrastructures, higher levels of urbanization, and greater manufacturing capacities. China is now the world's largest steel producer by more than 1000 million tons of steel in 2023. China provides more than 50% of the world's steel (Source: World Steel Association). Annual global steel production is approximately 1.9 billion tons in 2023 (2023 Data, World Steel Association), reflecting its substantial scale and impact on the global economy. The demand for steel continues to rise due to its various applications in construction, automotive, machinery, and other sectors. With all the benefits of the steel industry comes the fact that steel production is a significant contributor to greenhouse gas emissions, generating an estimated 1.5–3 tons of carbon dioxide for every ton of steel produced, with an average of 1.8 tons of CO_2 per ton of steel (Ho et al. 2013). This accounts for around seven percent of global greenhouse gas emissions. Coking coal is essential in the steel production process. It acts as a reducing agent that removes oxygen atoms from iron ore (Fe_3O_4 or Fe_2O_3), which is the primary raw material for steel production. This transformation results in the production of purified iron (Fe) for further processing. The production of steel from iron ore includes multiple processes, including coke production from coal, and combustion of coke to produce carbon monoxide used for iron ore reduction. Coke is a solid, carbon-rich material (C) produced by heating coal at high temperatures in the absence of oxygen with minor amounts of ash and other impurities. Compared with raw coal, coke burns at higher temperatures of 2900 °F compared to coal that burns at 637 °F. Coke burns much cleaner compared to coal and has much higher mechanical strength and porosity, which makes it a perfect choice to support a heavy column of ore in a blast furnace. These advantages are making coke a valuable fuel in steel production. Figure 2.15 shows an integrated steel mill in the Netherlands.

The United States is currently the world's fourth-largest producer of steel, with annual CO_2 emissions of 62 million tons from the steel industry. There are major

Fig. 2.15 Integrated steel mill in the Netherlands. By Jesper Schoen

challenges that the United States faces regarding the decarbonization of the steel industry, including the cost of implementing CCS in the steel industry, as it's estimated to cost $65 per ton of CO_2 captured, expressed in December 2018 prices (Hughes and Zoelle 2022). The adsorption of this extra cost is extremely difficult as steel prices are determined by competitive international markets rather than local production expenses, such as in China, with more than 50% share of steel production. This limits the U.S. steel industry's ability to pass decarbonization investments on to customers without losing market share. There are also technical difficulties regarding the implementation of the CCS technology, including CO_2 emissions from multiple sources, each with different CO_2 concentration, such as blast furnaces vs. reheating furnaces, the limitations of Amine-based solutions for carbon capture, and retrofitting existing steel facilities with carbon capture infrastructure (GAO-22-105,274).

2.4.1.1 The Al Reyadah Facility

The Al Reyadah facility, located in the United Arab Emirates, is one of the great examples in decarbonization of the steel industry and is the first industrial-scale CCS project in the Middle East and North Africa. The project is unique as it's a close collaboration between two major industries in the world, steel and the oil and gas industry. In the Al Reyadah facility, CCS technology with enhanced oil recovery (EOR) is combined to serve both purposes, increasing oil and gas production and battling environmental concerns associated with steel production (www.adnoc.ae).

This industrial-scale project is developed to capture 800,000 tons of carbon dioxide (CO_2) per year from the flue gas emissions coming from an Emirates steel production facility. The captured CO_2 is subsequently utilized in enhanced oil recovery (EOR) operations in the nearby oil fields operated by the Abu Dhabi National Oil Company. The primary goals of the project included the substantial reduction of the United Arab Emirates' carbon footprint to support the UAE's pledge to cut carbon emissions by 25% by 2030 and the implementation of EOR techniques to increase oil production.

The CO_2 is captured from the blast furnace using Amine solutions that can produce a CO_2 stream with 98% purity. The CO_2 is then compressed and transported to ADNOC at the Rumaitha oilfield via a 50-kilometer pipeline with an 8-inch diameter (Fathesha Sheikh 2021).

2.4.2 *The Cement Industry*

The worldwide cement production ranks among the most carbon-intensive industrial processes, with about 3–4% of global greenhouse gas emissions coming from chemical processes and combustion emissions during cement production. Annual CO_2 emissions from cement production exceed 2.4 billion tons annually (Ian Tiseo 2025). The United States annually emits more than 67 million tons of CO_2 from its cement production (EPA 430-F-21-004 October 2021). Each ton of cement generates approximately 800 kg of CO_2, with half originating from the chemical decomposition of limestone ($CaCO_3$) in kilns. This calcination reaction, which produces the essential calcium oxide (CaO) for cement, inherently releases CO_2 regardless of the energy source (Harju et al. 2006). The remaining emissions come from fossil fuel combustion, typically coal, required to reach the extreme temperatures (1400–1500 °C) needed for clinker production. The following equations show the CO_2 emission from chemical processes and combustion.

$$CaCO_3 \overset{900-1200^{o}C}{\rightarrow} CaO + CO_2$$

What makes cement particularly difficult to decarbonize is this dual emission source: core material and combustion. The reaction and decomposition of limestone, which is a core material for cement, means CO_2 release is chemically unavoidable. Although carbon capture technologies could potentially address 25–48% of emissions, their implementation remains at the pilot scale. In addition to inevitable CO_2 emissions due to the application of limestone, the CCS technology that can be adopted for the cement industry is also extremely expensive, with a predicted cost of \$45 to \$120 per ton of CO_2 (GAO-22-105,274). This estimate is \$62 to \$64 based on 2018 pricing (Hughes and Zoelle 2022). Figure 2.16 shows A cement plant owned by Asia Cement in Xincheng Township, Hualien County.

Figure 2.17 illustrates the geographical distribution of the steel and cement industrial sectors in the United States, along with the corresponding rates of CO_2 emissions measured in million metric tons per year ($MtCO_2$ /year). The map shows the concentration of steel and cement production facilities across the country and

Fig. 2.16 A cement plant owned by Asia cement in Xincheng Township, Hualien County. Photo by CEphoto, Uwe Aranas.

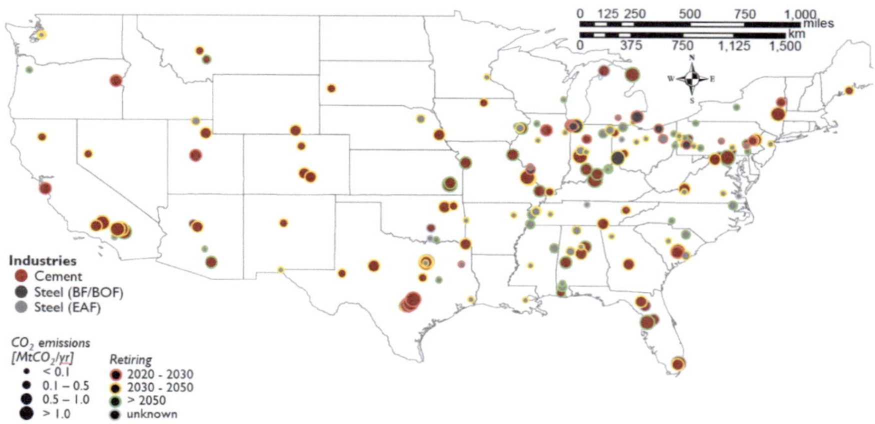

Fig. 2.17 Cement and steel facilities across the US carbon capture and storage US EPA, US DOE

provides an estimate of their combined CO_2 emissions, which account for approximately 135 $MtCO_2$/year. This presents the significant carbon footprint associated with these industries and the importance of addressing their emissions through mitigation strategies such as carbon capture and storage (CCS).

2.4.2.1 The Norcem Brevik Facility

The Norcem Brevik facility, operated by Heidelberg Materials, was recognized as the cement industry's first commercial Carbon Capture and Storage (CCS) initiative in 2020 (www.heidelbergmaterials.com). The construction project was completed in 2023, and carbon capture started in 2024, with full capacity expected to be 400,000

Fig. 2.18 Brevik CCS May 2021 by Eldart

tons of CO_2 annually. The amine-based solution CO_2 capture technology is used in the Norcem Brevik facility. The captured CO_2 is shipped to the terminal and then transferred via a 100 km offshore pipeline to be stored in a subsea saline aquifer. The total project cost about \$430 million; however, the government CO_2 tax of about \$200/ton of CO_2 made the project economically viable. Figure 2.18 shows the Norcem Brevik facility that serves as a hub for extensive research and innovation in the cement industry to reduce the carbon footprint of this industry.

2.4.3 The Chemicals Industry

Based on IEA (2023), the chemical and petrochemical industry is the third-largest industry subsector in terms of direct CO_2 emissions, with 0.9 to 1 billion tons of annual CO_2 emissions, after cement, with 2.4 billion tons of annual CO_2 emissions, and iron and steel, with roughly 2.6 billion tons of CO_2 emissions per year. Chemical production generates high-purity CO_2 streams, particularly in ammonia, hydrogen, and ethylene manufacturing. This inherent characteristic makes CCS implementation significantly more cost-effective, with capture costs as low as \$15–30 per ton compared to \$45 to \$120 per ton in the cement industry and cost \$40 to \$130 per ton in the iron and steel industry (GAO-22-105,274). The production of chemicals, such as fertilizers, continues to rely on fossil fuels as the primary resource. These fuels are used to create essential chemical compounds. Fertilizers are significant in global agriculture practices, making the fertilizer market large and globally influential. They are essential for improving crop yields and supporting food production. The process of producing fertilizers generates a substantial amount of CO_2 as an off-gassing. Large-scale CCS facilities have already been implemented in fertilizers and

Fig. 2.19 Oil refinery in Louisiana—an example of the chemical industry chemical By Jacinta Quesada

petrochemicals, such as Port Neal Nitrogen Complex in Louisiana (Fig. 2.19) and CO_2 capture and utilization at SABIC UNITED, Jubail, Saudi Arabia, with 500,000 tCO_2/year CCS capacity. Yara International, which is a Norwegian multinational company, operates as a major producer of ammonia and ammonia-derived nitrogen fertilizers.

2.4.3.1 The Koch Nitrogen Company

In an effort to address CO_2 emissions from the chemical sector, the Koch Nitrogen Company, which operates one of North America's largest nitrogen fertilizer plants in Enid, Oklahoma, planned to integrate Carbon Capture and Storage (CCS) into its facility (Fig. 2.20). This facility is specifically designed to capture and store CO_2 emissions from steam methane reforming (SMR) of natural gas. The project is expected to capture one million tons of CO_2 annually using amin solution technology, "Benfield process" applied to post-combustion with an objective of reaching 90% to 95% CO_2 purity. The captured CO_2 is compressed and planned to be transported via pipeline to either the depleted Golden Trend Field and Sho-Vel-Tum Field, where it is injected primarily into the Muddy Formation Sandstone for geologic storage and enhanced oil recovery (EOR) operations (Andrew Schaper 2021).

2.4.4 *Cost of Capturing CO_2 from Industrial Sources*

Table 2.1 provides a comparative analysis of carbon capture (CCUS) costs across key industrial sectors, measured in USD per metric ton of CO_2 captured based on 2018 pricing. It breaks down expenses into capital costs, fixed and variable operational and maintenance costs (O&M), and purchased energy inputs, while differentiating between greenfield (new) and retrofit (existing) facilities. The data reveals refinery

Fig. 2.20 Koch nitrogen company carbon capture and storage (CCS) facility in Enid, Oklahoma © Google Earth

hydrogen and iron/steel production emerging as the most capital-intensive applications, with the cost of capture (COC) being \$59.9 and \$65.9. Notably, retrofitting existing plants typically incurs slightly higher costs than greenfield projects.

2.4.5 Summary of Key Challenges

The steel industry is a significant contributor to global greenhouse gas (GHG) emissions. With over 1.9 billion tons of steel produced annually worldwide, the sector accounts for around seven to nine percent of total GHG emissions. The production of steel involves the use of coking coal, which releases large amounts of carbon dioxide (CO_2) during the manufacturing process. Despite efforts to optimize production and improve energy efficiency, the reliance on coking coal as an essential ingredient in steelmaking makes it challenging to reduce the carbon footprint of the industry. However, the implementation of carbon capture and storage (CCS) technologies and the exploration of alternative low-carbon steelmaking processes offer promising opportunities for emission reduction in the future.

The cement industry is a significant contributor to global GHG emissions, with an annual production of over 4 billion tons of cement. Cement manufacturing accounts for approximately six to eight percent of total CO_2 emissions worldwide. The main source of emissions in the cement production process is the chemical reaction that occurs when limestone (calcium carbonate) is heated in the cement furnace, resulting in the release of CO_2. This process, known as calcination, is essential to cement production and cannot be eliminated. However, efforts are being made to reduce emissions through carbon capture technologies. These efforts aim to mitigate the environmental impact of the cement industry and promote more sustainable construction practices.

Table 2.1 Cost of capturing CO_2 from industrial sources measured in USD per metric ton of CO_2 captured based on 2018 pricing (Hughes and Zoelle, 2022)

Case	Capital costs	Fixed O&M costs	Variable O&M costs	Purchased power/ Natural gas	Greenfield COC	Retrofit COC
Ammonia	6.1	3.9	2.7	6.3	19	19
Ethanol	14.1	9.2	1.7	6.8	31.8	32
NGP	6.2	3.4	1.5	5	16.1	16.2
Refinery hydrogen 90% capture	22.8	15.6	5.3	16.2	59.9	61.7
Refinery hydrogen 99% capture	21.3	14.4	5.1	16.5	57.3	58.9
Cement 90% capture	22.8	11.1	6.1	22.6	62.7	64.3
Cement 99% capture	21.8	10.6	5.9	22.6	60.8	62.4
Iron/steel 90% capture	28	9.5	5.7	22.6	N/A	65.9
Iron/steel 99% capture	27.8	9.3	5.6	22.6	N/A	65.4

The chemical industry is a major participant in global greenhouse gas (GHG) emissions because of its extensive energy use and dependence on fossil fuels as raw materials. Being the leading industrial energy consumer and the third-largest industry subsector with direct CO_2 emissions, the chemical industry releases substantial amounts of CO_2 during its production processes. Chemical production, including the manufacturing of fertilizers and other chemical compounds, often utilizes fossil fuels as resources, resulting in CO_2 emissions as byproducts. To address these emissions, initiatives such as carbon capture and storage, process optimization, and the development of renewable feedstocks are being explored. These efforts aim to reduce the carbon intensity of the chemical industry and promote the transition to more sustainable and low-carbon production methods. The combined impact of these three industries (Iron and Steel; Cement; chemical and petrochemical) accounts for 6 to 6.6 billion tons of CO_2 per year, equivalent to 16–20% of global CO_2 emissions.

2.4.6 Economic Implications

To effectively address global warming, it is important to prioritize the reduction of CO_2 emissions. While renewable energy sources are essential, focusing solely on their expansion may not be sufficient to meet the energy demands of all countries and achieve substantial emission reductions. One of the limitations of achieving 100% renewable energy lies in the lack of a cost-effective, large-scale, and long-term energy storage method. Except for hydropower, which has significant storage capacity, and fossil fuels, there is currently no widely available solution that can store energy at the scale required by an entire country's energy system.

Some countries have been successful in decarbonizing their electricity production through hydropower, nuclear energy, or a combination of both. However, many other countries do not have access to these options and must continue relying on conventional power stations.

These power stations utilize low-cost energy storage in the form of coal piles, as well as extensive pipelines for oil and natural gas, which have high carbon content. Considering concerns regarding the CO_2 emissions and climate change, CCUS technologies become critical for countries heavily reliant on conventional energy sources. CCUS enables the capture and storage of CO_2 emissions from fossil fuel-based power generation and industrial processes, helping to reduce their environmental impact and contribute to global emission reduction goals. Combining renewable energy expansion with CCUS deployment and energy mix policies, countries can expect a more balanced and effective approach to mitigating CO_2 emissions and addressing climate change, while meeting the energy demands.

2.5 Conclusion

This chapter has provided a comprehensive exploration of Carbon Capture and Storage (CCS) technologies, their classifications, real-world applications, and the challenges and opportunities they present. By categorizing CCS facilities into Carbon Negative, Carbon Neutral (or Near Carbon Neutral), and Carbon Positive systems, we established a framework for evaluating their environmental impact and effectiveness in decarbonization strategies. We demonstrated how these technologies operate across different industrial contexts.

A critical insight from this chapter is that CCS is not a single solution but rather a range of technologies that must be designed to fit specific industrial needs. Hard-to-decarbonize sectors like steel, cement, and chemical manufacturing face unique technical and economic challenges. However, projects such as Norcem Brevik in cement production and Al Reyadah in steel manufacturing illustrate how CCS can be integrated into industrial processes, even though challenges related to cost, scalability, and infrastructure persist.

Fig. 2.21 images designed by Freepik

The chapter also highlighted the rapid evolution of CCS technologies, from early projects like Sleipner to developments in bioenergy with CCS (BECCS) and direct air capture (DAC). However, scaling these solutions requires overcoming significant challenges, including high capital costs, environmental impacts, and regulatory uncertainties.

Looking ahead, the future of CCS relies on policy support, technological innovation, and cross-sector collaboration. In reaching net-zero emissions, CCS is a necessity, particularly in industries like power generation, steel, cement, and chemicals (Fig. 2.21).

References

J. Bartels, T. Vogt, V. Bertsch, Development of a strategic site selection framework for the green hydrogen economy: the example of Germany. Energy Convers. Manag. **342**, 120086 (2025). https://doi.org/10.1016/j.enconman.2025.120086

T.R. Carr, B.J. Carney, B. Panetta, Evaluating direct deep-use geothermal potential in the Appalachian Basin. Emerg. Energy: Build. Futur. Soc. Explor. (2024)

A. Cavanagh, R.S. Haszeldine, The Sleipner storage site: Capillary flow modeling of a layered CO_2 plume requires fractured shale barriers within the Utsira formation. Int. J. Greenh. Gas Control. **21**, 101–112 (2014). https://doi.org/10.1016/j.ijggc.2013.11.017

R.A. Chadwick, B.P. Marchant, G.A. Williams, CO_2 storage monitoring: leakage detection and measurement in subsurface volumes from 3D seismic data at Sleipner. Energy Procedia, **63**, 4224–4239. 1876-6102. (2014). https://doi.org/10.1016/j.egypro.2014.11.458. (https://www.sciencedirect.com/science/article/pii/S1876610214022735)

H. Chalmers, J. Gibbins, Carbon capture and storage: More energy or less carbon? J. Renew. Sustain. Energy **2**(3), 031006 (2010). https://doi.org/10.1063/1.3446897

Drax Group. Drax sets ambition to become world's first carbon negative company. (2019)

P. Ferreira-Aparicio, M.J. Benito, J.L. Sanz, New trends in reforming technologies: from hydrogen industrial plants to multifuel microreformers. Catal. Rev. **47**(4), 491–588 (2005). https://doi.org/10.1080/01614940500364958

Z.C. Guo, Z.X. Fu, Current situation of energy consumption and measures taken for energy saving in the iron and steel industry in China. Energy **35**(11), 4356–4360 (2010). https://doi.org/10.1016/j.energy.2009.04.008

J.A. Harju, M.D. Jensen, E.N. Steadman, J.A. Sorensen, E.M. O'Leary, 2006 IEEE cement industry technical conference, in *The plains CO_2 reduction (PCOR) partnership—identifying CO_2 sequestration opportunities for the cement industry in the central interior of North America*, (2006), pp. 1–7. https://doi.org/10.1109/CITCON.2006.1635716

K. Hnottavange-Telleen, E. Chabora, R.J. Finley, S.E. Greenberg, S. Marsteller, Risk management in a large-scale CO_2 geosequestration pilot project, Illinois, USA. Energy Procedia **4**, 4044–4051 (2011). https://doi.org/10.1016/j.egypro.2011.02.346

M.T. Ho, A. Bustamante, D.E. Wiley, Comparison of CO_2 capture economics for iron and steel mills. Int. J. Greenh. Gas Con. **19**, 145–159 (2013). https://doi.org/10.1016/j.ijggc.2013.08.003

S. Hughes, A. Zoelle, *Cost of Capturing CO_2 from Industrial Sources (Report No. DOE/NETL-2022/3300)* (National Energy Technology Laboratory, 2022)

IEA, in *A Vision for Clean Cooking Access for all*, ed. by Licence: CC BY 4.0, (IEA, Paris. https://www.iea.org/reports/a-vision-for-clean-cooking-access-for-all, 2023)

IEAGHG, The status and challenges of CO_2 shipping infrastructures (2020). https://ieaghg-publications.s3.eu-north-1.amazonaws.com/Technical+Reports/2020-10+The+Status+and+Challenges+of+CO2+Shipping+Infrastructures.pdf

I. Iyegbekedo, E. Fathi, T.R. Carr, F. Belyadi, Machine learning-driven quantification of CO_2 plume dynamics at Illinois Basin Decatur project sites using microseismic data. Energies **17**, 4421 (2024). https://doi.org/10.3390/en17174421

S. Lee, H.S. Kim, J. Park, B.M. Kang, C.-H. Cho, H. Lim, W. Won, Scenario-based techno-economic analysis of steam methane reforming process for hydrogen production. Appl. Sci. **11**, 6021 (2021). https://doi.org/10.3390/app11136021

D. Manzo, R. Thai, H.L. Thu, G.K. Venayagamoorthy, Fuel cell technology review: types, economy, applications, and vehicle-to-grid scheme, Sustain. Energy Technol. Assess. **75**, 104229 (2025) ISSN 2213-1388. https://doi.org/10.1016/j.seta.2025.104229

O. Massarweh, M. Al-khuzaei, M. Al-Shafi, Y. Bicer, A.S. Abushaikha, Blue hydrogen production from natural gas reservoirs: a review of application and feasibility. J. CO_2 Util. **70**, 102438 (2023). https://doi.org/10.1016/j.jcou.2023.102438

S. McDonald, Illinois Industrial Carbon Capture & Storage Project Eliminating CO_2 Emissions from the Production of Bio Fuels—A "Green" Carbon Process, U.S. Department of Energy (2017). [Online]. Available: https://www.energy.gov/sites/prod/files/2017/10/f38/mcdonald_bioeconomy_2017.pdf

S. Moazzem, M.G. Rasul, M.M.K. Khan, A review on technologies for reducing CO_2 emission from coal-fired power plants, in *Clean Energy for Sustainable Development*, ed. by M.M.K. Khan, (2012), pp. 27–46). IntechOpen. https://doi.org/10.5772/31876

C.K. Preston, *The Carbon Capture Project at Air Products' Port Arthur Hydrogen Production Facility* (IEAGHG, 2018)

G. Sallie, Illinois State Geological Survey (ISGS), Illinois Basin - Decatur Project (IBDP) Seismic Data, July 7, 2021. Midwest Geological Sequestration Consortium (MGSC) Phase III Data Sets. DOE Cooperative Agreement No. DE-FC26-05NT42588 (July 2021). https://doi.org/10.18141/1854142

A. Schaper, A survey of active U.S. carbon capture projects (2021).https://oaktrust.library.tamu.edu/server/api/core/bitstreams/8ef77e1b-5c49-45fc-924f-4088eff1be21/content

M. Schwander, Carbon capture and storage: an industry viewpoint. Sustain. Technol., Syst. Policies **2012**(1) (2012). https://doi.org/10.5339/stsp.2012.ccs.10

F. Sheikh, Abu Dhabi international petroleum exhibition and conference, in *Commercialization of Al Reyadah—World's 1st Carbon Capture CCUS Project from Iron and Steel Industry for Enhanced Oil Recovery CO_2-EOR*, (Abu Dhabi, United Arab Emirates, 2021). https://doi.org/10.2118/207676-MS

R.M. Stephens, *Quest CCS Project RAM Study—Final Report* (Shell Global Solution, 2010) https://open.alberta.ca/dataset/46ddba1a-7b86-4d7c-b8b6-8fe33a60fada/resource/a2b53125-0fbf-4eda-a314-005fe2931dd8/download/ramfinalreport.pdf

I. Tiseo, *Cement Industry Emissions Worldwide—Statistics & Facts* (Statista, Energy & Environment, Emissions, 2025)

Chapter 3
Regulatory and Economic Frameworks for CCS

3.1 Chapter Overview and Objectives

3.1.1 *Overview*

In this chapter, we will explore different regulatory approaches to Carbon Capture and Storage (CCS), including the Carbon Tax System, Cap-and-Trade System, and Voluntary Carbon Markets. Each of these mechanisms offers advantages and disadvantages in incentivizing and regulating CCS activities. Throughout the chapter, we will explore topics such as CCS regulation policies and frameworks, their environmental implications, legal and economic considerations, international collaboration in CCS regulation, and the outlook for CCS regulation in the context of achieving climate change goals. This overview provides a foundation for understanding the various regulatory approaches and classifications in CCS, which are essential for promoting the adoption of CCS technologies and mitigating climate change.

3.1.2 *Learning Objectives*

Upon completion of this chapter, you will be able to:

1- Apply different regulatory approaches for Carbon Capture and Storage (CCS).
2- Evaluate the advantages and disadvantages of each regulatory approach in promoting CCS adoption and reducing greenhouse gas emissions.
3- Analyze the importance of CCS regularization in mitigating climate change and achieving global emission reduction targets.

E. Fathi and F. Belyadi, *Subsurface CO_2 Sequestration*,
https://doi.org/10.1007/978-3-032-28296-5_3

3.2 Regulatory Frameworks for CCS

3.2.1 Overview of CCS Regulations

The regularization of CCS is critical for addressing climate change and reducing greenhouse gas emissions. There are various regulatory mechanisms and policy approaches that can be used to encourage and regulate carbon capture and storage (CCS) including the Carbon Tax system, Cap-and-Trade system, Voluntary Market, Renewable Portfolio Standards (RPS), Low Carbon Fuel Standards (LCFS), Technology-specific incentives, Emission Performance Standards (EPS), Public Procurement Policies, and Mandates for CCS deployment. These mechanisms, are often used in combination to create a regulatory environment that encourages the development and deployment of CCS technologies to mitigate climate change. The effectiveness of these approaches depends on the specific context, political motivations, and the level of support from various stakeholders. The key approaches that mainly practiced include the carbon tax system, Cap-and-trade system, and Voluntary markets. However, other mechanisms also practiced limiting the use of fossil fuels and promote renewable resources in a different scale.

3.3 Economic Mechanisms for Carbon Reduction

3.3.1 The Carbon Tax System

A carbon tax is a penalty applied on businesses and individuals based on the amount of greenhouse gas emissions they produce. The amount of emission is typically measured per ton of emission. A carbon tax serves as a form of "pollution tax: that targets carbon intensive industries and application of fuels like coal, oil, gasoline, and natural gas. This mechanism enforces the companies to improve their energy efficiency and reduce emissions by making use of fossil fuels more expensive through taxation. Obviously, companies transfer part of this extra cost to the consumers by increasing the prices. Therefore, the carbon tax also encourages consumers to shift towards cleaner energy sources by increasing the costs of gasoline and electricity. The primary goal of a carbon tax is to reduce emissions and drive the transition to a low-carbon economy.

A carbon tax is a government set price on carbon emissions in specific sectors, however, eventually part of the cost will pass on from businesses to consumers. The carbon tax objective is to increase the cost of greenhouse gas emissions such that consumption declines and lowers fossil fuel demand. This practice is encouraging companies to invest and develop environmentally friendly alternatives. Carbon taxes have been adopted in various countries globally, with Finland and Poland being the first to implement it in 1990 at a rate of $73.02 per ton of carbon. Sweden and Norway followed in 1991, with Norway's tax being one of the world's strictest at $69.00 per ton of CO_2 in gasoline and Denmark in 1992. Currently 39 carbon taxes have been fully implemented and enforced worldwide including such as those in Mexico 2014, Chile and Colombia 2017, Argentina 2018, Canada and South Africa in 2019, and Uruguay in 2022 to just name the few (World Bank Group 2025, https://carbonpricingdashboard.worldbank.org/).

3.3.1.1 Advantages and Disadvantages and Challenges of Carbon Tax System

Tackling climate change with mechanisms like the carbon tax is critical in the global effort to cut down on greenhouse gas emissions. Although this method has its benefits, like setting a clear price on carbon emission, it comes with its own set of complexities. Next, we take a closer look at the pros, cons, and challenges of the carbon tax system to understand how it contributes in guiding societies toward reducing the use of fossil fuels and increasing the application of renewable energies.

Advantages of Carbon Tax System

Carbon tax system has several benefits including providing a clear price tag on carbon emissions. By implementing emission pricing, it encourages both businesses and individuals to actively reduce their carbon emissions. This system also helps the market to naturally find the most cost-effective methods for emission reduction (Bretschger and Grieg 2024) and gives the businesses flexibility in choosing strategies that works best for their operations.

Governments implementing a carbon tax can also generate revenue through this mechanism. Carbon revenue for governments worldwide through carbon taxes are estimated to be US$29.3 billion (World Bank Group 2025). This revenue can then be allocated to various initiatives, including funding renewable energy projects, supporting climate adaptation measures, or even returning funds to citizens through rebates like Canada Carbon Tax system that issues checks sent to Canadian households. In comparison to some alternative regulatory mechanisms, a carbon tax is relatively straightforward to administer, therefore, policymakers find it an attractive option due to its simplicity. Moreover, businesses benefit from the predictability offered by a carbon tax, allowing them to anticipate the costs associated with emitting carbon. This predictability helps them in long-term planning and decision-making to reduce their fossil fuel use and increase the renewable portion of their energy requirements.

Disadvantages of Carbon Tax System

Despite its benefits, the carbon tax system has its own disadvantages. One significant concern regarding the carbon tax system is disproportionately impact on low vs high income individuals and big vs small industries (Shei et al. 2024). Smaller industries face difficulties digesting the extra cost associated with carbon tax, therefore, experiencing higher increase in production costs. This situation can potentially affect their competitiveness in the global market. In this scenario, industries might relocate to regions with less restricted regulations to cut costs or simply pass the expenses to the consumers. This will result in further burden for low-income individuals who might not be able to afford these extra costs on goods and services. On the other hand, industries heavily reliant on carbon intensive processes may also resist carbon taxes due to fears of increased operational costs and potential negative effects on their competitiveness.

Challenges of Carbon Tax System

Implementing and maintaining an effective carbon tax system presents certain challenges. One significant challenge is to come up with the optimal tax rate by the governments that can balance between emission reduction goals and potential economic impacts of implementing the carbon tax (Hänsel et al. 2022). If the tax rate is set too low, it might not drive sufficient changes in behavior and emission reduction as companies still see paying the carbon tax to be cheaper than emission reduction act. Conversely, if the tax rate is set too high, it could potentially harm economic growth. A quick review of the carbon tax in $US per ton of CO_2 equivalent shows significant discrepancy in prices as it ranges between \$0.76/tonCO_2 in Ukraine to \$167.7 /ton of CO_2 in Uruguay (World Bank Group).

Political resistance is another problem faced by carbon tax systems (Ewald et al. 2022). Industries, especially those with significant carbon emissions, may oppose such measures due to concerns about their economic impacts. Achieving a global consensus on carbon pricing is a complex task. Different countries have diverse economic structures and emissions profiles, leading to issues of competitiveness and fairness in the global market. The successful implementation of a carbon tax system also requires robust monitoring and enforcement mechanisms. These are critical to ensure compliance with the established regulations.

3.3.2 Cap-and-Trade Systems

Cap and Trade is a market-based approach to emission reduction, where a declining cap with time is set on total carbon emissions allowed, and permits allocated to entities. These permits can be traded in a market, allowing entities to buy and sell emissions allowances.

Cap and trade, also called Emission Trading System (ETS), is a regulatory program implemented by governments to set a limit, or cap, on the total emissions of specific chemicals, such as carbon dioxide, resulting from industrial activities. Supporters of the cap-and-trade system suggest that it can serve as a better alternative to a carbon tax system to mitigate CO_2 emissions while minimizing economic burdens on industries and individuals (Hua et al. 2024). Cap-and-trade programs in the energy sector are designed to gradually decrease pollution levels by providing companies with incentives to invest in cleaner alternatives. The program works by permitting companies allowances for a certain number of emissions, which can be bought, sold, or traded among participants. The gradual reduction of available allowances over time encourages emission reduction and promotes the adoption of cleaner technologies and practices.

Under a cap-and-trade system, the government initially establishes a fixed number of permits, which represents a cap on the allowable carbon dioxide emissions. These permits, also known as allowances, are divided among the participating companies. Each allowance allows a company to emit one ton of emissions. The government can distribute these allowances to companies in various ways, such as free allocation or through an auction. Companies that exceed their allocated allowances are subject to taxes or penalties for non-compliance. As time progresses, the total number of allowances issued by the government decreases, gradually tightening the cap on emissions. This reduction in available allowances creates an incentive for companies to seek more cost-effective alternatives and invest in cleaner technologies. Companies that successfully reduce their emissions below their allocated allowances can sell or trade their unused credits to other companies that may be exceeding their allowances. This trading mechanism allows for flexibility and efficiency in achieving emission reduction targets. Additionally, companies have the option to bank their unused allowances for future use. The cap-and-trade system encourages emission reductions by creating a market for allowances and providing financial incentives for companies to find ways to reduce their carbon dioxide emissions. Currently 36 cap-and-trade or Emission Trading Systems (ETS) are implemented around the world, among them California's cap and trade implemented since 2012, and European unions are the most famous examples.

3.3.2.1 Advantages and Disadvantages and Challenges of Cap-and-Trade Systems

Advantages of Cap and Trade

Cap and trade incentivize companies to invest in cleaner technologies to avoid purchasing costly permits, this results in promoting the companies in investing in more cleaner alternatives of fossil fuels (Li et al. 2025). It encourages companies to fund research into alternative energy resources and the development of renewable energy solutions. The system can lead to more rapid reductions in emissions, since companies that achieve emission cuts faster can sell their unused allowances to other companies generating more resource for investing in their renewable resources. Cap and trade serve as a revenue source for the government through the auctioning of emissions credits, providing funds for infrastructure, social programs, clean technology investments, or addressing budget deficits. In 2023 the share of governments from cap and trade is estimated to be US$75 billion (World Bank Group, 2025). The system benefits taxpayers by supplementing government resources with the income generated from the auctioning of permits. As an example, the UK ETS system uses the auctioning as the primary mechanism of allowance distribution. Regional Greenhouse Gas Initiative (RGGI, U.S. Northeast) also distributes 100% of allowances via auctioning.

Disadvantages of Cap and Trade

One disadvantage of Cap and trade is the potential for an overproduction of emissions if the government sets the allowable levels of emissions too high. In that case companies may continue using fossil fuels till up to the maximum limit each year, rather than making significant efforts to transition to cleaner energy sources. This could slow down the overall progress towards achieving environmental goals and reducing greenhouse gas emissions. Another challenge is that emissions credits, as well as penalties and fines for exceeding the cap limit, are often cheaper options compared to investing in cleaner technologies and resources. This cost differential may discourage companies from making substantial investments in transitioning to cleaner energy solutions, as they may find it more financially advantageous to simply purchase emissions credits or pay fines. Monitoring and determining the precise amount of emission can be a difficult task for many industries. This lack of accurate monitoring systems creates a potential gap, making it easier for businesses to manipulate their emissions reports. This challenges the integrity and effectiveness of the cap-and-trade system, as it relies heavily on accurate reporting for proper enforcement. The existence of different standards and maximum caps for emissions in each country adds complexity to the cap-and-trade system. It creates challenges in achieving global emission reduction targets and can lead to inequalities among countries. Unifying these standards and caps becomes critical for ensuring fairness and effectiveness in reducing emissions on a global scale. Lastly, the cap-and-trade system can be administratively complex and resource intensive. It requires robust monitoring, reporting, and verification mechanisms, as well as oversight to prevent fraud and ensure compliance. Implementing and managing such a system can be demanding for governments, requiring significant resources and expertise to effectively regulate and enforce the cap-and-trade framework.

Challenges for Cap and Trade

Setting appropriate emission caps is one of the major challenges (Al-Zawaidah et al. 2025). Governments face the challenge of accurately determining the optimal emission caps for industries. If caps are set too mildly, it may not effectively drive emission reductions, while overly strict caps can impose economic burdens on industries. The lack of reliable emissions data is the second most important challenge as the availability and reliability of emissions data can be limited, making it difficult to accurately measure and monitor emissions levels. This can reduce the effectiveness of cap-and-trade systems in achieving emission reduction targets. The third challenge is the international consensus as climate change and mitigation of CO_2 emissions is a global problem. Achieving consensus among different countries or regions on emissions targets and caps can be challenging. Varying priorities, interests, and policy approaches make it difficult to establish consistent cap-and-trade systems on a global scale. This makes the whole system inefficient in reaching its maximum capacity. This discrepancy can also promote reallocation of resources from most restricted jurisdictions to less intensive ones which does not help in the overall emission reduction goal. A study by World Bank Group (https://carbonpricingdashboard.worldbank.org/) shows the price in US$ for equivalent ton of CO_2 that has dramatic change in the cap-and-trade system between countries from 61 cents in Indonesia to $61.3 in European Union. There are also uncertainties associated with long-term effects of cap and trade (Wei et al. 2025). Predicting the long-term effectiveness and efficiency of such systems over extended periods of time are subject to uncertainties and can be influenced by various factors. Finally, while cap-and-trade systems aiming to reduce emissions, they can result in increased prices for fossil fuels such as oil, coal, and natural gas. This can impact the economy and potentially pose challenges for certain industries. Balancing the transition to alternative energy sources with economic considerations is a complex task that requires detailed studies to identify the system that cap and trade will be applied.

3.3.2.2 Does Cap and Trade Really Work?

Cap and trade have proven successful in reducing greenhouse gas emissions and achieving environmental goals worldwide. European nations have incorporated this approach since 2005, with the EU Emissions Trading System. China is actively moving towards a national cap program, with many cities and provinces already adopting carbon caps since 2021. In the United States, eleven states participated in the Regional Greenhouse Gas Initiative (RGGI), a cap-and-trade program launched in 2009. RGGI has significantly reduced emissions in the power sector and serves as a model for other regions considering similar initiatives. Indonesia and Australia are two newer countries that implemented cap and trade starting 2023. Cap and trade cover 19% of global greenhouse gas emissions while the carbon tax only covers 6%. The carbon tax was implemented since 1990 and cap and trade implemented later in 2005 (World Bank Group).

These examples demonstrate that when cap-and-trade programs are well designed and supported by robust policies, they can drive emission reductions and provide economic incentives for transitioning to cleaner energy sources. However, their success depends on careful planning, accurate emissions data, effective enforcement, and continuous monitoring and adjustments to ensure they effectively address climate change.

3.3.3 Carbon Markets

3.3.3.1 Overview of Carbon Markets

Carbon markets serve as platforms where carbon credits, which represent emission reductions, are traded. These markets are required for providing a financial incentive for reducing greenhouse gas emissions. They enable entities to buy and sell carbon credits, promoting a market-based approach to emission reduction. Governments and political entities, such as the EU, the UK, and the state of California, have established carbon markets as part of their strategies to mitigate climate change. These markets impose emission reduction targets on specific industry sectors, encouraging them to actively participate in emissions trading.

In addition to compliance schemes, there is a growing trend of voluntary participation in carbon markets. Various sectors, beyond those subject to mandatory regulations, voluntarily pledge to offset their emissions by engaging in carbon market activities. This voluntary engagement expands the scope of emission reduction efforts and allows a wider range of industries and organizations to contribute to addressing climate change. One distinguishing feature of voluntary carbon markets is their flexibility and lack of geographic constraints. Unlike compliance markets that are limited to specific regions or political boundaries, voluntary carbon credits can be traded globally. This flexibility provides opportunities for businesses and organizations from any sector or location to participate in emission reduction initiatives through voluntary carbon market. It promotes a more inclusive approach to sustainability that enables a broader range of entities to invest in emission reduction projects. Carbon markets create a mechanism for incentivizing and rewarding emission reductions. By facilitating the buying and selling of carbon credits, the voluntary carbon market contributes to global efforts in mitigating climate change and provides a platform that attracts more entities to get involved. As shown in Fig. 3.1 selling the carbon credit obtained from CO_2 emission reduction can help in advancement of renewable energies such as solar and wind to make them competitive with fossil fuels.

Fig. 3.1 Cartoon showing reinvesting the carbon credit in renewable energy and afforestation. Designed by Freepik

3.3.3.2 Carbon Credit (Carbon Offsets)

Carbon credits are a market-based system for reducing greenhouse gas emissions. They are certificates or permits that represent the right to emit one ton of carbon dioxide or other greenhouse gases. They represent units of measurement that allow organizations or individuals to offset their carbon footprint by purchasing credits generated through emission reduction projects. These credits are generated from projects that prevent or remove emissions, and individuals or organizations can purchase and apply them to offset their own emissions or meet carbon neutrality/net zero targets. Figure 3.2 shows an example of an individual generating carbon credit by planting trees and selling that to oil and gas production companies and generating revenue.

Carbon credits, also called carbon offsets, enable the owner to emit a designated quantity of carbon dioxide or other greenhouse gases. They are environmental commodities that derive value from reducing, removing, or capturing a ton of CO_2 or equivalent GHG before it enters the atmosphere, Fig. 3.3. The carbon credit mechanisms are distinguished based on their sources in international, independent, or governmental credits. The international credits are issued under international treaties and governed by international organizations mainly under United Nations such as Paris agreement or the Carbon Offsetting and Reduction Scheme for International Aviation (CORSIA) where airlines buy credits to meet international aviation emission targets. Independent credits are implemented by independent non-governmental entities such Verra that certifies forest and CCS projects such as Microsoft's mangrove credit or American Carbon Registry (ACR) that is a private carbon crediting program in both global compliance and voluntary markets. The governmental credits, however,

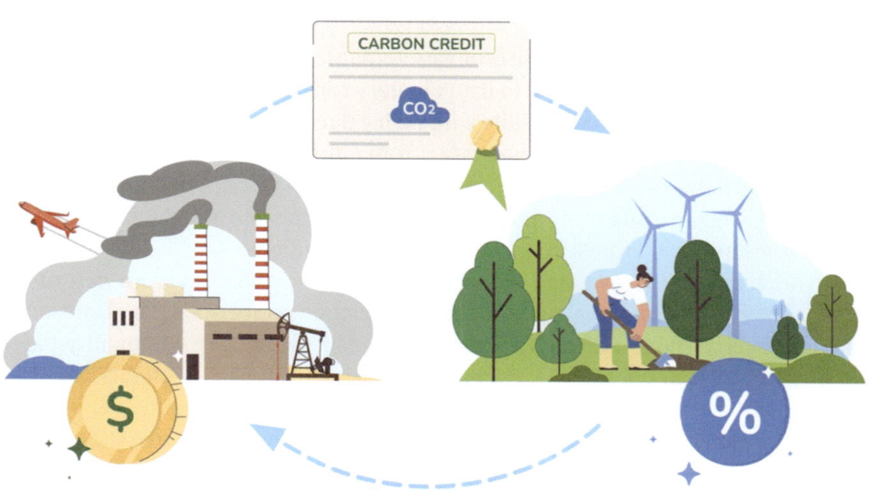

Fig. 3.2 Emission reduction/removal projects generate credits for purchase by organizations to offset emissions Designed by Freepik

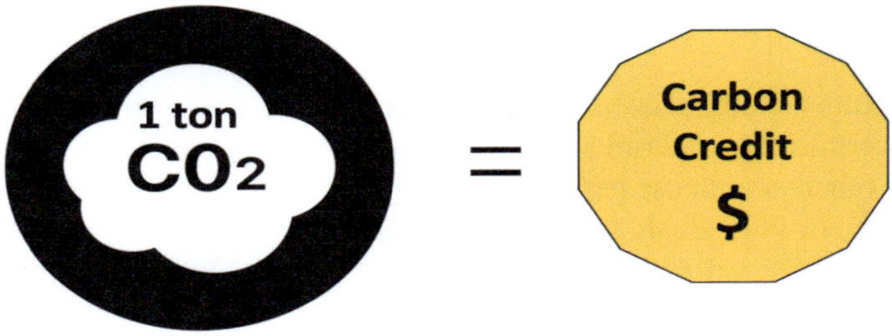

Fig. 3.3 carbon credit modified and designed by Freepik

can be managed by group of regional, national (e.g., China's certified emission reduction program or Australian carbon credit unit scheme) or subnational governments like California's Compliance Offsets. There are 1 regional, 17 national and 17 subnational crediting mechanisms that are already implemented worldwide. In total, $5.7 billion credits are issued from which $3.3 billion is issued via international mechanism while $1.9 billion is the share of independent mechanism and $450 million is issued by governments. The maximum allowance is issued in 2021 that is worth $506 million. Since then, the trend has been decreasing and reached to $372 million in 2023. East Asia and pacific with $676 million has been the major credit issuance region since 2018 followed by Latin America and Caribbean and South Asia. North America takes the fourth place in this ranking (World Bank Group, https://carbonpricingdashboard.worldbank.org/).

3.3.3.3 Voluntary Carbon Markets

The voluntary carbon market consists of various participants and processes that facilitate the trading of carbon credits. In the voluntary carbon market companies or individuals voluntarily buy carbon credits with no legal requirements. Examples include the Microsoft direct air capture purchases from Clime works credits to meet their net-zero emission goals or Shell's Forest offsets that implement nature-based solutions by investing in different emission reduction projects such as reducing emissions from deforestation and forest degradation (REDD) credits.

In voluntary carbon markets, project developers form the upstream part of the market. They initiate and implement emission reduction projects, ranging from large-scale industrial projects like high volume hydro plants to smaller community-based initiatives like clean cookstoves. Project developers issue carbon credits based on the emission reductions achieved by their projects. The downstream market includes end buyers, including tech companies like Apple, Google, and Microsoft or airlines like Delta, and oil and gas majors like Shell. In 2023, a report by Ecosystem Marketplace which is a voluntary carbon market shows Microsoft, Shell and Delta as top buyers in the voluntary carbon market. These buyers voluntarily participate in the market to offset their carbon footprint and meet sustainability goals. To bridge the gap between supply and demand, brokers and retail traders are the required elements. Retail traders purchase significant quantities of carbon credits directly from project developers. They bundle these credits into portfolios, typically consisting of hundreds to thousands of equivalent tons of CO_2, and sell them to end buyers, usually with a commission. Brokers, on the other hand, buy carbon credits from retail traders and market them to end buyers, also earning a commission. Standard organizations which are often non-governmental organizations (NGOs) are fundamental in the voluntary carbon market. They certify that specific projects meet their stated objectives and demonstrate the verified volume of emission reductions. These standards ensure the credibility and integrity of the carbon credits being traded. Together, these participants and processes create a functioning marketplace where project developers generate carbon credits, brokers and retail traders facilitate the trading, and end buyers purchase credits to offset their emissions. Standards organizations provide assurance and oversight to maintain the quality and reliability of the carbon credits in the market.

Based on the information provided by Trove Intelligence via Carbon Market Watch, India is a significant contributor to the voluntary carbon market. It is reported that nearly 20% of voluntary carbon offsets used globally come from India. This highlights the country's active participation in emission reduction projects such as Gujarat wind farms and its commitment to addressing climate change. Furthermore, the data indicates that India, along with the United States and China, has a considerable number of offsets still available on the market. This suggests that these countries have a substantial potential for further participation in the voluntary carbon market and can significantly contribute to driving emission reductions such as China CCER program. It's important to note that the dynamics of the voluntary carbon market can evolve over time, and the contributions of the countries may shift as market conditions and climate policies change. New reports by Ecosystem Marketplace 2023 shows that Turkey, Brazil, and Colombia are also developing hubs for wind and REDD projects that can significantly participate in the carbon market.

India, ranked as the world's third-largest emitter of greenhouse gases, has achieved notable success in earning carbon credits or emission reduction certificates (CERs). This has been accomplished through investments in low-carbon intensive technologies, the adoption of renewable energy, and efforts in forest conservation. India's renewable energy capacity exceeds 46.3% of the nation's total installed power generation capacity using solar, wind, hydro, and biomass. The revenue generated from selling these carbon credits in the international market has been substantial for India. Despite these accomplishments, India still holds a considerable surplus of CERs that it plans to sell. However, a significant challenge for India in carbon trading revolves around disagreements over trading rules. One major issue is the concern about double counting, where both the buyers and sellers place claim to the same carbon credit. To illustrate this issue, consider a country or company that sells a credit earned from constructing a solar park, the buyer uses the purchased credit to offset their carbon emissions. However, if the seller also includes the same credit in their emission reduction calculations, it is considered double counting. Critics argue that allowing double counting enables countries to sidestep meaningful emission reductions, thereby undermining the effectiveness of carbon trading mechanisms. Addressing the issue of double counting and establishing transparent and robust rules for carbon trading is now required for India to be able to meaningfully contribute to the carbon market. It would ensure the credibility and integrity of carbon markets while encouraging countries to genuinely reduce their emissions. India's experience highlights the complexities and challenges associated with carbon trading and emphasizes the need for international cooperation and consensus on trading rules to facilitate effective emission reduction efforts globally.

Demand for voluntary carbon credits has risen in recent years till 2022!

Figure 3.4 by Ecosystem Marketplace, illustrates the trend of carbon credits demand from 2016 to 2024. The x-axis represents time, starting from 2016 and ending in 2024, while the y-axis represents the market size in million ton of equivalent CO_2 and market share in percentage between 2016 and 2024. Each carbon credit is typically equivalent to 1 metric ton of CO_2 emission reduced or avoided. From 2016 to 2022 the market size grows from 40 million ton of CO_2 to 500 million to of CO_2 when it reaches its peak. This upward trend indicates a growing recognition and interest in reducing and offsetting carbon emissions using carbon credits. It suggests an increasing commitment from businesses and organizations to act against climate change and support projects that contribute to emission reductions or avoidance. However, sharp decline has been observed after 2022 mainly due to integrity concerns such as Verra controversies, where a significant concern raised regarding low integrity and overall effectiveness of carbon credits especially related to avoid deforestation. To rebuild the trust in the carbon market few measures included the shifting the focus to more tangible projects such as direct air capture and biochar and 2024 reforms with high transparency in project definition, monitoring, and verification.

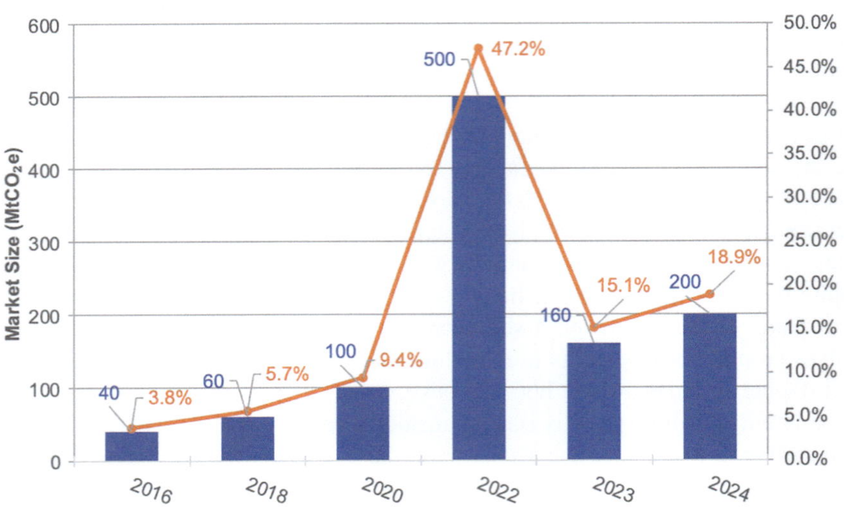

Fig. 3.4 Demand for voluntary carbon credits between 2016 and 2024 (Ecosystem Marketplace)

"The carbon market is expected to become the world's biggest commodity market and looking to be a trillion- dollar industry very soon." ~ New York Times.

The statement in The New York Times suggests that the carbon market has the potential to become the largest commodity market in the world and is on track to become a trillion-dollar industry soon. This projection reflects the growing importance and scale of carbon markets as a mechanism to address climate change and reduce greenhouse gas emissions. Several factors contribute to this expectation. Firstly, there is a growing global agreement and urgency to tackle climate change, with countries, businesses, and individuals increasingly recognizing the need to reduce carbon emissions. As a result of these developments the demand for carbon credits, which represent emission reductions, is expected to rise significantly. Secondly, governments around the world are implementing regulatory frameworks and policies to incentivize emission reductions and promote the use of carbon markets. This creates a favorable environment for the growth of the carbon market and attracts participation from a wide range of industries and sectors. Furthermore, as the global awareness of climate change and sustainability increases, businesses are seeking ways to demonstrate their environmental responsibility and achieve carbon neutrality. Many companies are voluntarily participating in carbon markets to offset their emissions, which further drives the demand for carbon credits. Considering these factors, experts predict that the carbon market will experience exponential growth in terms of volumes and market value. As the world transitions to a low-carbon economy, the carbon market is predicted to be critical in emission reductions and driving investments in clean technologies. However, this requires proven integrity and building trust that can bring the credibility of the carbon market to the higher level as it stands today.

3.3.3.4 Carbon Market Exchange

Carbon trading, also known as carbon emissions trading, is the process of buying and selling credits that permit companies or entities to emit a specific amount of carbon dioxide. It operates through marketplaces where these credits are traded. Among which followings are the most common Markets.

1- European Union (EU) Emissions Trading System (EU ETS) stablished in 2005 and operated by European Energy Exchange (EEX) is a central European carbon market exchange that operates in 30 countries including all 27 European Union member states, Iceland, Lichtenstein, and Norway. It specializes in electric power and heat generation, industrial manufacturing, and aviation. EU ETS is working based on cap-and-trade regulations. The EU ETS allowances are sold in auction and tradable. BY 2023 the EU ETS contribution to emission reduction is estimated to drop the European power and industrial plant emissions by 47% compared to 2005 (European Commission).

2- California Cap and Trade program operated by California Air Resources Board (CARB) established in 2012 covers power, industrial, and fuel distribution. This includes 80% of the state greenhouse gas emissions. The California Cap and Trade market is linked to Quebec's market and has price floor of $24.70 / ton of CO_2 in 2024 (Thuerbach and Johnson 2024).
3- Regional Greenhouse Gas Initiative (RGGI, U.S. Northeast) covers 11 north-eastern and mid-Atlantic states launched in 2009, is also based on Cap-and-Trade regulations. RGGI currently covers power plants only and sets regional cap on CO_2 emissions. The allowances are auctioned and generated revenue is reinvested in clean energy programs (Ruth et al. 2008).
4- China National ETS operated by Shanghai Environment and Energy Exchange was stablished in 2021 also covering power plants, but it is expected to expand to cement and steel industries. It is estimated that ETS covers 5.2 billion tons of CO_2 emission equivalent to 40% of China's emissions (International Carbon Action Partnership), Liu et al. 2022.
5- Korea Emissions Trading Scheme (K-ETS) operated by Korea Exchange (KRX) covers steel and petrochemical industries that started operation in 2015. The allowances are allocated in hybrid as combination of free, fixed benchmarking and auctioning with the Cap of 567.1 million tons of equivalent CO_2 in 2024 (International Carbon Action Partnership), Suk et al. 2018.
6- Voluntary Carbon Markets such as CBL (Xpansiv), Air Carbon Exchange (ACX), and CME Group are among major trading platforms that facilitate credit trading. These platforms work with voluntary carbon market registries such as Verra, Gold Standard and Climate Action Reserve (CAR). In this collaboration chain registries issue carbon credits while platforms put these credits as part of nature-based N-GEO or tech-based C-GEO contracts enabling their trading by selling them to clients. The registry like Verra (VCS) was providing 40% of credits before encountering the integrity issues. Gold Standard applies more restricted rules than Verra and produce credits through clean water and solar while Climate Action Reserve (CAR) focused on forestry and methane capture credits.

These platforms contribute to the functioning and development of carbon trading markets, enabling participants to trade emissions allowances and move toward a low-carbon economy. In terms of the contribution European Union through EU Emission Trading System is one of the major players by largest carbon market covering 40% of EU emissions and around 8 billion tons of CO_2 trading volume. European Union, China, and the United States control more than 80% of global carbon trading. China with national ETS is the world largest compliance Market covering emissions with roughly 200 million tons of CO_2 per year but does not trade internationally. In voluntary markets, top buyers are the United States tech companies such as Microsoft, airlines and oil and gas majors with more than 40% of voluntary market demand.

3.3.4 Other Carbon Regulatory Mechanisms

The Clean Development Mechanism (CDM), established under the Kyoto Protocol, allowed developed countries to invest in emission reduction projects in developing nations, offsetting their own emissions. These projects generated certified emission reduction (CER) credits contributing to Kyoto Protocol targets. The mechanism followed rigorous public registration and approval processes overseen by Designated National Authorities and the CDM Executive Board to ensure integrity. From 2006 till 2020, 7837 individual projects registered by CDM and 2.45 billion Certified Emission Reductions (CERs) issued (2025 UNEP-CCC). In 2020 the Kyoto Protocol was replaced by Paris Agreement, therefore no new project was registered by CDM after 2020 and only ongoing projects continued till 2025.

Emission Performance Standards (EPS) enforced by the United States Environmental Protection Agency (EPA) identifies the maximum amount of CO_2 emissions allowed per unit of electricity produced, requiring power facilities exceeding these limits to adopt emission reduction technologies, such as CCS. These standards are imposed by states to control greenhouse gas emissions from power stations. These standards will be reviewed and updated every five years by assessing the average emissions output of new natural gas turbines, focusing on commercially available turbines in the United States. The standards apply to baseload generation units with a capacity factor of 60% or greater, restricting utilities from entering long-term contracts or investing in facilities that exceed the CO_2 emissions standard. These regulations are applicable to all utilities in the state, with exemptions for renewable and nuclear-powered electricity and long-term commitments with the Bonneville Power Administration (BPA). BPA provides hydropower generated in the Columbia River Basin to communities throughout the Northwest. EPS also implements Public Procurement Policies where the governments prioritize the application of low carbon emission products such as the European Union Green Public Procurement.

Renewable Portfolio Standards (RPS) and Low Carbon Fuel Standards (LCFS), exploring how regions set standards to promote the use of cleaner energy sources and technologies. In some places, there are rules that set a certain amount of energy must come from clean sources. Also, there are standards for using fuels with less emissions. The Renewable Portfolio Standards (RPS) has practiced in some states in the United States where the % of energy consumption is mandated to be from renewable sources. This aims to reduce the application of fossil fuels and reducing CO_2 emissions. Like renewable portfolio standards (RPS), the state of California provides credits for low carbon fuels under "Low Carbon Fuel Standards (LCFS)" to support CCS in hydrogen and ethanol production. This plan is started by California Air Resources Board (CARB) to cut down on pollution from transportation fuels in California.

Governments can incorporate CCS into public procurement policies, encouraging public institutions to choose goods and services produced with lower carbon intensity, including those with CCS. Governments can also provide funding for research and development initiatives focused on advancing CCS technologies, making them more cost effective and efficient. These mechanisms are often used in combination to create a regulatory environment that encourages the development and deployment of CCS technologies. The effectiveness of these approaches depends on the specific context, political will, and the level of support from various stakeholders.

They also must address the costs, reliability and proper monitoring to ensure the scalability of CCS projects.

3.4 Government Support for CCUS: Subsidies, Grants, and Financial Incentives

Subsidies, grants, and financial incentives are widely used by governments around the world to promote the adoption of Carbon Capture, Utilization, and Storage (CCUS) technologies across various industrial sectors. This is because climate change is an urgent global challenge, and CCUS is recognized as one of the most effective and immediate solutions for reducing industrial emissions and achieving net-zero targets.

CCUS is already established and proven as a viable technology. However, large-scale implementation of CCUS faces significant challenges, especially the high capital and operating costs, which make it less accessible to industries and private companies. To overcome these challenges, governments have introduced a mix of funding strategies to lower the financial risk of CCUS investments and to encourage private-sector involvement in research, development, and deployment of these technologies on an industrial scale. These government-supported financial incentives initially targeted the most emission-intensive industries, such as energy production, power generation, cement, steel, and chemical manufacturing.

The most common financial incentives for CCUS projects include:

1- Grants and direct funding,
2- Tax credits,
3- Operational subsidies and price guarantees,
4- Public procurement, and.
5- Public funding for strategic infrastructure.

Many governments offer non-repayable grants to support CCUS projects at different stages, from early research and development (R&D) to full-scale deployment. For example, the U.S. Department of Energy (DOE) defines four phases of CCS grants: (I) integrated CCS pre-feasibility, (II) storage complex feasibility, (III) site characterization and permitting, and (IV) construction (U.S. Environmental Protection Agency [EPA], 2022). These grants often fund breakthrough R&D, pilot and feasibility studies, demonstration projects to prove technical and economic viability, and capital costs for building large-scale CCUS facilities. Authors of this book have been involved in multiple U.S. federal funded projects for developing and advancing CCUS projects including Citronelle, Alabama, that served as a primary testing for the U.S. Department of Energy's National Risk Assessment Partnership (NRAP) program, Illinois Basin-Decatur CCS Project and Wyoming CarbonSAFE Projects as part of DOE-NETL SMART initiatives, and the Tri-State CCS Hub in Ohio, West Virginia and Pennsylvania.

In addition to grants, governments also use tax credits and other fiscal incentives to stimulate private investment. Tax relief mechanisms are a popular way to lower the cost of CCUS. For instance, the U.S. 45Q tax credit provides a per-ton incentive for CO_2 that is captured and either permanently stored or used in other applications. Similar policies exist elsewhere, such as Norway's CO_2 tax exemptions for stored emissions and Canada's investment tax credits (ITCs) for CCUS equipment.

Governments offer operational subsidies and price guarantees to address the higher operating costs of CCUS-equipped facilities compared to traditional plants, for example, a conventional coal-fired power plant versus the one equipped with post-combustion carbon capture. These may include guaranteeing a fixed carbon price or offering premiums for low-carbon products, which help bridge the gap between the market price of CO_2 and the actual cost of capturing and storing it.

Governments can also increase the demand for CCUS by purchasing low-carbon products from industries that integrate CCUS, by requiring CCUS in public infrastructure projects, and by supporting markets that use captured CO_2, such as enhanced oil recovery (EOR).

In some countries, CCUS networks are treated as critical public infrastructure, much like highways or energy grids, because of their importance in national decarbonization strategies. For example, the Norwegian government co-finances CO_2 transport and storage hubs (as in Norway's Longship Project), and the UK funds shared CO_2 pipelines and storage sites. This approach lowers investment risk and enables economies of large-scale projects by connecting multiple projects through shared infrastructure.

3.5 Corporate Social Responsibility (CSR) and its Role in Advancing CCUS for Sustainable Development

In addition to the government's role in accelerating CCUS through incentives, industries especially high-emission sectors carry a significant responsibility. This responsibility is best understood through Corporate Social Responsibility (CSR), which encourages companies to go beyond regulatory compliance. Corporate Social Responsibility (CSR) reflects a company's commitment to doing business in a way that is ethical, sustainable, and beneficial to society. It goes beyond just making a profit by focusing on protecting the environment and ensuring long-term balance between the economic and environmental aspects of the projects. When it comes to Carbon Capture, Utilization, and Storage (CCUS), CSR becomes even more important because companies using this technology must align their operations with global climate goals and minimize environmental risks, (Fig. 3.5).

CCUS is a key technology in addressing climate change because it can significantly reduce industrial CO_2 emissions. However, it must be deployed carefully to

Fig. 3.5 corporate social responsibility (CSR) Designed by Freepik

avoid causing new environmental problems and to ensure that the carbon reductions are real and meaningful. CSR helps companies integrate transparency and ethical practices into their CCUS projects. To be responsible, companies need to report their results honestly and allow independent verification to confirm their claims. Companies also have a duty to use an advanced monitoring technology to ensure that CO_2 stays securely stored over the long term.

A complete evaluation of the entire CCUS process in terms of the emissions and CSR core values from capturing and transporting CO_2 to using or storing it is also essential. This includes analyzing energy use, land impacts, and potential long-term risks to ensure that the overall environmental footprint is minimized.

Engaging with stakeholders including local communities, regulators, and environmental groups is another central part of CSR in CCUS. Open communication and addressing public concerns help build trust and earn the community support for sustainability of the CCUS operations.

Finally, CCUS projects guided by CSR should support global climate agreements, such as the Paris Agreement, and contribute to net-zero goals. Companies should also push to move beyond simply meeting minimum requirements and have CSR, a central part of corporate strategy.

3.6 Case Studies

Cap-and-Trade Programs: A Global Effort to Cut Emissions.

In 2005, the European Union (EU) pioneered the first global cap-and-trade program, known as the EU Emissions Trading System (EU ETS). This system places restrictions on emissions from various sectors and enables the trading of emission allowances. The EU set a goal to achieve a 21% reduction in emissions from covered sectors by 2020 as a success measure in terms of lowering greenhouse gas emissions.

In the United States, during the Obama administration, a clean energy bill with a cap-and-trade system was proposed in the Congress. The American Clean Energy and Security Act aimed to reduce carbon emissions. While it passed in the House of Representatives, it faced challenges in the Senate due to political issues and a lack of agreement. California established its cap-and-trade program in 2013, known as the California Cap-and-Trade Program, initially focusing on about 400 businesses, including power plants and large industries. The program limits emissions and allows for the trading of allowances. The program achieved its goal of reducing greenhouse gas emissions to 1990 levels by 2020. This is equivalent to 431 million metric tons of carbon dioxide equivalent. They have reached this target three years ahead of schedule, thus demonstrating the effectiveness of cap and trade in driving emission reductions. After that early success California aimed to reduce the emission by 40% and 85% more by 2030 and 2045 respectively (www.lao.ca.gov October 24 2023).

In January 2020, Mexico initiated a unique cap-and-trade pilot program, marking the first of its kind in Latin America. This program is designed to simulate a fundamental policy tool outlined in Mexico's General Law of Climate Change and aligned with the Paris Agreement, with the aim of being fully operational by 2018. By committing to a 22% reduction in greenhouse gas emissions by 2030, Mexico seeks to drive emission reductions and promote the adoption of cleaner technologies through cap and trade. The program covers key sectors like transportation, aviation, electricity generation, and various industries, featuring the voluntary participation of 70–120 entities with the goal of addressing 70 million tons of carbon dioxide equivalent (tCO_2 e).

These examples illustrate the diverse global implementation of cap-and-trade programs, highlighting their potential to reduce greenhouse gas emissions and facilitate the transition to a low-carbon economy.

3.7 Conclusion

This chapter has examined the regulatory and economic mechanisms that govern Carbon Capture and Storage (CCS). Carbon taxation, cap-and-trade systems, and voluntary carbon markets and how each policy influences CCS adoption is discussed.

The carbon tax system, with its price certainty, offers simplicity but faces challenges in setting optimal tax rate, wide discrepancy in prices globally and political resistance. In contrast, cap-and-trade programs provide flexibility through market-driven carbon pricing but risk setting appropriate emission caps and the lack of reliable emissions data. Meanwhile, voluntary carbon markets introduce more flexibility and innovation and involve wide participation from major companies to individuals, however, suffer from lack of enforcement and accountability and plausibility of the credits generated.

Fig. 3.6 images Designed by Freepik

A key takeaway is that no single policy is sufficient. Effective CCS regulation requires multilevel strategies such as government subsidies (e.g., the U.S. 45Q tax credit) and public-private partnerships. International treaties like the Paris Agreement, and COPs can help to regularize carbon accounting across the world. However, discrepancies persist where developing nations often lack infrastructure to participate effectively in carbon markets. Case studies of national policies (Canada's carbon tax, California's cap-and-trade) illustrated how well-designed structures can lead to great success in reducing the greenhouse gas emission moving toward net zero emissions. Looking ahead, the evolution of CCS regulation must address three gaps: (1) standardization of CCS regularization, (2) integration of CCS into broader net-zero policies, and (3) global effective contribution from all the countries around the world (Fig. 3.6).

References

H. Al-Zawaidah, M. Kammerer, D.M. Mitrano, K. Waldschläger, Opportunities and challenges of a cap-and-trade system for plastics. Environ. Sci. Technol. **59**(4), 1887–1896 (2025). https://doi.org/10.1021/acs.est.4c04931

L. Bretschger, E. Grieg, Carbon taxes, CO_2 emissions, and the economy: the effects of fuel taxation in the UK. Energy Policy **195**, 114359 (2024). https://doi.org/10.1016/j.enpol.2024.114359

J. Ewald, T. Sterner, E. Sterner, Understanding the resistance to carbon taxes: drivers and barriers among the general public and fuel-tax protesters. Resour. Energy Econ. **70**, 101331 (2022). https://doi.org/10.1016/j.reseneeco.2022.101331

M.C. Hänsel, M. Franks, M. Kalkuhl, O. Edenhofer, Optimal carbon taxation and horizontal equity: a welfare-theoretic approach with application to German household data. J. Environ. Econ. Manag. **116**, 102730 (2022). https://doi.org/10.1016/j.jeem.2022.102730

J. Hua, K. Wang, J. Lin, Y. Qian, Carbon tax vs. carbon cap-and-trade: Implementation of carbon border tax in cross-regional production. Int. J. Prod. Econ. **274**, 109317 (2024). https://doi.org/10.1016/j.ijpe.2024.109317

X. Li, B. Jiang, Y. Bai, Manufacturers' decision making on emission reduction based on blockchain technology—Analysis under the government's carbon cap-and-trade policy. J. Clean. Prod. **499**, 145224 (2025). https://doi.org/10.1016/j.jclepro.2025.145224

J. Liu, J. Hou, Q. Fan, H. Chen, China's national ETS: global and local lessons, Energy Rep., **8**(Supplement 6), 428–437 (2022). ISSN 2352-4847. https://doi.org/10.1016/j.egyr.2022.03.097

M. Ruth, S.A. Gabriel, K.L. Palmer, D. Burtraw, A. Paul, Y. Chen, B.F. Hobbs, D. Irani, J. Michael, K.M. Ross, R. Conklin, J. Miller, Economic and energy impacts from participation in the regional greenhouse gas initiative: a case study of the state of Maryland, Energy Policy **36**, 6, 2008, 2279–2289, ISSN 0301-4215. https://doi.org/10.1016/j.enpol.2008.03.012

C.-H. Shei, J.C.-E. Liu, I.-Y.L. Hsieh, Distributional effects of carbon pricing: An analysis of income-based versus expenditure-based approaches. J. Clean. Prod. **465**, 141446 (2024). https://doi.org/10.1016/j.jclepro.2024.141446

S. Suk, S. Lee, Y.S. Jeong, The Korean emissions trading scheme: Business perspectives on the early years of operations. Clim. Pol. **18**(6), 715–728 (2018). https://doi.org/10.1080/14693062.2017.1346499

K. Thuerbach, N. Johnson, Analyzing California carbon allowance prices: a case for the rate-adjusted Price floor. Commodity Insights Digest (2024) bayes-cid.com

J. Wei, Y. Wang, Y. Liu, Installation and pricing decision in a two-channel supply chain under carbon cap-and-trade mechanism. Sustain. Futur. **10**, 100876 (2025). https://doi.org/10.1016/j.sftr.2025.100876

Chapter 4
Fundamentals of Combustion and CO_2 Capture Technologies

4.1 Chapter Overview and Objectives

4.1.1 Overview

Combustion is widely used in industries and power generation contributes significantly to carbon dioxide (CO_2) emissions. To address this negative environmental impact of combustion, several CO_2-capturing technologies have been developed, including pre-combustion capture, oxyfuel combustion, and post-combustion capture, to prevent CO_2 emissions release to the atmosphere (Ahmed et al. 2020). These technologies aim to capture and store CO_2 emissions from combustion processes, reducing their impact on climate change. In this chapter, we will explore each technology in detail, examining their mechanisms, efficiency, scalability, and challenges. The chapter also discusses the concept of CCS-readiness in combustion technologies, with a special focus on innovations that facilitate CO_2 capture from existing combustion systems.

By the end of the chapter, you will have a comprehensive understanding of various carbon capture technologies and their capacity to contribute toward meeting the goals set by the Paris Agreement.

4.1.2 Learning Objectives

Upon completion of this chapter, you will be able to:

1. Explain the concept of combustion and its role in CO_2 emissions.
2. Explain the principles and mechanisms behind pre-combustion capture, oxyfuel combustion, and post-combustion capture technologies.

E. Fathi and F. Belyadi, *Subsurface CO_2 Sequestration*,
https://doi.org/10.1007/978-3-032-28296-5_4

3. Analyze the advantages and limitations of each CO_2 capturing technology in terms of efficiency, cost, and scalability.
4. Evaluate the potential impact of implementing these technologies in reducing CO_2 emissions and mitigating climate change.

4.2 Introduction to Combustion and CO_2 Capture Technologies

Industrial activities require a significant amount of energy that is traditionally provided through combustion of fossil fuels. As of 2025, fossil fuels support more than 70% of energy consumption. However, this dependency on combustion has a substantial environmental impact due to the significant amount of CO_2 emissions. Therefore, understanding combustion and its different types, such as complete and incomplete combustion, becomes essential in finding ways to reduce emissions or find mitigation strategies. This chapter is significantly important as we explore the role of combustion in various industries and power generation, and its contribution to carbon dioxide (CO_2) emissions. It is also critical to gain knowledge on different CO_2-capturing technologies and their capacity and readiness to remediate CO_2 emissions.

Throughout this chapter, we will review three critical CO_2 capture technologies including pre-combustion capture, oxyfuel combustion, and post-combustion capture. These innovative approaches, which aim to capture and store CO_2 emissions from combustion processes, are required in minimizing CO_2 emissions and their negative impact on our environment (Fig. 4.1).

Fig. 4.1 Combustion designed by Freepik

4.3 Fundamentals of Combustion

Combustion is a fundamental process through which the primary energy stored in fuels such as fossil fuels, biomass or hydrogen is converted into a usable energy form such as heat and light. It involves a chemical reaction between a fuel and an oxidizer, commonly atmospheric oxygen. This reaction also generates reaction products such as CO_2, water or other pollutants. For example, when methane (CH_4), the main component of natural gas, undergoes combustion with oxygen, the resulting products are carbon dioxide (CO_2) and water (H_2O), while when hydrogen (H_2) reacts with oxygen (O_2), the only byproduct is water (H_2O).

In most cases, combustion involves hydrocarbons, and the required oxygen is typically sourced from the air, which consists of approximately 79% nitrogen and 21% oxygen. To prevent the release of CO_2 into the atmosphere during combustion, an additional step is needed. This involves the use of specialized technologies that can selectively separate CO_2 from the combustion gases while leaving nitrogen, oxygen, and water unaffected. This process ensures that the captured CO_2 can be properly managed, such as through carbon capture and storage (CCS) techniques, to prevent its release into the environment.

Understanding the details of combustion and the associated emissions helps to explore strategies to optimize the CO_2 capture technologies and minimize the emissions. The development and implementation of effective carbon capture technologies are necessary for achieving significant reductions in greenhouse gas emissions.

4.3.1 *Complete Combustion*

In Fig. 4.2, we can observe the process of complete combustion of fossil fuel where a hydrocarbon reacts with an excess amount of oxygen (O_2). The result is the generation of heat and light, along with the formation of water (H_2O) and carbon dioxide (CO_2). The following equation shows the complete combustion of propane with excess oxygen where 1 mol of propane (C_3H_8) combusted with 5 moles of oxygen results in 3 moles of CO_2, 4 moles of water, and heat.

$$C_3H_8 + 5O_2 \rightarrow 3CO_2 + 4H_2O + Heat$$

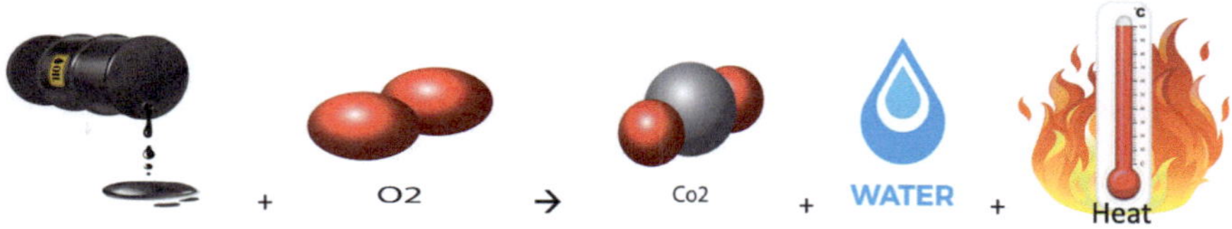

Fig. 4.2 Complete combustion of hydrocarbon with excess oxygen modified (Designed by Freepik)

If different fossil fuels such as methane or gasoline are used as fuel different amount of oxygen is required for full combustion. The amount of CO_2 emissions will also differ as the fuel type changes. Following are some examples for complete combustion of methane, gasoline and ethanol.

$$CH_4 + 2O_2 \rightarrow CO_2 + 2H_2O + Heat \rightarrow \text{Methane combustion}$$

$$2C_8H_{18} + 25O_2 \rightarrow 16CO_2 + 18H_2O + Heat \rightarrow \text{Gasoline combustion}$$

$$C_2H_5OH + 3O_2 \rightarrow 2CO_2 + 3H_2O + Heat \rightarrow \text{Ethanol combustion}$$

In the case of hydrogen combustion, the only reaction product would be water and heat as follows:

$$2H_2 + O_2 \rightarrow 2H_2O + Heat \rightarrow \text{Hydrogen combustion}$$

In hydrogen combustion in addition to enough hydrogen, high temperatures are required to make sure the full breakdown of fuel molecules, and it is essential to provide good mixing of fuel with oxygen.

4.3.2 Incomplete Combustion

If there is an insufficient supply of oxygen, incomplete combustion occurs, leading to the formation of carbon monoxide (CO), pure carbon particles, other pollutants, and water (H_2O). This is not the case if hydrogen is used as a fuel, where low oxygen will result in unburned hydrogen, which might just be wasted. In the following equation, we can see an example of incomplete combustion due to lack of sufficient oxygen. Initially, propane (C_3H_8) reacts with an insufficient amount of oxygen (3/2 O_2), resulting in the formation of three molecules of carbon monoxide (3CO) and four molecules of water (4 H_2O). This is an incomplete combustion reaction, and can occur when there is a lack of oxygen supply.

To address the incomplete combustion, propane is then reacted with an increased amount of oxygen ($2O_2$) in the second part of the equation. This results in the formation of three molecules of carbon (3C) and four molecules of water ($4H_2O$). This reaction shows the consequences of insufficient oxygen, leading to the production of carbon particles instead of carbon dioxide. This illustrates the importance of ensuring complete combustion, where sufficient oxygen is available to minimize the production of harmful byproducts such as carbon monoxide. It highlights the need for proper control of oxygen supply in combustion processes to maximize energy efficiency and reduce emissions.

$$C_3H_8 + 3\frac{1}{2}O_2 \rightarrow 3CO + 4H_2O$$

$$C_3H_8 + 2O_2 \rightarrow 3C + 4H_2O$$

4.3.3 *Combustion Natural Gas Versus Coal*

Within the different sources of fuel for power generation, coal-fired and gas-fired power plants are fundamental in providing electricity. These two technologies significantly differ in terms of their CO_2 emission intensity. Coal-fired plants are historically associated with much higher carbon dioxide emissions compared to gas-fired power plants. Figure 4.3 shows a gas-fired power plant in Duisburg-Huckingen, including 2 units generating 320 megawatts of electrical power. This plant utilizes blast furnace gas, coke oven gas, and uses natural gas as its fuel source. On the right, Figure 4.3 shows a coal-powered facility equipped with three generating units, each capable of producing 750 megawatts of electricity with a smokestack emitting smoke into the air. The Navajo Generating Station, near Page, Arizona, burns approximately 25,000 t of coal daily to meet energy demands till 2019. The visual comparison of the two images highlights the considerable difference in the amount of CO_2 emissions.

4.3.3.1 Environmental Impact of Coal and Gas-Fired Power Plants

Table 4.1 illustrates the approximate byproducts generated after combustion in coal-fired and gas-fired power plants with and without CCS technologies implemented. In coal-fired power plants without CCS technology, approximately 13.5% of the byproducts are CO_2, while in gas-fired power plants, it accounts for around 4.0%. The primary byproduct in both cases is nitrogen (N_2), with coal-fired power

Fig. 4.3 On the left, Heizkraftwerk Mitte (Natural Gas combined heat and power plant), and right The Navajo Generating Station, near Page, Arizona. (left) By Georg Slickers, (right) By Alan Stark

plants producing 68.1% N_2 and gas-fired power plants producing 74.3% N_2. Other byproducts include water (H_2O), oxygen (O_2), and argon (Ar).

Table 4.1 provides insight into the composition of byproducts, coal emits 3 times more CO_2 in compared to gas; it also produces more water. The gas plants use more air (O_2) for efficiency therefore more N_2 is produced. The amount of argon is negligible in both cases. It is also important to note that gas-fired power plants produce 80% less nitrogen oxide and near zero sulfur dioxide with no ash that takes 5–15% of coal mass in the coal-fired power plant. When CCS Technology is implemented a significant reduction in CO_2 emissions from both gas-fired and coal-fired power plants was observed. In coal-fired and gas-fired power plants the CO_2 emission reduction can reach somewhere between 80–90%, with coal-fired power plants showing slightly higher reduction due to more concentrated CO_2 emissions in exhaust.

In Table 4.2, we compare the CO_2 emissions per kilowatt-hour (kWh) of electricity generated from both coal-fired and gas-fired power plants. Without carbon capture and storage (CCS) technology, coal-fired power plants produce higher CO_2 emissions, ranging from 800 to 1000 kg/kWh. In contrast, gas-fired power plants emit lower CO_2, ranging from 350 to 400 kg/kWh. When CCS is implemented, the CO_2 emissions from coal-fired power plants can be reduced to approximately 100 kg/kWh, while gas-fired power plants with CCS can achieve emissions of around 40 kg/kWh. This shows around a ten-fold reduction in CO_2 emissions from the most aggressive sources of environmental concern. In summary, Table 4.2 demonstrates that coal-fired power plants emit higher levels of CO_2 compared to gas-fired power plants, with CCS playing a significant role in reducing CO_2 emissions for both types.

Table 4.1 byproducts generated after combustion in coal-fired and gas-fired power plants

	CO_2%	H_2O %	O_2%	N_2%	Ar%
Coal-fired plant	13.5	15.2	2.4	68.1	0.8
Gas-fired plant	4.0	8.7	12.1	74.3	0.9
Coal-fired plant with CCS	1–4	16–18	2–3	75–80	0.9
Gas-fired plant with CCS	0.5–1.5	10–12	12–13	80–85	1

Table 4.2 CO_2 emissions with and without CCS comparison in gas- and coal-fired plants

CO_2 Emissions (kg/MWh)		
	Without CCS	With CCS
Coal-fired plant	800–1000	~100
Gas -fired plant	350–400	~40

4.4 Overview of CO_2 Capture Technologies

4.4.1 Overview and Technologies

CO_2 capture technologies such as pre-combustion, oxyfuel combustion, and post-combustion capturing technologies are fundamental in reducing greenhouse gas emissions by capturing and storing CO_2 that would otherwise be released into the atmosphere. They provide various approaches to mitigate CO_2 emissions across industries such as power generation, cement production, steel, and chemical manufacturing. Scaling up the implementation of these technologies is required for meeting climate goals and transitioning toward a low-carbon future. Figure 4.4 shows the emitted CO_2 from power plants that is collected using direct air capture technology and stored underground as one mitigation plan for fighting against power plant emissions. However, other technologies like pre-combustion, post-combustion, and oxyfuel might result in more economical and efficient emission reduction strategies to reduce power plants' negative environmental impacts.

Pre-combustion capture technology works by capturing CO_2 before the combustion process. It does this by converting fossil fuels into synthesis gas (syngas), composed of hydrogen and carbon monoxide (Kumar and Ravikumar 2010). The CO_2 is then separated from the syngas, leaving behind hydrogen as a clean fuel. This method is commonly employed in integrated gasification combined cycle (IGCC) power plants and is known for its high efficiency in CO_2 capture.

In oxyfuel combustion, fuel is burned in an oxygen-rich environment instead of air, creating a flue gas mainly consisting of carbon dioxide and water vapor. The CO_2 can be separated from the water vapor using various technologies like condensation or

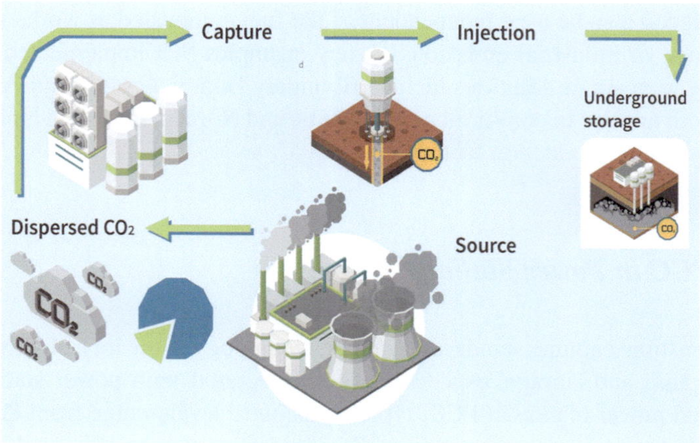

Fig. 4.4 Schematic of CCS technology using direct air capture technology and underground storage (Designed by Freepik)

absorption. This separation results in a concentrated CO_2 stream that can be captured and stored.

Post-combustion capture involves capturing CO_2 from the flue gases released during the combustion process. Different techniques, such as absorption using solvents, adsorption onto solid materials, or membrane-based separation, are used to separate CO_2 from the flue gas. Separated CO_2 then can be compressed and transported for storage or utilization (Ahmed et al. 2020).

4.5 Post-Combustion Capture (PCC)

4.5.1 Overview and Technologies

Post-Combustion capture is a CO_2 capture technology that separates carbon dioxide from flue gases after combustion. Various techniques can be employed for post-combustion capture, including CO_2 absorption using solvents, membrane technology, or distillation. CO_2 absorption with solvent technology uses amine solutions, a group of chemicals mixed with water, to selectively capture CO_2 from flue gases like nitrogen. Amine solvents have been used since the 1930s and are widely utilized in the oil and gas industry for CO_2 capture (Soo et al. 2024). In some cases, solvent technology is employed to separate natural gas mixed with high concentrations of CO_2, enabling purification for commercial use (Mohd Pauzi et al. 2022). The PCC has a significant advantage over other CO_2 capture technologies, as it can be added to existing facilities without major modifications in the existing infrastructure. This means that the technology can be implemented in existing power plants. This advantage makes PCC a viable option for reducing carbon emissions from already existing facilities. It can also be used independent of the fuel combusted; it works with coal, gas, biomass, or industrial emissions. A few examples that implemented the PCC technology with amine solutions include Boundary Dam in Canada and Petro Nova in USA, both applied to coal-fired power plants, and Norcem Brevik in Norway, that is a major cement production factory.

4.5.2 PCC in Power Stations

Post-combustion capture stands as a major technology with lots of flexibility in carbon capture and storage, especially when integrated with power stations such as coal-fired power plants. In PCC, typically capture levels range from 80 to 90%, which indicates a significant portion of CO_2 emissions can be effectively captured and prevented from entering the atmosphere. However, it is possible to achieve even higher capture levels of 98–99% at higher cost (Karayil et al. 2024). Once the CO_2 has been effectively separated from the nitrogen, oxygen, and water molecules in the flue gas, the remaining exhaust gas is released into the atmosphere through the power plant stack. The captured CO_2 is then compressed by increasing its pressure to reach the supercritical state and then transported for geological storage. Compression enables the CO_2 to be efficiently transported to suitable storage sites. It is important to note that the capture process and the subsequent compression of carbon dioxide require energy input, which is derived from the power plant's electricity and heat generation. This is one of the challenges and limitations of PCC as the process of CO_2 separation is very energy intensive and might take 20–30% of power plant output (Hou et al. 2022). This will result in high cost of CO_2 capture. The effective PCC needs to achieve high CO_2 capture rates, high compression efficiency, and optimum energy consumption from the power plant. Currently, the most common CO_2 capture technology used in power plants with PCC technology is the application of amine solutions. The cost of amine solutions, solvent degradation during CO_2 desorption, and reactions with nitrogen oxide and sulfur dioxide are other challenges that impact the cost of implementing PCC.

4.5.2.1 The Los Medanos Energy Center by Calpine

As stated earlier, one of the challenges of CO_2 separation using an amine solution is the extremely high energy consumption of this process. Efforts have been made to come up with lower energy and higher CO_2 capturing capacity using amine solutions. Application of Advanced Amine-Based Solvents (AABS) is one of these efforts. In a collaboration between Linde industrial gases and an engineering company with BASF chemical company, a new advanced amine solution for CO_2 capturing was developed. The solution is designed for natural gas combined cycle (NGCC) power plants. Other solvents have been developed with higher CO_2 capture efficiency and more cycle times, such as ION ICE solvent. Figure 4.5 shows The Los Medanos Energy Center by Calpine, a commercially operational natural gas combined cycle (NGCC) power plant that serves as the location for testing ION Clean Energy's solvent technology at an engineering scale.

4.5.3 Ideal Solvent for PCC

When implementing carbon capture using chemical solvents at an industrial scale, it is essential to be able to process massive volumes of flue gases. This is not possible if enough contact between the flue gasses and the solvent is not achieved. To accomplish this, specific engineering devices are designed to efficiently facilitate the interaction between millions of cubic meters of gases and an equivalent amount of solvent. In addition to the contact between flue gasses and solvent the reaction rate between these two should be high enough for effective separation, therefore another key factor for successful carbon capture is fast reaction kinetics and mass transfer. This enables the solvents to effectively absorb and capture CO_2 from the gas stream.

Fig. 4.5 The Los Medanos energy center by Calpine, Nathan Fine, DE-FE0031950, DOE, NETL, August 13, 2021

Another important aspect is the high adsorption capacity of the solvents as the goal is to minimize the cost of CO_2 capture by minimizing the required solvent volume (Mota-Martinez et al. 2017). Solvents with a high adsorption capacity can capture a significant amount of CO_2, allowing for smaller equipment sizes and lower solvent flow rates, resulting in more efficient carbon capture processes. Utilizing solvents with good reaction enthalpy properties can lead to reduced energy consumption during the carbon capture process. Optimizing the thermodynamic characteristics of the solvents will result in energy requirement minimization that can make the overall process more cost effective. High thermal stability and low solvent degradation are major considerations for carbon capture using CO_2 absorption with solvent technology as CO_2 desorption from the solvent mainly drives with applying temperature to break the chemical bonds between the solvent and CO_2. Solvents that exhibit these properties can withstand high temperatures and chemical reactions without significant degradation and this will result in reduced solvent waste and longer operational lifetimes of solvents.

The cost of solvents is a significant factor in large-scale carbon capture operations. Utilizing solvents with low costs can contribute to the economic viability of the carbon capture process, making it more feasible for implementation on an industrial scale. Finally, it is important to note that the chosen solvents should have no negative environmental impact. The selection of environmentally friendly solvents ensures that the carbon capture process itself does not introduce additional harm to the environment.

In Fig. 4.6 the animation simplifies the use of a solvent in the Post-Combustion Capture (PCC) process. This process, outlined in a recent study featuring EEMPA, a solvent developed by Pacific Northwest National Laboratory, demonstrates how carbon is captured from flue gas emissions in power plants. EEMPA (depicted in

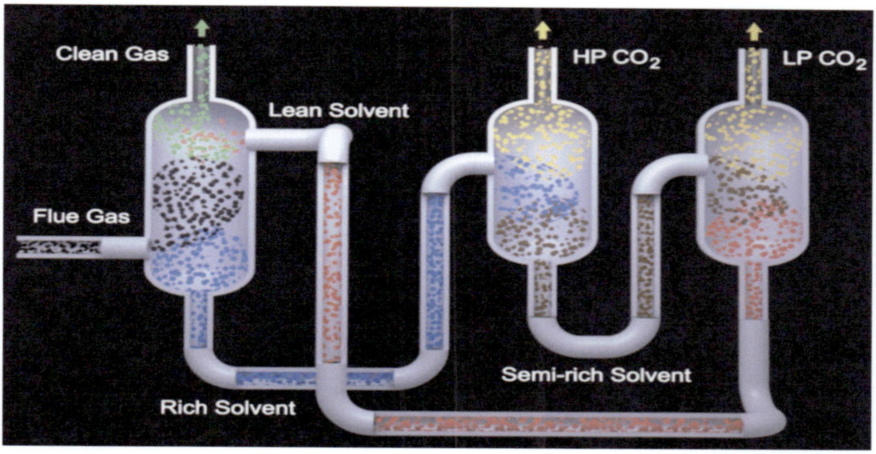

Fig. 4.6 The animation showcases the post-combustion capture (PCC) process utilizing a solvent. [*Source* https://www.pnnl.gov/sites/default/files/media/image/CleanedCO_2b.gif]

red) enters the reactor (on the left) from the top, interacting with the flue gas (shown in black) entering from the bottom to absorb carbon dioxide. The clean gas (without CO_2) is then removed from the top of the reactor (shown in green). As a saturated CO_2 solvent (displayed in blue), EEMPA undergoes carbon dioxide removal in high- and low-pressure tanks using heat stream. Semi-rich and lean solvents are prepared to reenter the carbon dioxide absorber. The cycle completes by reintroducing the stripped solvent to the carbon dioxide absorber, restarting the process. (Animation by Michael Perkins | Pacific Northwest National Laboratory) [Source: https://www.pnnl.gov/sites/default/files/media/image/CleanedCO_2b.gif].

4.5.4 PCC with Solvents

To achieve the high contact area between the solvent and flue gases in the post-combustion capture process, the structured packing commonly used in the chemical industry is used. In structured packing, a repeating geometric pattern is used that provides multiple channels for both gas and liquid to flow through and get in contact. These structured packing elements are installed within large vessels with diameters and heights of tens of meters. This configuration allows for the efficient capture of millions of tons of carbon dioxide from combustion gases (Zhao et al. 2011). Figure 4.7 shows the structured packing used for distillation columns like those used in the PCC process.

The process of post-combustion capture involves a continuous loop of the chemical solvent, adsorbing and desorbing CO_2 from flue gases. On one side, the solvent

Fig. 4.7 Structured packing for distillation column, photo taken by Luigi Chiesa

absorbs carbon dioxide from the combustion gases, while on the other side, the CO_2 is desorbed from the solvent as detailed in Fig. 4.6.

The amine-based solvents are the most common chemical solvents used in a continuous loop, capturing and releasing carbon dioxide. Typically, monoethanolamine (MEA) is used, which is highly effective in capturing 85–90% of the CO_2 from the flue gas emitted by coal-fired power plants (Du et al. 2021). However, the reaction between CO_2 and MEA leads to the formation of a carbamate salt, limiting its loading capacity to 0.5 mol of CO_2 per mole of the amine. The viscosity of MEA's aqueous solution is similar to that of water, but MEA is a relatively volatile compound, which means that there is a tendency for partial vaporization of the amine and subsequent release into the environment. This means the amine solution needs to be added to the closed loop to compensate for loss of amine solution due to vaporization. Moreover, the process of desorbing the CO_2 from the solvent requires a significant amount of energy to break the stable carbamate bond (approximately 85 kJ mol^{-1}). To overcome these problems—lower CO_2 capacity and higher energy requirements—different solvents are studied for use in the PCC process, such as secondary and tertiary amines (Gómez and Gutiérrez 2025).

Secondary amines, such as diethanolamine (DEA), and tertiary amines, such as triethanolamine (TEA) and N-methyl-diethanolamine (MDEA), have been extensively studied for their potential use in carbon capture (Chen et al. 2015; de Ávila et al. 2015). These secondary and tertiary amines exhibit a slower reaction rate and require less energy for regeneration; however, they show higher capacity to adsorb CO_2 (Sartori and Savage 1983). The reason behind this behavior lies in the relatively low stability of the carbamate salt, which results in higher concentrations of free amine in the solution, influencing their performance.

Researchers have also explored alternative solvent classes, including blended solutions, biphasic solvents, poly(ethylene glycol) ethers, ionic liquids, and deep eutectic solvents, as potential options for carbon capture (Li et al. 2024). Some commercially deployed solvents, such as Shell's Cansolv technology, claim impressive results with an energy requirement of only 2.3 gigajoules per ton of CO_2 (Singh and Stéphenne 2014). Similarly, Fluor's Econoamine and MHI's hindered amine KS-1 demonstrate comparable performance in this regard (Davison 2007). Scientists at the Pacific Northwest National Laboratory have developed an innovative and cost-effective carbon capture and conversion system. This breakthrough has significantly reduced the cost of capturing CO_2 to $39 per metric ton, making it the most affordable system to date. The process involves extracting flue gas from power plants, utilizing a PNNL-patented solvent to remove CO_2, and subsequently converting the captured CO_2 into industrially useful methanol (Brendan Bane and PNNL 2023).

4.5.4.1 Technology Centre Mongstad (TCM) in Norway

Figure 4.8 displays the Technology Centre Mongstad (TCM) in Mongstad, Norway, launched in 2012 and recognized as the world's largest open access test center for carbon capture technologies. It is used for conducting engineering-scale tests under real flue gas conditions for post-combustion CO_2 capture technology (Koeijer et al. 2011). The flue gas is sourced from the Mongstad refinery and gas power plant. Some newly designed non-amine solvents, such as ION clean energy, are tested in this facility (Erik Meuleman 2022). The lack of a permanent CO_2 storage site limited the TCM facility from performing full commercial testing; however, the facility is a great example of a path toward decarbonizing the major pollutant power industry.

Fig. 4.8 The national energy technology laboratory partners with technology centre mongstad (tcm) in Norway. *Source* NETL.doe.gov (Jak Tanthana, DE-FE0032218, 2023)

4.5.4.2 Gas Sweetening

A procedure similar to post-combustion capture can also be applied to remove H_2S (hydrogen sulfide) as well as CO_2 in a process called gas sweetening, where both H_2S and CO_2 are removed from the sour gas stream. Sour gas is a natural gas or any other gas mixture that contains H_2S and often CO_2. Amine-based solvents are commonly used in an absorption-regeneration loop for carbon dioxide and hydrogen sulfide capture. The process involves several steps as shown in Fig. 4.9 including:

1. Absorption: The industrial gas (e.g., sour gas), containing carbon dioxide, is introduced into an absorber. Inside the absorber, the amine-based solvent reacts with carbon dioxide and hydrogen sulfide and captures both gases from the gas stream.
2. Transfer to Regenerator: The solvent, now enriched with CO_2 and H_2S, is transferred to a regenerator. In the regenerator, the temperature is raised to approximately 120 degrees Celsius. This elevated temperature facilitates the carbon dioxide and hydrogen sulfide desorption from the amine solution.
3. Regeneration: As the temperature increases, the amine solvent releases the captured carbon dioxide and hydrogen sulfide. This separation allows for restoring the amine solution to be reused in the closed loop.
4. The removed H_2S and CO_2 are then passed through a condenser and extracted as acid gas.
5. Fresh Solvent and Cycle Restart: After the regeneration process, the solvent, now free from CO_2, is considered "fresh" and can be recycled back to the absorber. This enables the solvent to begin the absorption process again, continuing the cycle of capturing carbon dioxide from the industrial gas.

4.5.5 PCC with Cryogenic Distillation

Cryogenic carbon capture process is a modern technology that involves a multi-step process to separate and recover CO_2 from combustion gases through cryogenic distillation. This technique is highly effective in removing over 90% of CO_2 from combustion gases (He et al. 2023). The process begins by cooling the combustion gas to extremely low temperatures, causing the CO_2 to solidify. This solidified CO_2 is then separated from other light gases and compressed at high pressures. Through a series of compression, refrigeration, and separation steps, the components of the gas mixture can be separated (Shen et al. 2022; Oyewole et al. 2023). What makes cryogenic distillation unique is its chemical-free operation, which eliminates the need for chemicals such as solvent supply and the risk of secondary pollution during solvent production or evaporation. This process offers a high CO_2 recovery rate, typically reaching 90–95% of the flue gas. However, the cryogenic distillation process is significantly energy intensive due to the need for extremely low temperatures and high pressures. It is estimated to consume around 600–660 kWh per ton of

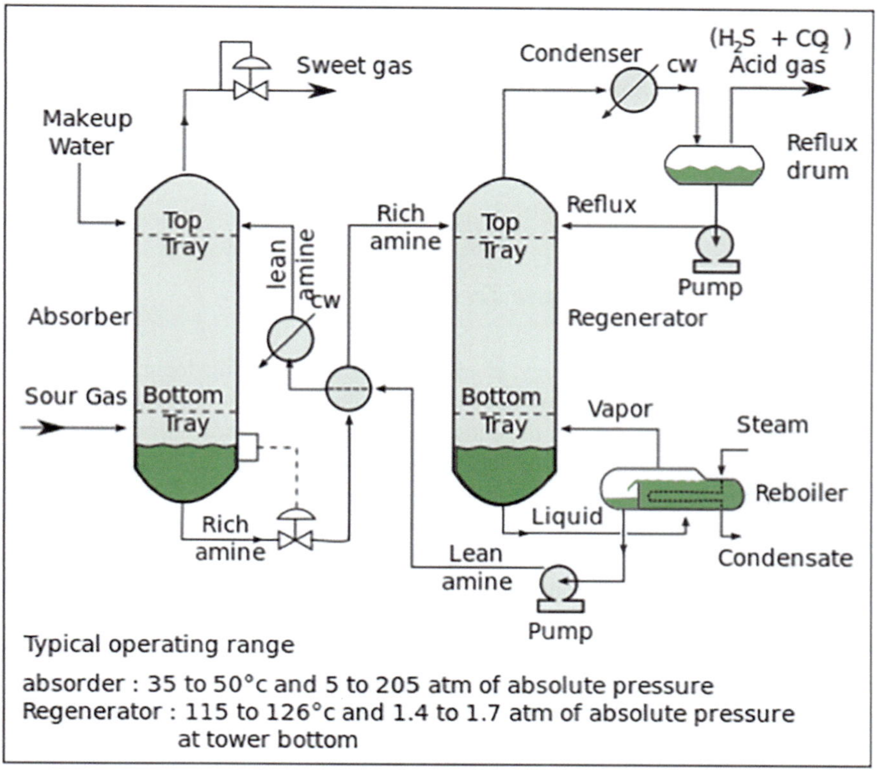

Fig. 4.9 Diagram illustrating the typical amine treating process employed in petroleum refineries, natural gas processing plants, and various industrial facilities. (Attribution: Raminagrobis)

CO_2 recovered in liquid form (Leung et al. 2014). Despite its energy requirements, cryogenic distillation offers advantages in terms of CO_2 transport. The captured CO_2 is obtained in liquid form at low pressure, making transportation easier and more efficient. This eliminates the need for compressors typically used in other carbon capture methods like solvent or membrane separations. Although cryogenic carbon capture technology is still relatively young, it is actively being developed and holds promise for large-scale implementation. Currently, there are only a few operational CCS projects utilizing cryogenic distillation, such as Cryocap™ and Linde & BASF's Cryogenic CO_2 Purification (Air Liquide 2022; Linde plc. 2019).

The Post-Combustion Capture (PCC) with Cryogenic Distillation and Oxygen production begins with an air compressor that pressurizes atmospheric air. The compressed air then passes through the heat exchanger, where it is cooled to lower temperatures. The cooled air passes through a separator, where CO_2 is solidified and separated from the stream. The air then enters an expansion chamber, where pressure and temperature further drop. In the distillation column, the air is further cooled to cryogenic temperatures. The components of air, such as nitrogen, oxygen, and argon, separate based on their different boiling points. The oxygen, which has been separated and purified, is collected in an oxygen container. It is then prepared for transportation and application in different industries. While capturing CO_2 from flue gases, the system also facilitates the production and separation of oxygen from the air. This combination demonstrates the potential for simultaneous CO_2 capture and oxygen production.

4.5.6 PCC with Membrane Technology

PCC with membrane technology is evolving as a promising technology where membranes act as filters that allow specific molecules to pass through while blocking others. The selectivity of gases depends on factors such as molecular size, affinity to the membrane material, and molecular weight (Rezaei et al. 2023). Membrane technology has already found extensive applications in various liquid separations, including micro-filtration, ultra-filtration, reverse osmosis, forward osmosis, desalination, and medical applications (Van der Bruggen 2018). Its flexibility and efficiency make it an attractive option for carbon capture technology. The membrane area required for CO_2 removal depends on various parameters, such as the purity of the gas stream and the temperature. To achieve desired separations while minimizing membrane area, CO_2 removal is typically performed using one or two membrane stages. The gas stream from the first stage is further processed in the second stage to recover additional carbon dioxide. Figure 4.10 by Ji and Zhao (2017) (illustrates the feed gas passing through the membrane, which consists of selective top layers, interlayers, and substrates. This illustration represents the selective permeation of gases through the membrane and highlights the role of different membrane layers in achieving the desired separation. Membrane technology offers a selective approach for CO_2 removal in the PCC process. Polymeric or inorganic membranes are used to capture carbon dioxide from gas mixtures like those from coal gasification or steam reforming. Inorganic membranes are especially beneficial in cases where elevated temperatures or the presence of corrosive gases are very common, such as CO_2 removal from integrated gasification combined cycle (IGCC) power plants. Even though inorganic membranes are very costly, their extended lifespans and consistent performance under harsh environments justify their initial cost. Furthermore, inorganic membranes are shown to be more efficient and effective in carbon dioxide capture from the gas stream in IGCC power plants. Other options, which are cheaper and are readily available, such as polymeric membranes, are facing major limitations when it comes to CO_2 capture in large-scale power production due to their poor high-temperature stability. Even though the CO_2 capture using a commercially available membrane technology offers an innovative solution for various power plant off gases, it also faces a major drawback due to the higher energy penalties imposed on power generation efficiency compared to conventional chemical absorption processes.

Figure 4.11 shows size sieving, surface diffusion, solution diffusion, and facilitated transport mechanisms involved in separation mechanisms in PCC with

Fig. 4.10 (Left) Utilizing membrane separation for binary gas mixtures. (Right) The diagram illustrates the asymmetric structure of a membrane coated on a substrate, showcasing the technology discussed in "Membrane Separation Technology in Carbon Capture" by Ji and Zhao (2017)

membrane technology. In the size-sieving separation mechanism, the membrane acts as a physical barrier with pores of specific sizes. Smaller molecules, such as CO_2, can pass through the pores, while larger molecules are retained. This mechanism relies on the molecular size difference between gases to achieve separation. The size-sieving separation mechanism is simple to implement and relatively energy efficient, but it is limited by the size difference between gases, and pore blockage may occur over time.

Surface diffusion separation mechanism occurs when gas molecules selectively adsorb and desorb on the surface of the membrane. This mechanism exploits differences in the affinity of gases for the membrane surface, allowing certain gases, like CO_2, to diffuse across the membrane while others are denied. The surface diffusion separation mechanism is selective based on gas affinity for the membrane surface and can operate under mild conditions. However, it has relatively slow diffusion rates and is subject to inefficiency with time due to adsorption pollution.

In solution diffusion, gas molecules dissolve in the membrane and then diffuse through it. The driving force for separation is the difference in solubility and diffusion rates of gases within the membrane material. This mechanism relies on the preferential dissolution and transport of CO_2 through the membrane. The solution diffusion separation mechanism is effective for gases with varying solubilities and allows for separation based on both solubility and diffusion rates. However, it is energy intensive due to the need for dissolving and desorbing gases and is limited by the solubility of gases in the membrane.

Facilitated transport involves the use of specialized molecules, called carriers, within the membrane. These carriers selectively bind to CO_2 molecules and assist their transport across the membrane. This mechanism enhances the permeability of CO_2 and improves the separation efficiency by leveraging the specific interaction between the carrier molecules and CO_2. Finally, facilitated transport is highly selective and efficient for CO_2 separation and can operate under mild conditions. However, it may require periodic replacement of carrier molecules and is sensitive to impurities that may interfere with carrier function (Ji and Zhao 2017).

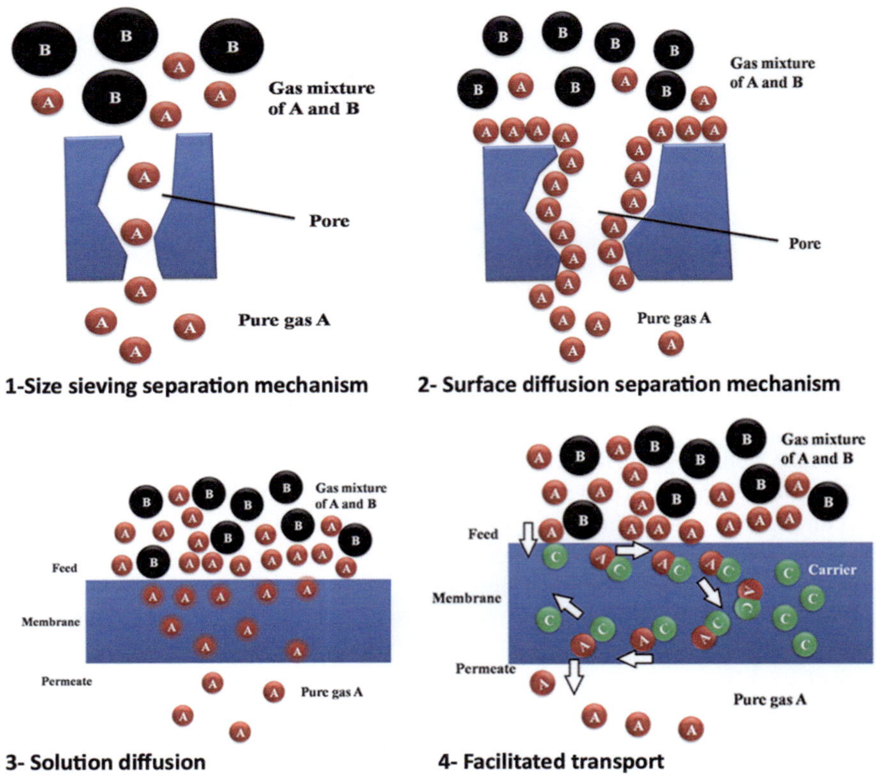

Fig. 4.11 Mechanisms including size sieving, surface diffusion, solution diffusion, and facilitated transport in PCC with membrane technology Source: Membrane Separation Technology in Carbon Capture by Ji and Zhao (2017)

When comparing different membrane separation mechanisms, we consider factors such as energy efficiency, selectivity, process complexity, and applicability. In terms of energy efficiency, size sieving and surface diffusion are generally more energy efficient, while solution diffusion may require more energy. The efficiency of facilitated transport depends on the characteristics of the carrier. Moving on to selectivity, both facilitated transport and surface diffusion offer high selectivity. Size seiving is selective based on size difference, while solution diffusion achieves selectivity based on solubility. Considering process complexity, size sieving is straightforward, while solution diffusion and surface diffusion are moderately complex. Facilitated transport, involving specialized carriers, may be more complex. Regarding applicability, both size sieving and surface diffusion are flexible and applicable in various scenarios. Solution diffusion is effective for gases with different solubilities, and facilitated transport is highly effective in CO_2 separation. This analysis helps in understanding the strengths and limitations of each separation mechanism for diverse applications. Each mechanism has its own set of advantages and disadvantages, and the choice depends on the specific requirements and conditions of the separation process. However, for CO_2 removal in PCC, the most common technique, especially in large-scale industrial applications, is amine-based absorption, as discussed earlier. Other capture techniques, such as cryogenic distillation, adsorption, and membrane separation, are also used in specific applications, but amine-based absorption remains a dominant method for large-scale CO_2 capture from industrial sources. The choice of technique often depends on factors such as the concentration of CO_2 in the flue gas, the specific industry, and economic considerations.

4.6 Oxyfuel Combustion

4.6.1 Overview and Technologies

Oxyfuel combustion is a promising technology for carbon dioxide capture in power plants equipped with CCS technology. This involves burning fuel with almost pure oxygen instead of air. The main goal here is to generate a flue gas with a high concentration of CO_2 and water vapor, which can then undergo separation through dehydration and low-temperature purification processes (Tak et al. 2010; Fernández-Miranda et al. 2016; Stanger et al. 2015; Cho et al. 2010).

In comparison to air-fired combustion, oxyfuel combustion produces a significantly smaller volume of flue gas, containing approximately 70 percent CO_2 by volume. This reduced gas volume simplifies the removal of pollutants like sulfur oxide (SOx), nitrogen oxide (NOx), mercury, and particulates from the flue gas (Buhre et al. 2005). Additionally, since nitrogen is eliminated from the air during oxyfuel combustion, the production of NOx is significantly lowered, contributing to improved air quality. Oxyfuel combustion has the advantage of producing a concentrated CO_2 stream that facilitates the subsequent capture and storage. This technology in power plants can effectively reduce greenhouse gas emissions and mitigate their environmental impact. Economic development and implementation of oxyfuel combustion can further advance the feasibility of large-scale CCS deployment.

Oxyfuel combustion is a power generation process that involves several key units working together to enable CO_2 capture. These units are necessary for efficient and effective operation of the oxyfuel combustion system. The Air Separation Unit (ASU) plays a major role in producing high-purity oxygen required for the combustion process. It separates oxygen from atmospheric air, typically utilizing technologies such as cryogenic distillation or membrane separation. The produced oxygen is then supplied to the combustion unit. In a Boiler or Gas Turbine unit, the fuel (e.g., coal or natural gas) is burned with nearly pure oxygen instead of air in an atmospheric or pressurized condition, depending on the design. The combustion process generates heat, which is used to produce steam in a boiler or drive a gas turbine. This heat energy is converted into mechanical or electrical power, providing the primary output of the power generation system. For controlling the flame temperature in a boiler, a

Fig. 4.12 Alstom's 15-MWth Boiler simulation facility https://www.osti.gov/servlets/purl/1159325/

portion of the resulting flue gas is circulated back into the furnace/boiler. The flue gas processing unit is responsible for cleaning the flue gas and controlling its quality. It typically involves the removal of pollutants such as sulfur dioxide (SO_2), nitrogen oxides (NOx), mercury, and particulate matter. Various technologies like scrubbers, electrostatic precipitators, and fabric filters are employed to ensure compliance with environmental regulations and minimize the release of harmful emissions. The CO_2 Processing Unit (CPU) is the final stage of the oxyfuel combustion process, where the captured CO_2 undergoes purification for transport and storage. This unit removes impurities and moisture from the flue gas, resulting in a high-purity CO_2 stream ready for transportation to a storage site. Different separation techniques like dehydration, compression, and low-temperature purification are utilized to achieve the desired CO_2 purity levels. The purified CO_2 is then compressed and transported for permanent underground storage. Figure 4.12 shows 15 MWth Oxyfuel Pilot Plant: Alstom Boiler Laboratories, Windsor, CT, which was one of the earliest research facilities that focused on oxyfuel combustion technology from early 2000–2015 (Levasseur 2014). Currently, there is no large-scale commercial power plant that uses oxyfuel combustion, mainly due to a much higher energy penalty and much cheaper options such as PCC with amine solutions.

4.6.2 Advantages and Disadvantages of Oxyfuel Combustion

Advantages of Oxyfuel combustion include high CO_2 capture efficiency, where Oxyfuel combustion has the potential to capture up to 100% of the CO_2 emissions produced during the combustion process. This makes it an effective technology for reducing greenhouse gas emissions.

By using oxygen instead of air for combustion, oxyfuel combustion eliminates the formation of nitrogen oxides (NOx), which are significant air pollutants. This results in cleaner exhaust gases and improved air quality. However, it comes with major disadvantages compared to post-combustion and pre-combustion capture technologies. Oxyfuel combustion requires the development of specialized turbines that can operate efficiently with high-purity oxygen. The design and engineering of such turbines can be complex and costly, which introduces a major challenge for the implementation of oxyfuel technology in high-efficiency gas power plants. The process of air separation, which is necessary to obtain high-purity oxygen for combustion, is also energy-intensive. It requires significant energy input to separate oxygen from air using methods such as cryogenic separation or membrane separation. This can increase the overall energy consumption of the oxyfuel combustion process.

4.6.3 Vattenfall Kraftwerk Schwarze Pumpe Pilot Plan

The White Rose project was one of the most impressive oxyfuel projects aiming to be the first large-scale implementation of oxyfuel combustion and CO_2 capture from a coal-fired power plant. The captured CO_2 was also planned to be stored offshore in the North Sea. In 2012 the project started, and in 2013 the engineering design was completed (Metcalfe et al. 2017). Unfortunately, the UK government canceled the funding for this project in 2015, two years after it completed the Front-End Engineering and Design studies. The major challenges that White rose faced were the high cost of the project that was not secured, and the lack of carbon pricing that led to unstable future of the project. Another notable project, which successfully tested various components of oxyfuel combustion, ranging from the boiler to the CO_2 purification unit, was Vattenfall's Schwarze Pumpe Pilot Plant, which was launched in 2008 and ended in 2014 with similar concerns as White Rose project. The project was the world's first Oxyfuel combustion and carbon capture project implemented in coal-fired power generation in Brandenburg, Germany (Vattenfall 2008) (Fig. 4.13).

4.6.4 Direct-Fired Oxyfuel Combustion

Direct-Fired Oxyfuel combustion is an advanced approach that combines two innovative technologies: supercritical carbon dioxide (sCO_2) and oxyfuel combustion (Sleiti et al. 2021).

sCO_2, or supercritical carbon dioxide, is carbon dioxide kept at a temperature and pressure above its critical point. At supercritical condition, CO_2 behaves like both gas and liquid, with high density and distinct fluid characteristics. sCO_2 density can significantly change with slight variations in temperature or pressure, making

Fig. 4.13 Vattenfall Kraftwerk Schwarze Pumpe Pilot Plant (Attribution: I, SPBer)

it an ideal medium for power generation. Oxy fuel combustion involves burning fuel with a nearly pure oxygen that results in a flue gas with a high concentration of CO_2. This enables the efficient capture of almost 100% of the CO_2 emissions produced during the combustion process. The flue gas heat is used to drive the turbine to generate the electricity and the remaining heat in the flue gas is captured in the recapturator and the flue gas is cooled down to extract the water and leave the pure CO_2 stream. At this point some of the pure CO_2 is compressed to the desired pressure and passed through the recuperator for preheating before being recycled to the combustor as the diluent. The remaining CO_2 is prepared for compression and storage. By combining sCO_2 technology and oxy-combustion, direct-fired sCO_2 power cycles offer a highly efficient and environmentally friendly solution for power generation (Indelicato et al. 2020). The utilization of sCO_2 as the working fluid, with its unique properties, enhances the overall system efficiency and allows for effective CO_2 capture. This innovative approach holds great promise for significantly reducing greenhouse gas emissions and advancing the development of sustainable and low-carbon energy systems.

Why sCO_2 Power Cycles?

sCO_2-based oxyfuel combustion offers many advantages compared to conventional oxy combustion, including higher efficiency power cycles, built-in capture technology, and more efficient power plant design. sCO_2 is a supercritical fluid, which allows sCO_2 power cycles to achieve higher thermal efficiencies compared to conventional power cycles. In fact, they can offer a significant improvement of + 3–+5 percentage points over supercritical steam cycles, especially in indirect coal-fired applications. The high fluid densities of sCO_2 contribute to enhanced heat transfer and overall system efficiency. The high density of sCO_2 compared to steam allows for the design of more compact turbomachinery, including compressors and turbines. The smaller size and reduced weight of these components make the power cycle equipment more compact, leading to a smaller plant footprint and lower capital costs. The compactness also facilitates the integration of sCO_2 power cycles with other energy systems, making them suitable for various applications. sCO_2 power cycles can be integrated with oxycombustion to efficiently capture CO_2 emissions. Burning fuel with nearly pure oxygen results in flue gas that consists mainly of CO_2 and water vapor. This facilitates the separation and capture of CO_2 for storage or utilization purposes. sCO_2 power cycles offer flexibility in terms of fuel sources, including coal, natural gas, biomass, and waste heat from industrial processes. Additionally, the modular nature of sCO_2 power cycles allows for scalability, enabling the construction of power plants with various capacities to meet different energy demands.

4.7 Pre-Combustion Capture

4.7.1 Overview and Technologies

Pre-combustion capture involves the removal of CO_2 from fossil fuels before they undergo complete combustion. One common method is through gasification processes, where a feedstock like coal is partially oxidized in the presence of steam and oxygen/air at high temperatures and pressures. This results in the formation of synthesis gas, or syngas, which consists of hydrogen, carbon monoxide, CO_2, and other gases like methane. To further refine the syngas, the water-gas shift reaction can be employed. This reaction converts CO and water (H_2O) into hydrogen (H_2) and CO_2, leading to the production of a gas mixture rich in H_2 and CO_2. The concentration of CO_2 in this mixture typically ranges from 15% to 50%. The captured CO_2 can then be separated, transported, and ultimately sequestered, while the H_2-rich fuel can be combusted for energy generation. The process of hydrogen combustion involves a chemical reaction with oxygen, typically from the air. This reaction generates water vapor and releases energy. The balanced chemical equation representing the combustion of hydrogen is as follows (g stands for gas or steam):

$$2H_2(g) + O_2(g) \rightarrow 2H_2O(g)$$

This chemical reaction is highly exothermic, meaning it releases a substantial amount of energy in the form of heat. When the combustion is complete, the only byproduct is water vapor. Pre-combustion capture produces clean energy as hydrogen-rich fuel while capturing CO_2. Compared to post-combustion technology, pre-combustion capture offers several advantages. Pre-combustion capture technology produces higher CO_2 concentrations compared to post-combustion methods (Saini 2017). Higher concentration of CO_2 allows more efficient carbon capture and removal before the hydrogen (H_2) is combusted. Therefore, concentrated CO_2 in the shifted synthesis gas stream makes pre-combustion capture more efficient in capturing CO_2 compared to post-combustion capture. However, the upfront expenses for the gasification process are frequently higher than those of conventional pulverized coal power plants. Pre-combustion carbon capture technologies available in the market typically use physical or chemical adsorption processes. The approximate cost for capturing CO_2 from an integrated gasification combined cycle (IGCC) power plant is \$60 per ton of CO_2 (House et al. 2011). Pre-combustion capture provides a promising solution for capturing CO_2 emissions; however, the high-cost considerations related to the gasification process are a major barrier.

Figure 4.14 shows the pre-combustion capture technology process including air separation, gasifier, CO_2 capture, combustion unit, and steam turbine. The process begins with air separation, where air is separated into its component gases, primarily nitrogen and oxygen. This is typically done using methods such as cryogenic distillation or membrane separation. The fossil fuel, such as coal, goes through a gasification process. In a gasifier, the coal reacts with steam and oxygen/air under high temperature and pressure, resulting in the production of synthesis gas or syngas. The syngas produced in the coal gasifier contains impurities and particles. These impurities, such as ash and solid contaminants, need to be removed through a particle removal step to make it suitable for further processing. The cleaned syngas then enters a shift reactor, where a water-gas shift reaction takes place. This reaction converts carbon monoxide (CO) and water (H_2O) into hydrogen (H_2) and CO_2, resulting in a gas mixture rich in hydrogen and CO_2. The gas mixture from the shift reactor may contain sulfur compounds, which need to be removed to prevent sulfur emissions. Various techniques, such as scrubbing or catalytic processes, are employed to remove sulfur compounds from the gas. The gas, now primarily composed of hydrogen and CO_2, enters a CO_2 adsorber unit. Physical solvents with higher CO_2 adsorption capacity are used to separate the CO_2. The captured CO_2 is then released from the adsorbent material, and the lean solvent is captured to recirculate for further CO_2 adsorption. This step involves applying heat or changing pressure conditions to facilitate the desorption of CO_2 from the adsorbent material. The captured CO_2 is then compressed and transported for permanent storage underground. The hydrogen-rich gas, on the other hand, is combusted in a gas turbine, where it produces power by driving the turbine. The high-temperature exhaust gases from the gas turbine are utilized for heat recovery in a heat exchanger (Koornneef et al. 2010). In the heat recovery process, the high-temperature exhaust gases are cooled, and the heat is transferred to a working fluid, often water, to generate steam. The steam can be used for various purposes, such as electricity generation or heating. Excess air is also added to the combustion process to optimize the combustion efficiency. Pre-combustion capture technology enables the removal and capture of CO_2 from fossil fuels before combustion.

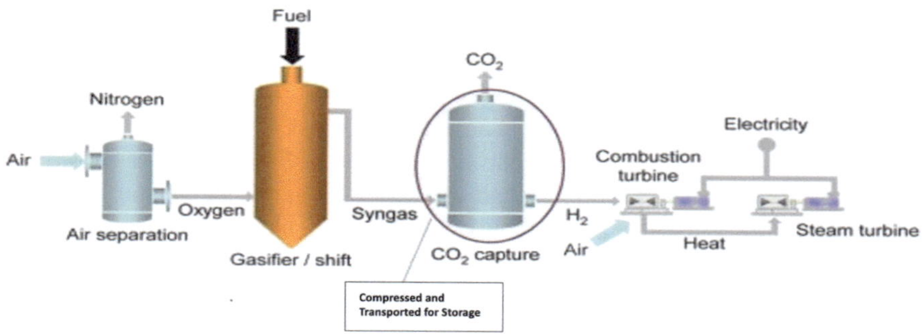

Fig. 4.14 Pre-combustion capture technology (Carbon Sequestration Technology and Program Plan 2007) https://netl.doe.gov

4.7.2 Advantages and Disadvantages of Pre-Combustion

Advantages of the pre-combustion capture include higher CO_2 concentration, making it easier to capture and separate. This technology also offers higher efficiency since the concentration of CO_2 in the shifted synthesis gas allows for efficient capture before combustion, reducing energy penalties. The process of pre-combustion capture can also produce hydrogen as a byproduct, which can be utilized as a clean fuel. Pre-combustion capture reduces the release of greenhouse gases into the atmosphere by capturing CO_2 before combustion.

The pre-combustion technology also faces disadvantages, including high capital costs, as the base gasification process required for pre-combustion capture is generally more expensive than traditional power plants, adding to the overall cost of implementation. The pre-combustion technology is also more complex than post-combustion capture, as it involves multiple process steps and equipment, requiring sophisticated engineering and operational expertise. The technology is primarily applicable to fossil fuels like coal and natural gas, limiting its use in other energy sources. However, with advancements in technology and reduced costs, the pre-combustion technology becomes more compatible with post-combustion capture technology with time.

4.8 CCS-Readiness in Combustion Technologies

Post-Combustion Capture (PCC) offers the advantage of being easily added to existing power stations, and the concept of CCS-Readiness emphasizes the importance of designing new power stations to facilitate future implementation of CCS technology. Considering CCS-Readiness during the location, design, and engineering phases of new power stations, the facility can be prepared to ensure access to geological storage that allows smooth integration of CCS without significant additional costs later when the resources are ready to do so.

In the U.S.A the department of energy and global CCS institute published a technology level readiness of different technologies in 9 levels and 4 categories, including concept, laboratory, pilot, and full scale. The concept category includes (1) the basic principle and (2) technology concept. The laboratory category includes (3) proof of concept, (4) component validation, and (5) system validation. The third category is pilot with (6) pilot-scale system validation and finally the full scale that includes (7) full-scale demonstration of actual system, (8) Actual system completed and proven to work, and (9) Actual system completed and proven to work under expected conditions. Based on the above, post-combustion gets the highest-level technology readiness level of 9, which means that post-combustion can be adopted to existing facilities; pre-combustion has also reached level 9 readiness in some cases, such as hydrogen production, but is still level 7 in the case of coal-fired power generation. Oxyfuel combustion is less developed and not prepared for full demonstration under various conditions and has achieved level 7 of technology readiness (GAO-22-105,274).

Gas separation technologies depending on the application, either solvent, sorbent, or membrane, have shown different levels of technology readiness from 1 to 9. However, cryogenic separation only reached as high as level 6, which is pilot-scale system validation in specific conditions (GAO-22-105,274).

The United Kingdom has taken a practical approach by publishing technical guidelines on CCS-Readiness, which serve as a framework for large new fossil fuel and biomass power stations. Following these guidelines ensures that the necessary infrastructure and requirements are in place to accommodate future CCS implementation in these power generation facilities. The idea comes from industrial-scale successful CCS projects implemented in old power plants such as The Boundary Dam facility in Saskatchewan, Canada. This facility is the world's leading power station incorporating Carbon Capture and Storage (CCS) into power generation. The Boundary Dam 3 CCS Facility (BD3) is the first fully integrated CCS operation on a global scale within a coal-fired power plant. BD3's fully integrated system includes the Carbon Capture Facility, Transportation to Enhanced Oil Recovery, CO_2 Storage in an aquifer, Carbon Capture Test Facility, and Emissions Control Research Facility. Boundary Dam Unit #3 CCS Facility, with 110 megawatts (MW) of power capacity, achieves remarkable environmental milestones, with up to a 90% nominal reduction in CO_2 emissions. During the third quarter of 2023, the CCS facility demonstrated exceptional operational efficiency, reaching 98.3%. It successfully captured 217,366 tons of CO_2 during this period, with a daily average capture rate of 2674 tons while operating. These accomplishments result in an emissions intensity of 407 tons of carbon dioxide per gigawatt-hour, aligning with the current federal carbon tax threshold (SASKTODAY.ca staff).

Another example of successful application of PCC is the Petra Nova, which was the largest power station with Carbon Capture and Storage (CCS) until its closure in 2020. The Petra Nova is in Thompsons, Texas, southwest of Houston. The operational history of the Petra Nova CCS project serves as a model for integrating commercial-scale post-combustion carbon capture technology into a pre-existing dual-fired power plant. The facility resumed operations in September 2023 after a three-year shutdown; it is designed to capture approximately 90 percent of carbon dioxide (CO_2) from a 240 MW flue gas slipstream. Petra Nova utilizes a proven carbon capture process employing a high-performance solvent for efficient CO_2 absorption and desorption. The captured CO_2 undergoes compression and travels via an 80-mile pipeline to Hilcorp's West Ranch oil field near Vanderbilt, Texas. Here, it facilitates enhanced oil recovery (EOR) and sequestration, amounting to an annual total of around 1.4 million metric tons. Successful CCS facilities like Boundary Dam and Petra Nova can enhance our understanding of the practical implementation and potential benefits of CCS in power generation, that can justify the implementation of theCCS-Readiness concept to all planned fossil fuel and biomass power stations.

4.9 Conclusion

This chapter has provided a detailed examination of combustion processes, and the carbon capture technologies designed to mitigate their environmental impact. Combustion, which is a foundation of industrial and power generation activities, is the major contributor to global CO_2 emissions. However, as demonstrated throughout this chapter, advanced carbon capture technologies such as post-combustion capture (PCC), oxyfuel combustion, and pre-combustion capture can significantly reduce these emissions.

A key insight from this chapter is that each capture technology has distinct advantages and limitations depending on the application. Post-combustion capture is adaptable to existing power plants but faces challenges in energy efficiency and solvent management. Oxyfuel combustion, while a highly efficient carbon capture process, requires substantial energy for oxygen production and poses operational complexities. Meanwhile, pre-combustion capture offers high capture rates but involves high capital costs and infrastructure demands.

The chapter also highlighted the importance of the CCS-Readinessconcept that suggests designing combustion systems with the flexibility to integrate carbon capture technologies in the future. This approach is critical for industries such as natural gas and coal power generation. Technological innovations, such as advanced solvents, cryogenic distillation, and membrane separation, are pushing the boundaries of capture efficiency and scalability. However, challenges remain, including energy penalties and high operational costs. Looking ahead, the successful deployment of these technologies relies on policy support, continued R&D investment, and cross-industry collaboration (Fig. 4.15).

Fig. 4.15 Images designed by Freepik

References

R. Ahmed, G. Liu, B. Yousaf, Q. Abbas, H. Ullah, M.U. Ali, Recent advances in carbon-based renewable adsorbent for selective carbon dioxide capture and separation—a review. J. Clean. Prod. **242**, 118409, ISSN 0959-6526 (2020). https://doi.org/10.1016/j.jclepro.2019.118409

B. Bane, Scientists unveil least costly carbon capture system to date. Pacific Northwest National Laboratory. (2023). https://www.pnnl.gov/news

B.J.P. Buhre, L.K. Elliott, C.D. Sheng, R.P. Gupta, T.F. Wall, Oxy-fuel combustion technology for coal-fired power generation. Prog. Energy Combust. Sci. **31**(4), 283–307 (2005). https://doi.org/10.1016/j.pecs.2005.07.001

I. Campello Gómez, C. Gutiérrez, Desorption strategies in CO_2 capture technologies: Novel approaches and future perspectives, Journal of Environmental Chemical Engineering, **13**(2), (2025), 116109, ISSN 2213–3437. https://doi.org/10.1016/j.jece.2025.116109

S. Chen, S. Chen, Y. Zhang, H. Chai, L. Qin, Y. Gong, An investigation of the role of N-methyldiethanolamine in non-aqueous solution for CO_2 capture process using ^{13}C NMR spectroscopy. Int. J. Greenh. Gas Con. **39**, 166–173 (2015). https://doi.org/10.1016/j.ijggc.2015.05.011

S.-H. Cho, S.K. Park, T.S. Kim, et al., Performance analyses of oxy-fuel power generation systems including CO_2 capture: Comparison of two cycles using different recirculation fluids. J. Mech. Sci. Technol. **24**, 1947–1954 (2010). https://doi.org/10.1007/s12206-010-0623-x

J. Davison, Performance and costs of power plants with capture and storage of CO_2. Energy **32**(7), 1163–1176 (2007). https://doi.org/10.1016/j.energy.2006.07.039

S.G. de Ávila, M.A. Logli, J.R. Matos, Kinetic study of the thermal decomposition of monoethanolamine (MEA), diethanolamine (DEA), triethanolamine (TEA) and methyldiethanolamine (MDEA). Int. J. Greenh. Gas Con. **42**, 666–671 (2015). https://doi.org/10.1016/j.ijggc.2015.10.001

G. de Koeijer, J. H. Borch, M. Drescher, H. Li, Ø. Wilhelmsen, J. Jakobsen, CO_2 transport–Depressurization, heat transfer and impurities. Energy Procedia, **4**, 3008–3015, ISSN 1876-6102 (2011). https://doi.org/10.1016/j.egypro.2011.02.211

Y. Du, T. Gao, G.T. Rochelle, A.S. Bhown, Zero- and negative-emissions fossil-fired power plants using CO_2 capture by conventional aqueous amines. Int. J. Greenh. Gas Con. **111**, 103473 (2021). https://doi.org/10.1016/j.ijggc.2021.103473

N. Fernández-Miranda, M.A. Lopez-Anton, T. Torre-Santos, M. Díaz-Somoano, M.R. Martínez-Tarazona, Impact of oxy-fuel conditions on elemental mercury re-emission in wet flue gas desulfurization systems. Environ. Sci. Technol. **50**(13), 7247–7253 (2016). https://doi.org/10.1021/acs.est.6b00210

T. He, Z. Liu, H. Son, T. Gundersen, W. Lin, Comparative analysis of cryogenic distillation and chemical absorption for carbon capture in integrated natural gas liquefaction processes. J. Clean. Prod. **383**, 135264 (2023). https://doi.org/10.1016/j.jclepro.2022.135264

R. Hou, C. Fong, B.D. Freeman, M.R. Hill, Z. Xie, Current status and advances in membrane technology for carbon capture. Sep. Purif. Technol. **300**, 121863 (2022). https://doi.org/10.1016/j.seppur.2022.121863

K.Z. House, A.C. Baclig, M. Ranjan, E.A. van Nierop, J. Wilcox, H.J. Herzog, Economic and energetic analysis of capturing CO_2 from ambient air. Proc. Natl. Acad. Sci. **108**(51), 20428–20433 (2011). https://doi.org/10.1073/pnas.1012253108

G. Indelicato, P.E. Lapenna, R. Concetti, M. Caputo, M. Valorani, G. Magnotti, F. Creta, Numerical investigation of high pressure CO-diluted combustion using a flamelet-based approach. Combust. Sci. Technol. **192**(11), 2028–2049 (2020). https://doi.org/10.1080/00102202.2020.1811243

G. Ji, M. Zhao, Membrane separation technology in carbon capture, recent advances in carbon capture and storage. InTech, Mar. 08, (2017). https://doi.org/10.5772/65723

A. Karayil, A. Elseragy, A.M. Aliyu, An assessment of CO_2 capture technologies towards global carbon net neutrality. Energies **17**(6), 1460 (2024). https://doi.org/10.3390/en17061460

J. Koornneef, A. Ramirez, T. van Harmelen, A. van Horssen, W. Turkenburg, A. Faaij, The impact of CO_2 capture in the power and heat sector on the emission of SO_2, NOx, particulate matter, volatile organic compounds and NH_3 in the European Union. Atmos. Environ. **44**(11), 1369–1385 (2010). https://doi.org/10.1016/j.atmosenv.2010.01.022

H. Kumar, S. Ravikumar, 2010 international conference on recent advances in space technology services and climate change, in *An approach of CO_2 capture technology for mitigating global warming and climate change—an overview*, (2010), pp. 364–371. https://doi.org/10.1109/RSTSCC.2010.5712870

D.Y.C. Leung, G. Caramanna, M.M. Maroto-Valer, An overview of current status of carbon dioxide capture and storage technologies. Renew. Sust. Energ. Rev. **39**, 426–443 (2014). https://doi.org/10.1016/j.rser.2014.07.093

A. Levasseur, Recovery act: Oxy-combustion technology development for industrial-scale boiler applications. Task 4—testing in Alstom's 15 MWth boiler simulation facility (report no. DOE-NETL-40763). U.S. Department of Energy (2014). https://doi.org/10.2172/1159325

Q. Li, X. Huang, N. Li, T. Qi, R. Wang, L. Wang, S. An, Energy-efficient biphasic solvents for industrial CO_2 capture: absorption mechanism and stability characteristics. Energy **293**, 130710 (2024). https://doi.org/10.1016/j.energy.2024.130710

R. Metcalfe, K. Thatcher, G. Towler, A. Paulley, J. Eng, Sub-surface uncertainty and contingency planning for the endurance CO_2 store of the White rose project, UK. Energy Procedia **114**, 4313–4320 (2017). https://doi.org/10.1016/j.egypro.2017.03.1578

E. Meuleman, Validation of Transformational CO_2 Capture Solvent Technology with Revolutionary Stability (Apollo), in *U.S. Department of Energy National Energy Technology Laboratory Carbon Management Project Review Meeting, United States*, (2022)

M.M. Mohd Pauzi, N. Azmi, K.K. Lau, Emerging solvent regeneration technologies for CO_2 capture through offshore natural gas purification processes. Sustainability **14**(7), 4350 (2022). https://doi.org/10.3390/su14074350

M.T. Mota-Martinez, J.P. Hallett, N. Mac Dowell, Solvent selection and design for CO_2 capture—how we might have been missing the point. Sustain. Energy Fuels **1**(10), 2078–2090 (2017). https://doi.org/10.1039/C7SE00404D

K.A. Oyewole, O.B. Okedere, K.O. Rabiu, K.O. Alawode, S. Oyelami, Carbon dioxide emission, mitigation and storage technologies pathways. Sustainable. Environment **9**(1) (2023). https://doi.org/10.1080/27658511.2023.2188760

S. Rezaei, A. Liu, P. Hovington, Emerging technologies in post-combustion carbon dioxide capture & removal. Catal. Today **423**, 114286 (2023). https://doi.org/10.1016/j.cattod.2023.114286

D. Saini, Engineering aspects of geologic CO_2 storage. Spring (2017). https://doi.org/10.1007/978-3-319-56074-8

G. Sartori, D.W. Savagebr, Industrial & engineering chemistry fundamentals **22**(2), 239–249 (1983). https://doi.org/10.1021/i100010a016

M. Shen, L. Tong, S. Yin, C. Liu, L. Wang, W. Feng, Y. Ding, Cryogenic technology progress for CO_2 capture under carbon neutrality goals: A review. Sep. Purif. Technol. **299**, 121734 (2022). https://doi.org/10.1016/j.seppur.2022.121734

A. Singh, K. Stéphenne, Shell Cansolv CO_2 capture technology: achievement from first commercial plant. Energy Procedia **63**, 1678–1685 (2014). https://doi.org/10.1016/j.egypro.2014.11.177

A.K. Sleiti, W. Al-Ammari, S. Ahmed, J. Kapat, Direct-fired oxy-combustion supercritical-CO_2 power cycle with novel preheating configurations—Thermodynamic and exergoeconomic analyses. Energy **226**, 120441 (2021). https://doi.org/10.1016/j.energy.2021.120441

X.Y.D. Soo, J.J.C. Lee, W.-Y. Wu, L. Tao, C. Wang, Q. Zhu, J. Bu, Advancements in CO_2 capture by absorption and adsorption: a comprehensive review. J. CO_2 Util. **81**, 102727 (2024). https://doi.org/10.1016/j.jcou.2024.102727

R. Stanger, T. Wall, R. Spörl, M. Paneru, S. Grathwohl, M. Weidmann, G. Scheffknecht, D. McDonald, K. Myöhänen, J. Ritvanen, S. Rahiala, T. Hyppänen, J. Mletzko, A. Kather, S. Santos, Oxyfuel combustion for CO_2 capture in power plants. Int. J. Greenhouse Gas Control **40**, 55–125 (2015). https://doi.org/10.1016/j.ijggc.2015.06.010

S.H. Tak, S.K. Park, T.S. Kim, et al., Performance analyses of oxy-fuel power generation systems including CO_2 capture: comparison of two cycles using different recirculation fluids. J. Mech. Sci. Tech., **24**, 1947–1954 (2010). https://doi.org/10.1007/s12206-010-0623-x

B. Van der Bruggen, Microfiltration, ultrafiltration, nanofiltration, reverse osmosis, and forward osmosis, in *Fundamental Modelling of Membrane Systems*, ed. by P. Luis, (Elsevier, 2018), pp. 25–70. https://doi.org/10.1016/B978-0-12-813483-2.00002-2

Vattenfall. (2008). Vattenfall inaugurates world's first pilot for a coal-fired power plant with CO_2 capture [Press release]. https://group.vattenfall.com/press-and-media/press-releases/2008/vattenfall-inaugurates-worlds-first-pilot-for-a-coal-fired-power-plant-with-co2-capture

X. Zhao, K. H. Smith, M. A. Simioni, W. Tao, S. E. Kentish, W. Fei, G. W. Stevens, Comparison of several packings for CO_2 chemical absorption in a packed column. International Journal of Greenhouse Gas Control, **5**(5), 1163–1169 (2011). https://doi.org/10.1016/j.ijggc.2011.07.006

Chapter 5
CO_2 Transport: Monitoring, Surveillance, and Leak Detection Technologies

5.1 Chapter Overview and Objectives Overview

5.1.1 Overview

This chapter provides an overview of the latest advancements in CO_2 Transport, Monitoring, Surveillance, and Leak Detection Technologies. We will discuss various monitoring techniques that enable the continuous observation and measurement of critical parameters along the transport infrastructure, such as flow rates, pressure, temperature, and composition. The chapter highlights the significance of leak detection technologies, emphasizing their role in early detection and prevention of CO_2 leakage. We examine industry standards, particularly API requirements, that govern CO_2 transport and emphasize the importance of robust leak detection and surveillance systems to prevent and mitigate any possible leaks. Additionally, it delves into the concept of surveillance, and integrity of CO_2 transport systems. Advanced surveillance systems and strategies will be discussed, highlighting their role in identifying potential risks and enhancing overall safety.

5.1.2 Learning Objectives

Upon completion of this chapter, you will be able to:

1. Evaluate the importance of monitoring, surveillance, and leak detection technologies in the context of CO_2 transport.
2. Analyze various monitoring techniques used to measure critical parameters along the CO_2 transport infrastructure.
3. Explain the concept of surveillance and learn about the integrity of CO_2 transport systems.

E. Fathi and F. Belyadi, *Subsurface CO_2 Sequestration*,
https://doi.org/10.1007/978-3-032-28296-5_5

4. Evaluate different leak detection technologies employed to detect and prevent CO_2 leakage during transport.
5. Explain the significance of these technologies in enhancing the safety, efficiency, and sustainability of CO_2 transport practices.

5.2 Introduction to CO_2 Transport

Transporting captured carbon dioxide from different sources to a destination for various purposes such as underground storage or utilization is the critical step in successful implementation of CCUS. The efficiency of this process depends on the state of CO_2 being transferred such as gas, liquid or supercritical, the means of transportation (e.g., pipeline, ship, trucks, or train), proper monitoring, and compliance with regulatory standards.

Each mode of CO_2 transport has its advantages and disadvantages. Factors such as distance, volume, infrastructure availability, and regulatory requirements are necessary in determining the most suitable transportation technique. It is essential to include different CO_2 transportation strategies including the construction of an extensive CO_2 pipeline network, specialized CO_2 tanks for shipping, and the optimization of railroads and trucks to support global carbon capture and storage (CCS) projects. This diversified approach provides flexibility and practical solutions that are essential in the lack of widespread CO_2 pipeline networks.

Trucks and trains serve as efficient connectors between local facilities and short to medium-distance transport of CO_2 to markets. They offer flexibility in reaching locations that may not have direct pipeline access. While truck and rail transport are suitable for smaller quantities and short distances, they are predominantly utilized at project sites. However, compared to pipelines and ships, they have limited capacity and may be more costly for large-scale transportation. With the anticipated large-scale adoption of CCS technology in different industries, truck and rail transports are expected to have limited roles. Much of the load therefore is expected to be carried by ship and pipeline transportation as viable alternatives on a global scale.

CO_2 can be transported either as a gas, supercritical CO_2 or as a liquid, with supercritical CO_2 pipelines being the preferred option for new infrastructure. The most common state to transport the CO_2 is at supercritical state, when it's compressed above its critical temperature and pressure of 31.1 °C and 73.8 bar (87.9 °F and 1070.4 psi). Supercritical CO_2 (sCO_2) is dense like liquid but has viscosities like gas and that's why we call it liquid like fluid. At this condition CO_2 is very favorable for pipeline transportation due to high flow efficiency. Liquid CO_2 is more favorable for shipping where it requires low temperatures of −20 to −55 °C (−4 to −67 °F) and moderate pressures of 6–20 bar (87–290 psi). (U.S. Japan CO_2 Shipping Feasibility Study: Screening Assessment, 2025). The CO_2 transport in the gaseous phase is not common as its not efficient and requires high pressure compression. The CO_2 purity for transportation depends on the means of transportation, application, and specific region. However the purity of CO_2 should be above 95% for pipelines and more than 99% for liquified CO_2 shipping (https://thundersaidenergy.com/downloads/ccs-what-CO2-purity-for-transport-and-disposal/).

5.3 Methods of CO_2 Transport

5.3.1 Pipeline Transportation

The safe transport of CO_2 is a critical aspect to consider when implementing carbon capture and storage (CCS) projects. Pipelines are the most used and cost-effective methods of CO_2 transportation for large scale CCS projects. They offer a continuous, reliable, and efficient option of transporting large volumes of CO_2 over long distances. Currently, there are approximately 6000 to 8000 miles of pipelines for transporting CO_2 exist worldwide including recently constructed, expanded, and soon to be commissioned. The United States has around 5300 miles of existing CO_2 pipelines which are mainly used for transporting CO_2 for enhanced oil recovery projects. Since 2021 additional CO_2 pipelines are proposed in Midwest including Summit Carbon solution to build 2000 miles of CO_2 pipeline, Wolf Carbon Solutions with 280 miles of CO_2 pipeline and 392 miles of natural gas pipeline conversion to CO_2 by Tallgrass Energy (P.W. 1944) IN12269). CO_2 pipelines are designed to ensure the safe and secure transfer of CO_2 from the capture site for storage or utilization facilities.

However, still there is a critical need to significantly broaden the CO_2 pipeline network to achieve meaningful greenhouse gas reduction through CCS. A recent study proposes that such a network could expand approximately 66,000 miles of pipeline by the year 2050, requiring an estimated $170 billion in new capital investment (Larson et al. 2021). The construction of a national CO_2 pipeline network has raised safety concerns due to the potential hazards of high concentrations of CO_2 to human health. These safety considerations may impact nearby communities and potentially delay the widespread deployment of CCS. Ensuring the safe operation of CO_2 pipelines involves implementing rigorous monitoring, inspection, and maintenance protocols to detect and mitigate any potential leaks or accidents. Comprehensive Uncertainty and Contingency Plannings along with emergency response plans following regulatory guidelines are required for safe CO_2 transportation. This process demands careful consideration of pipeline design, protection against corrosion, leak monitoring, and precautions against overpressure, particularly in populated areas.

The high strength carbon steel pipeline is commonly used for CO_2 transport, but the steel grade depends on CO_2 purity and contents like water and H_2S. In some cases, internal coating of pipeline is also applied to avoid corrosion. The size of CO_2 pipelines also varies depending on the flow capacity but commonly 10" to 24" (inches) for onshore, and up to 36" for high-capacity pipelines are implemented. The pipeline transportation of CO_2 also needs to be supported by the following: a dehydration unit to ensure the purity of CO_2 stream, a compression station where

Fig. 5.1 CO_2 pipeline https://netl.doe.gov/carbon-management/carbon-storage/transport

CO_2 is converted to the supercritical form and a monitoring system. Additional safety measures like the presence of pressure relief valves or flare system is also recommended. Major challenges regarding the development of the CO_2pipelines includes the extremely high capital cost of the pipelines ensuring mechanical integrity of the pipelines (the risk of rupture and leakage) and flow assurance problems in addition to regulatory constraints (Grant et al. 2018; Ozaki and Ohsumi 2011; Morgan et al. 2024) (Fig.5.1).

5.3.2 *Shipping of CO_2*

Ships are another mode of CO_2 transport, particularly for long-distance transport across oceans. The history of transporting liquid CO_2 in ships goes back to 1980s, where liquid CO_2 is provided for multiple purposes such as food industry especially carbonated beverages chemical and medical industries (Topham et al. 2014) Clarkson PLC49. CO_2 can be liquefied and stored in specialized tanks onboard ships, which provide a cost-effective option for large-scale transport. However, this method requires adherence to strict safety measures to prevent leakage and ensure environmental protection.

Specially designed ships for liquefied CO_2 transport offer a secure and reliable option for long-distance transportation. In Europe, small-scale shipments of liquefied CO_2 (approximately 1000 tons) are practiced. This includes transporting food-grade CO_2 between sources and coastal distribution terminals (Burger et al. 2024).

However, efficient collection from diverse industrial capture sites requires specialized ships. The primary challenge in CO_2 shipping is the absence of ships capable of transporting high volumes over extended distances. Key challenges include the need for high-pressure tanks for liquid CO_2 transfer. Additionally, factors such as annual and per-load CO_2 volumes, distance to the storage site, and the quality of the captured CO_2 in terms of impurities and condition impact tank requirements and the type of ship design. Currently four ships are capable of transporting between 1000 and 2000 m^3 equivalent of 1100–2200 tons of CO_2, commercially operated by Larvik Shipping in Norway. However, these ships are not implemented in CCUS projects (Splash247.com 2023; Parker 2024). As we discussed earlier the amount of CO_2 required to be transported are much larger than the current shipping capacity and there is an urgent need to have carriers than can accommodate much larger volumes such as 20,000 tons of CO_2 (Zero Emissions Platform and Carbon Capture and Storage Association 2024).

To maximize the efficiency of liquid CO_2 transportation using ships, the attention is on providing the pressure and temperature conditions where the liquid CO_2 has the highest density. However, this depends on CO_2 source volume, purity, transport distance, and utilization. Based on these, three shipping conditions of low, medium, and high-pressure conditions are investigated in the literature, as low pressure (6–10 bar and –50 °C, or 87 to 145 psi and −58 °F) is identified as the prospect as it acheives the highest CO_2 density (Zero Emissions Platform and Carbon Capture and Storage Association 2018; IEAGHG 2020). However, this narrow pressure condition is extremely hard to maintain during consider harsh shipping environments that could lead to major safety and risk concerns such as solid CO_2 formation or rapid vapor formation (Engel and Kather 2018; Wetenhall et al. 2017). The medium pressure condition is considered as 15 to 18 bar at –25 °C (217–261 psi and −13 °F) and high pressure of 34–45 bar at 0–25 °C (493–652 psi and 32–50 °F). Currently available liquid CO_2 ships are classified as medium pressure conditions, where there are no current low- or high-pressure ships practiced in the industry.

The major considerations for ship design are the state of liquid CO_2 being in low- or high-pressure category and the ship capacity. The state of liquid CO_2 impacts both the ship's capacity and scalability while the ship's capacity should be scaled to CO_2 capacity of the project (Bond 2024). Additional requirements like CO_2 transportation to liquification facility via pipelines, its capacity, temporary storage containers to accommodate any waiting time between the loadings and ship loading and off-loading infrastructure also need to be considered. Figure 5.2 outlines different liquid CO_2 ship loading and off-loading scenarios from and to offshore and onshore conditions. Shore to ship loading can be done using different techniques depending on the conditions such as loading liquid CO_2 tanks using loading arms or using flexible cryogenic hoses (Global Centre for Maritime Decarbonization 2024; Orchard 2019). Offshore loading can also be done using flexible cryogenic lading arms or a bow loading system through the connection point located at the front (Bow) of the ship instead of traditional loading from midship. Sometimes floating storage is also used as the buffer hub between the platform and the main carrier/ship. Also for offloading, depending on onshore or offshore, different strategies can be used as shown in Fig.5.2, such as

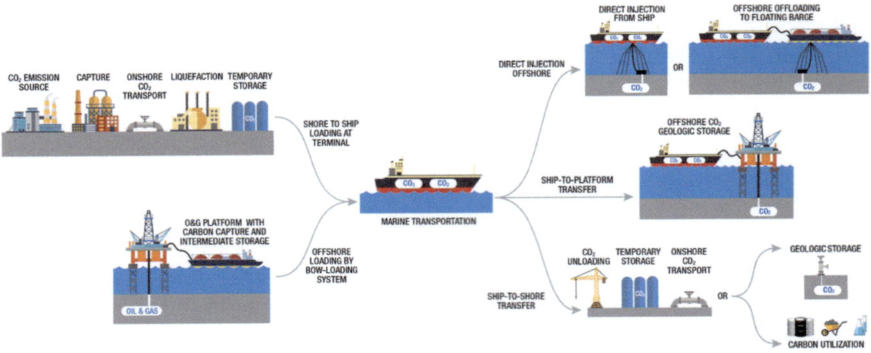

Fig. 5.2 Liquid CO_2 loading and off-loading configurations (NETL U.S. Japan CO_2 shipping feasibility study: screening assessment 2025)

direct injection to the storage site or transfer to the floating carrier or the platform in offshore and loading arms or using flexible cryogenic hoses to the shore.

The Society of International Gas Tanker and Terminal Operators (SIGTTO) has issued guidelines for designing the infrastructure needed to load and unload liquid CO_2 (SIGTTO 1997).

5.3.3 CO_2 Clusters and Transportation Networks

Efficient CO_2 transport from capturing facilities to storage sites require delicate planning and implementation. This can be done by using CO_2 clusters and transportation networks, where both CO_2 gathering facilities and storage sites are grouped and integrated through interconnected networks. CO_2 clusters play a critical role in the efficient transportation of carbon dioxide. As large-scale CCS projects are implemented, it becomes increasingly important to establish transportation networks that can effectively move CO_2 from sources to the storage locations. The size and magnitude of these CO_2clusters depend on the geographic characteristics of the region. Factors include proximity between sources and storage sites availability of suitable storage sites and existing infrastructure (Schultz et al. 2023). One significant advantage of clustering CCS facilities is the potential for reducing the unit cost per ton of CO_2 transported. The clustering increases the efficiency and reduces the cost associated with the CCS project by consolidating multiple sources, storages, and the necessary transportation infrastructure. Instead of each facility handling CO_2 independently, they share infrastructure for transport and storage, improving efficiency and reducing costs. This cost reduction is particularly important in making CCS more economically viable and helps in spreading the application of CCS. Moreover, CO_2 clusters and transportation networks offer reliability in the overall management and oversight of the CCS system. This significantly improves the monitoring and coordination of the project. The establishment of reliable and interconnected transport networks ensures a continuous and secure flow of CO_2 minimizing potential disruptions and ensuring safe and efficient operations. Few examples of using clustering in CCS projects include the Humber Zero (UK) Northern Lights (Norway) and Port of Rotterdam (Netherlands) Hong 2022. Humber Zero captures CO_2 from the largest industrial cluster of CO_2 emitting industries in the UK including refinery and power stations. Humber Zero is integrated into the broader East Coast Cluster to form the CCS infrastructure in the UK. Northern Lights also capture CO_2 emissions from multiple sources within Europe. Port of Rotterdam also captures CO_2 emissions from multiple industrial emitters in the Rotterdam port area and transports that for storage in underground depleted gas reservoirs Hong 2022. Figure 5.3 illustrates a schematic of CO_2 source clustering from various sources such as end users, power stations, and industries. This includes direct air capture and pre-and post-capture from biomass carbon capture, steel, cement, and chemical industries. The captured CO_2 is then transported through a pipeline to a cluster of CO_2 storage sites, including both onshore and offshore geological storage locations.

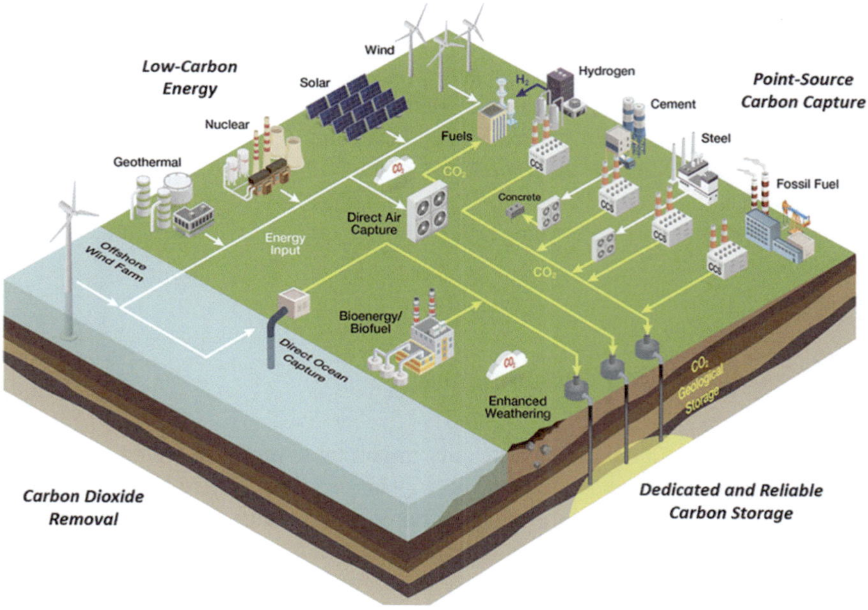

Fig. 5.3 CO_2 Source clustering and transport for comprehensive storage source. https://www.energy.gov/

5.4 Safety and Regulatory Considerations for CO_2 Transport

Transporting CO_2 through pipelines and ships is following rigorous safety standards and regulations, like other industrial activities. In fact, CO_2 transport is as safe as transport of hydrocarbons, natural gas, and oil, which are successfully managed using established safety protocols.

The safety of CO_2 pipelines falls under the jurisdiction of the Pipelines and Hazardous Materials Safety Administration (PHMSA) within the Department of Transportation (DOT), which holds legal authority over CO_2 pipeline safety. PHMSA establishes and enforces regulations (49 C.F.R. §190, 195–199) governing the construction, operation, maintenance, and spill response planning for CO_2 pipelines. Despite CO_2being classified as a Class 2.2 non-flammable gas under DOT regulations PHMSA currently applies safety requirements like those for pipelines transporting hazardous liquids like crude oil and anhydrous ammonia (Parfomak 1944). Currently PHMSA is updating the CO_2 transport standards to assure the safety measure regarding the emergencies and responses. It is important to note that PHMSA regulates CO_2 transport with pipeline only and is not involved in other non-pipeline transportations of CO_2. State level permits are also required to build the pipelines.

The safety of CO_2 pipelines is evident through the track record of safe transport and extensive experience of the industry to perform this transportation. Currently, there are 36 pipelines in the USA dedicated to transporting CO_2. These pipelines cover approximately 6500 kilometers and transport 50–60 million tons of CO_2 annually from natural sources and industrial facilities. The safety regulations applied and extensive experience in fluid transportation using pipelines resulted in secure and reliable transport of CO_2.

5.5 Leak Detection Technologies

Leak detection techniques are significantly important to ensure the safety and reliability of carbon capture and storage projects. This can be classified in leak detection in pipelines and in reservoirs through wellbores with poor cement job or broken or fraced caprocks. The leak detection in CO_2 transportation networks focuses on both the surface above the ground pipelines and underground or buried pipelines. Since the CO_2 transportation networks expand across large distances and through regions with rough topography part of the pipeline could be on the surface or buried under the ground. Here we look at the leak detection technique in pipelines in general and subsurface leak detection form underground CO_2 storage sites such as deep saline formations, depleted oil and gas reservoirs, etc.

5.5.1 *Pipeline Leak Detection Techniques*

As discussed earlier, pipeline networks are the most used and cost-effective method of CO_2transportation for large-scale CCS projects. These extensive networks are subject to a range of potential issues including defects damage or even rupture. These can occur due to factors such as regular wear and tear corrosion caused by reactive fluids sudden changes in pressure and temperature or unforeseen environmental disasters. One significant challenge is the age of the infrastructure. More than 50% of the major pipelines utilized in industries are over 30 years old making them more prone to failure (Hussain et al. 2024). This aging factor of pipeline network requires special attention to continuously monitor, detect, and mitigate possible leaks. Before the incident in Satartia, Mississippi on February 22, 2020, pipeline operators had reported only one injury and no fatalities caused by regulated pipelines over the past 20 years in the US. In the Satartia incident, the soil around the pipeline slid due to two months of rain, causing a pipe weld to break. This rupture resulted in the release of carbon dioxide, leading to the evacuation of more than 200 people, with at least 45 individuals hospitalized, according to the statistics from the Pipelines and Hazardous Materials Safety Administration (PHMSA).

Pipeline leaks not only introduce immediate dangers to human health and the environment but can also result in substantial economic losses. Given the extensive nature of these networks, conducting manual inspections alone is timely, costly, and impractical. Therefore, there is a growing need for advanced technologies and automated leak detection systems. These automated systems can rapidly identify and locate leaks that enables rapid response and minimize the potential impact on surrounding areas. Automated leak detection offers several advantages. It allows for real-time monitoring of the network, that enables early detection and quick mitigation. Additionally, automated leak detection systems can enhance the overall operational efficiency of the transportation networks by minimizing downtime and optimizing maintenance activities. Developing effective leak detection techniques and technologies that identify and allocate the possibility of single or multiple leaks is still a challenge. It requires continuous research and innovation to improve the sensitivity and accuracy of the detection methods, as well as to address the unique challenges due to diverse landscape and environmental conditions.

5.5.1.1 Internal Leak Detection Techniques

Internal pipeline integrity assessment techniques involve evaluating the internal condition of the pipelines using in-line inspection devices (Shawn 2015). These devices such as pipe inspection gauges or “pigs” travel through the pipeline and collect information regarding pressure flow rate and temperature using flow meters or pressure transducers. Pigs can detect anomalies or disturbances between the upstream and downstream sides of the pig that can be used to identify the leak location. They can also assess pipe wall thickness and geometry for deformation or erosion. However pigging can be costly due to pipeline downtime for inspection and the potential challenges of navigating complex topography or transporting corrosive fluids (Thiberville et al. 2020).

Figure 5.4 shows a leak detector pig pressure measurements with three pressure transducers P1 P2 and P3 at different positions (Head Middle and Tail) within the pig. The graph illustrates the pressure values plotted against the position along the pipeline. This pressure behavior then can be used in the analysis of fluid flow behavior and leak detection based on pressure variations. The pressure transducer P2 located at the center of the pig is particularly important for leak detection. As the pig moves through the pipeline it collects continuous pressure data from P2. Any significant change in pressure readings from P2 compared to expected values can indicate the presence of a leak. In a properly functioning pipeline section the pressure measured by P2 remains stable and consistent. However a leak disrupts the flow resulting in a pressure drop at that specific location. Monitoring the pressure readings from P2 and comparing them to the expected pressure profile helps operators identify anomalies that suggest a leak. A sudden and significant decrease in pressure measured by P2 indicates a potential leak in the corresponding section of the pipeline (Camerini et al. 2004). Pressure transducers P1 and P3 provide additional information regarding pressure conditions ahead of and behind the pig, respectively. By comparing the pressure readings from P1, P2, and P3, operators can analyze pressure differentials and patterns along the pig's path. Significant variations between these transducers can indicate the presence of leaks or other irregularities in the pipeline. Integrating the pressure data from P1, P2, and P3 completes the pressure profile within the pig. This will lead to effective leak detection and localization during pipeline inspections. Running the pig in the pipelines helps in prompt maintenance and repair scheduling that ensures the integrity and safety of the pipeline system. However, as mentioned before, the application of Pig can be costly as it results in interruption in CO_2 transport and consequently loss of revenue.

Visual inspection is an alternative internal pipeline assessment approach that deploys robotic rovers, floating platforms, or cameras inside the pipeline to identify defects or damages. However, this method can also be expensive and requires

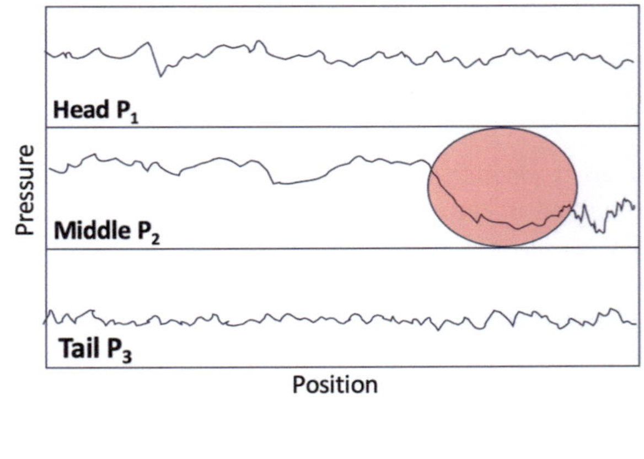

Fig. 5.4 Schematic of "pig" compartment and pressure measurements

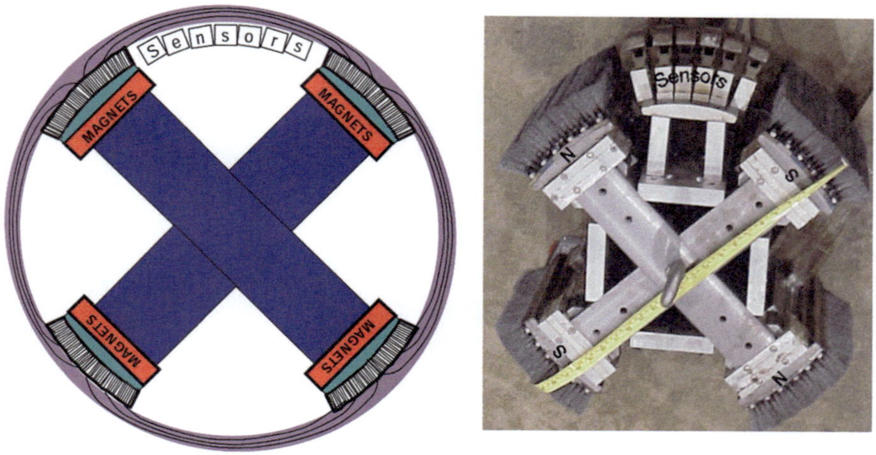

Fig. 5.5 A circumferential MFL magnetizer (https://www.netl.doe.gov/sites/default/files/2018-03/TSA41881-BatelleCrawlers.pdf)

taking the pipeline out of service during inspection. Other internal techniques include ultrasonic inspection, eddy current inspection, and magnetic flux leakage inspection. Magnetic flux leakage (MFL), Fig. 5.5 is commonly used and can detect various types of defects including cracks and corrosion (El Sherbeny et al. 2015). Ultrasonic testing (UT), Fig.5.6, utilizes sensors to detect ultrasonic waves and is effective in detecting high-pressure fluid escape or gas leakage (Al Kindi et al. 2025). However, UT signals require preprocessing techniques for signal enhancement and advanced methods like artificial neural networks to distinguish leak events accurately. Figure 5.7 shows the internal electronic part of ultrasonic leak detection pig that is used for recording the leak noise at the midpoint of the pig.

Despite advancements in internal pipeline assessment techniques, including MFL and UT, they share limitations with pigging, and the results obtained may have uncertainties and constraints. It is essential to quantify and evaluate the results to ensure accurate pipeline integrity assessment. Additionally, these techniques can result in a high rate of false alarms, meaning the leaks identified are not actually happening that introduces unnecessary inspection and additional cost which emphasizes the need for careful analysis and interpretation of the data. Ongoing research and development are necessary to improve the sensitivity, reliability, and efficiency of internal leak detection methods.

5.5.1.2 External Leak Detection Techniques

External leak detection techniques offer various approaches to actively monitor the potential of leakage in the pipelines. These methods involve human visual inspection, or the utilization of equipment placed outside the pipeline to identify signs of leaks.

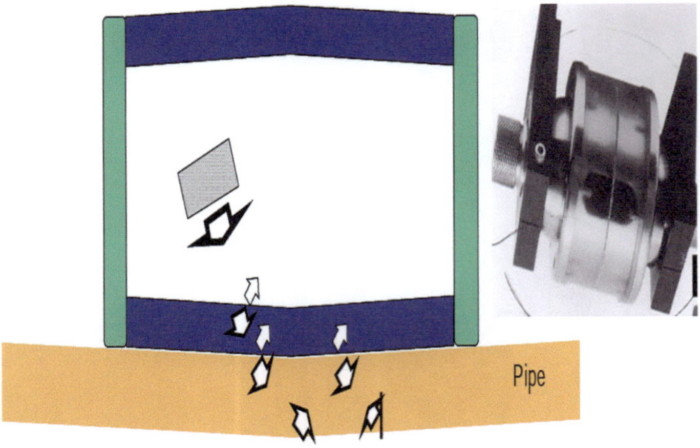

Fig. 5.6 Ultrasonic inspection for cracks. Implementing current in-line inspection technologies on crawler systems, June 2003. https://www.netl.doe.gov/sites/default/files/2018-03/TSA41881-BatelleCrawlers.pdf

Fig. 5.7 The internal electronic part of ultrasonic leak detection pig. *Source* Wilfra–Own work, Public Domain, 2009

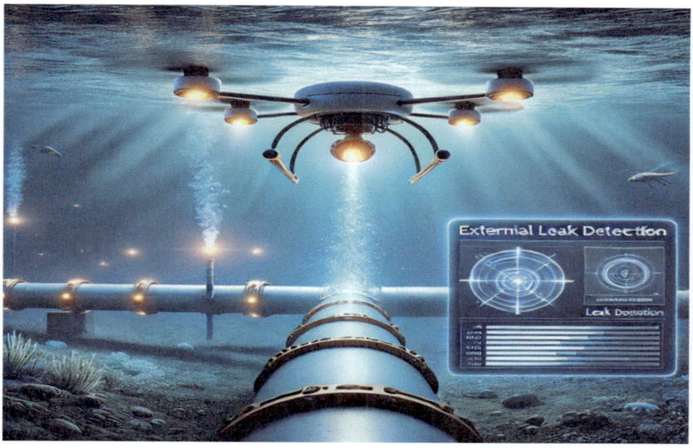

Fig. 5.8 Illustration of external leak detection techniques using drones for subsea pipelines. The drone employs advanced sensors to scan for pipeline leaks beneath the ocean surface created by OpenAI's DALL-E

For instance, drones and airplanes (Figs. 5.8 and 5.9) can be deployed to conduct aerial inspections, satellite imagery can detect changes in surface features or vegetation that can be result of pipe leaks, and thermal imaging can identify temperature anomalies along the pipeline caused by fluid leakage. Pressure transducers installed at different points along the pipeline can also detect pressure drops, which may suggest the presence of a leak.

In recent years more advanced real-time measurements have been introduced. Fiber optic cables installed near the pipeline enable liquid leak detection through collection acoustic (Distributed Acoustic Sensing DAS) temperature (Distributed Temperature Sensing DTS) or strain (Distributed Starin Sensing DSS) data while optical remote sensing techniques are used to detect gas emissions in the atmosphere (Chairunissa et al. 2021). Although these external techniques provide valuable information, they come with certain limitations. They require significant investment in terms of installation and monitoring costs and often struggle to accurately pinpoint the exact location of a leak within a reasonable timeframe. Achieving precise and timely leak localization remains a challenge for these methods. Despite these limitations, ongoing advancements in technology and research continue to enhance external leak detection techniques. Efforts are focused on improving accuracy, efficiency, and cost-effectiveness to ensure the effective monitoring and safety of CO_2 transportation networks.

The application of pressure transducers for pipeline leak detection has seen significant advancements in recent years. The Negative Pressure Wave (NPW) technique that incorporates pressure transducers and computational analysis has shown to be a promising approach for real-time and accurate leak identification and localization (Fathi et al. 2023). By analyzing pressure data and time delays associated with pressure drops the NPW technique can detect and quantify leaks with reasonable

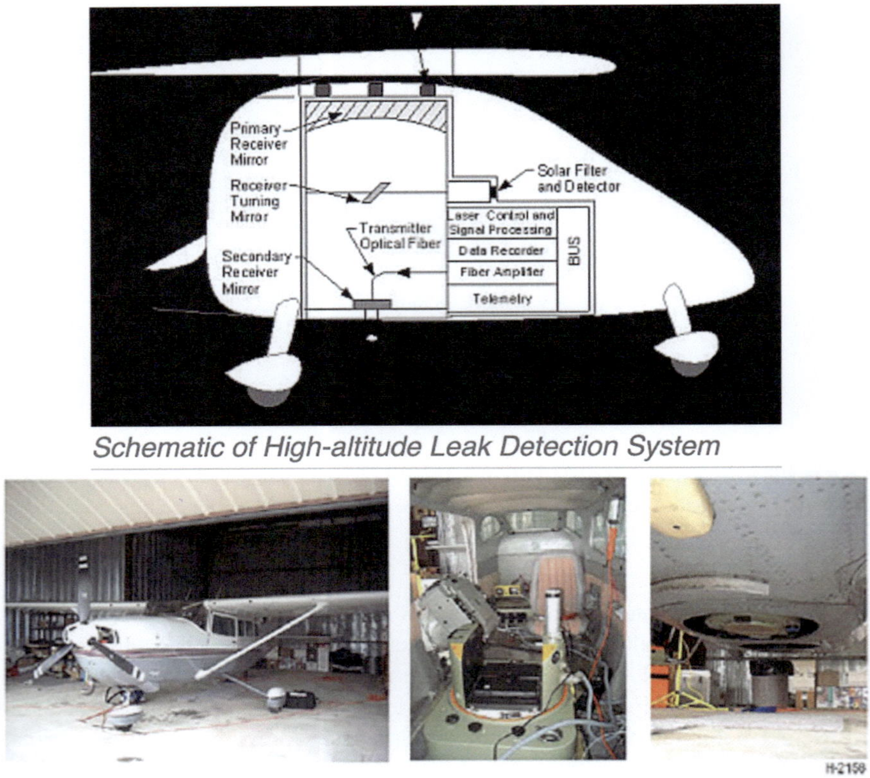

Fig. 5.9 Utilizing drones and airplanes for leak detection, featuring a Cessna 207 with installed photographic equipment in the cabin and an external view of the camera port. *Source*. https://netl.doe.gov/node/3200

accuracy (Reynolds and Kam 2019). It is a cost-effective method that relies on transducer data points and pump schedules or flow measurements. However denoising and filtering are necessary for precise pressure event detection. The NPW technique's success relies on factors such as the velocity of negative pressure waves which can be estimated by processing pressure signals from multiple transducers. It enables the detection of single or multiple leaks across the pipeline network. Integration of NPW with the supervisory control and data acquisition (SCADA) system can help in real time monitoring and leak detection. Physics-based numerical simulations of fluid flow in pipes can also enhance the technique's effectiveness if combined with machine learning algorithms providing more predictive capabilities by revealing the pressure behavior patterns that lead to leaks with different intensity and using that as a predictive tool when such patterns are identified during the real time monitoring of pressure behavior. Despite its advantages the NPW technique has limitations. Locating leaks accurately with low-resolution equipment and high noise-to-signal ratios may introduce uncertainty. It can pose challenges in complex network configurations and exhibits a high false alarm rate due to pump and valve operations. Researchers have addressed these challenges by applying appropriate filtering techniques and adaptive thresholding to improve data quality and reduce false alarms. However studies have mainly focused on simplified cases with single leaks and limited accuracy achieved with low-precision pressure transducers (Fathi et al. 2023). Continued research and refinement are necessary to overcome these limitations and optimize the NPW technique for various pipeline configurations, noise levels, and equipment resolutions.

5.5.2 *Advanced Technologies: Laser-Based, Infrared Cameras, and Leak Sniffers*

External leak detection techniques utilize various technologies to detect and identify leaks in different mediums. Three commonly used techniques are laser-based detection infrared cameras and "leak sniffers" that employ air sampling for analysis (Fahy 2023). Laser-based detection involves setting the laser to the absorption wavelength of the medium to be detected. When the laser beam encounters the medium, a portion of its energy is absorbed by the medium. By measuring the amount of energy absorbed, the quantity of leaked medium can be determined. This method is particularly effective for detecting leaks in gases or volatile substances where the absorption properties of the medium are known.

Infrared cameras are equipped with specialized filters that highlight specific infrared wavelengths. These cameras capture video footage, and by analyzing the selected spectrum of infrared wavelengths, leaks can be identified. When a leak occurs, the escaping medium can exhibit temperature variations compared to the surrounding environment. Infrared cameras can detect these temperature anomalies, allowing for the identification of potential leaks. In Fig. 5.10 the midwave infrared image captures a static frame from an optical gas imaging camera revealing a methane leak (Spidle 2020).

"Leak sniffers" are devices that draw in air samples from the surroundings to evaluate the concentration of the leaked medium (Appah et al. 2021). These samples are then analyzed in a separate unit to determine the presence and concentration of the leaked substance. Leak sniffers are commonly used for detecting gas leaks such as in industrial settings or natural gas pipelines. They provide direct measurements of the leaked medium concentration and are effective in identifying leaks in confined spaces or areas where visual inspection is challenging (Spidle 2020). Figure 5.11 shows handheld sensors, known as "sniffers," designed to identify and measure the presence of leaks.

These external leak detection techniques offer valuable tools for detecting and quantifying leaks in different mediums. Laser-based detection provides precise

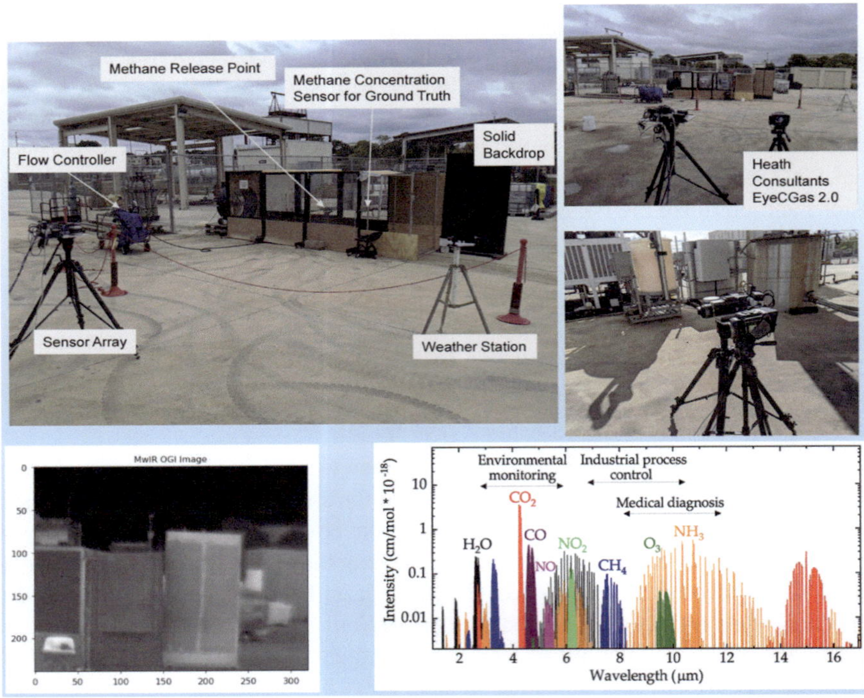

Fig. 5.10 The Midwave Infrared image captures a static frame from an optical gas imaging camera revealing a methane leak (Spidle 2020).https://netl.doe.gov/sites/default/files/netl-file/20VPRONG_28_Spidle.pdf

Fig. 5.11 Handheld sensors, known as "sniffers," designed to identify and measure the presence of leaks. https://netl.doe.gov/node/3138

measurements of leaked medium quantities, infrared cameras can identify temperature anomalies associated with leaks, and "leak sniffers" offer direct measurements of leaked medium concentrations and protection in various industries.

5.5.2.1 Fiber Optic Leak Detection

Monitoring pipelines with fiber optic cables is a promising technique for leak detection and localization. By installing fiber optic cables along the pipeline route it becomes possible to detect leaks by analyzing changes in light signals transmitted through the cables. The advantage of this method lies in its ability to provide accurate and real-time information about the location of the leak (Eisler and Glenn 2012). However, it is important to note that the use of fiber optic cables for leak detection is feasible only within certain limitations in terms of pipeline length. Installing fiber optic cables for monitoring extremely long pipelines can be costly and technically challenging. When installing fiber optic cables, careful consideration must be given to the nature of the medium being monitored. If the medium is gas, it is advisable to install the cables above the pipeline. Since gas tends to rise, positioning the cables above ensures efficient detection of gas leaks. On the other hand, for monitoring liquid pipelines, it is more logical to install the fiber optic cables below the pipeline. This placement allows for effective detection of liquid leaks as the fiber optic cables can capture any liquid that may seep downward. The accuracy and reliability of leak detection can be significantly improved by adapting the installation of fiber optic cables based on the specific characteristics of the medium being monitored, such as gas or liquid (Fig. 5.12).

The fiber leak detection technique involves the use of short pulses of laser light that are sent through optical cables, Fig. 5.13. These laser pulses travel along the cables and interact with the surrounding environment. When the laser light encounters a leak or any disturbance in the pipeline it scatters or reflects (Martin H. Weik 1997). The backscattered light containing valuable information about the pipeline conditions is collected by a signal processing unit. This unit is connected to an optical splitter which splits the incoming light into different channels for analysis. The signal processing unit receives the backscattered light and performs various calculations and measurements to identify any changes or anomalies in the signal. Three distinct fiber-optic measurements are collected: Distributed Acoustic Sensing (DAS) Distributed Temperature Sensing (DTS) and Distributed Strain Sensing (DSS) (Fathi et al. 2024a). By analyzing the characteristics of the backscattered light, such as intensity, wavelength, and time of flight, the presence and location of leaks can be determined. The signal processing unit compares the received signal with a baseline signal or predetermined thresholds to detect any variations that indicate a leak. This fiber leak detection technique offers several advantages, including real-time monitoring, high sensitivity, and the ability to cover long distances. It provides continuous surveillance of the pipeline by analyzing the backscattered light, allowing for early detection and prompt response to leaks or other abnormalities.

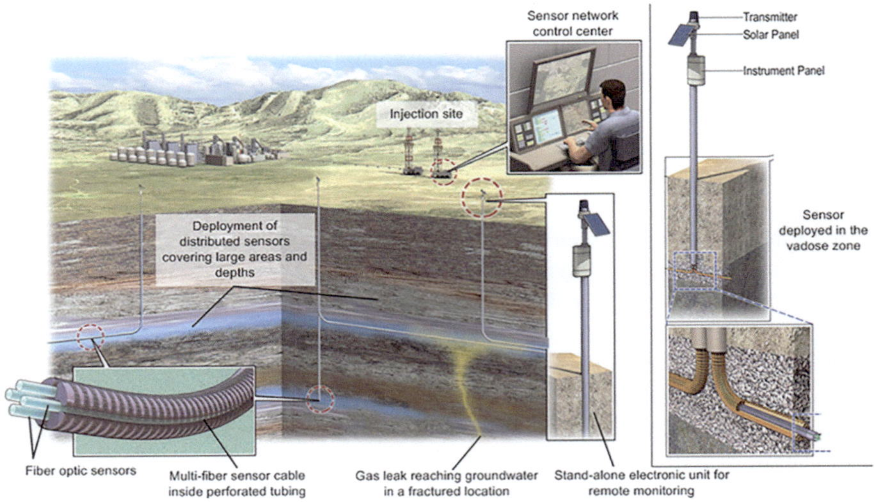

Fig. 5.12 Schematic shows the deployment of fiber optics for gas leak detection *Source.* https://netl.doe.gov/node/5872

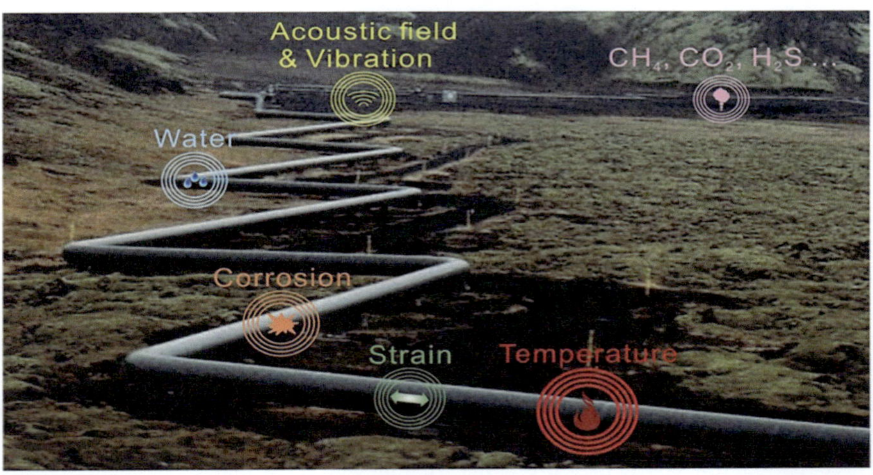

Fig. 5.13 Advanced sensors for real-time monitoring of natural gas pipelines, Dr. Ruishu Wright, FWP-1022424, Project No.1611133." 2021. NETL USDOE. https://netl.doe.gov/sites/default/files/netl-file/21CMOG_OG_Wright.pdf

5.5.2.2 Acoustic Leak Detection

Leak detection using acoustic signals relies on the fact that the flow of gases or liquids through a crack or hole in a pipeline generates distinctive acoustic signals (Brown et al. 2015). To detect leaks, acoustic sensors are installed outside the pipeline at

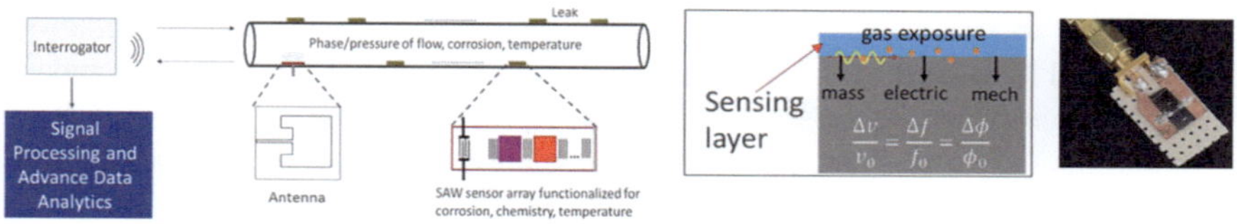

Fig. 5.14 Surface acoustic wave (SAW) Sensors source. https://netl.doe.gov/sites/default/files/netl-file/21CMOG_OG_Wright.pdf

various locations, Fig. 5.14. These sensors measure the noise levels along the pipeline. The collected noise data is used to create a baseline noise profile of the pipeline under normal operating conditions. Any deviations from this baseline profile are indicative of a potential leak. When the measured noise levels exceed the expected or predefined thresholds, a leak alarm is triggered, indicating the presence of a leak in the pipeline. Acoustic leak detection offers several advantages, such as non-invasiveness and the ability to monitor large sections of the pipeline. Analyzing the acoustic signals helps operators to promptly identify and locate leaks. This will result in timely actions to mitigate risks and prevent further damage. The technique requires careful calibration and optimization to minimize false alarms and achieve accurate leak detection. Additionally, factors like environmental noise, pipeline material, and distance between sensors can influence the sensitivity and effectiveness of the acoustic leak detection system. Regular maintenance and calibration are necessary to ensure reliable performance over time using acoustic leak detection.

5.5.2.3 Specialized Leak Detection Techniques

The Negative Pressure Waves (NPW) technique is an effective method for detecting and localizing leaks in pipelines. Leaks cause a significant pressure to drop in the pipeline which generates a propagating pressure wave over time (Reynolds and Kam 2019). This pressure drop is a key indicator of a leak event. When pressure transducers are well placed along the pipeline, they record the pressure drop at different times due to their varying distances from the leak. Analyzing the time delay between the pressure drop events recorded by the transducers can be used to detect and localize the leaks. The analysis focuses on identifying the differences in timing between pressure drop occurrences at different locations.

The Negative Pressure Waves (NPW) technique for leak detection in pipelines faces several challenges. High-resolution equipment required for accurate detection can be expensive making it a cost consideration for implementing the NPW technique (Fathi et al. 2023). Transducer data collected for leak detection is often overwhelmed by significant noise which can make it challenging to accurately identify and analyze pressure events associated with leaks. Filtering techniques are used to remove noise and isolate expected events can sometimes mask the presence of a leak leading to difficulties in detecting and pinpointing its location. The NPW technique encounters complexities in localizing leaks within the pipeline networks that have numerous branches and interconnected segments. The complex network structure can make it harder to precisely locate the source of a leak. The NPW technique involves multiple computational processes for real-time leak detection which can be computationally intensive. Managing these processes and ensuring accurate and timely detection may pose challenges in certain operational scenarios. In certain situations there might be an inadequacy of available leak data for a particular pipeline network. This limitation can impede the calibration of the workflow for a specific site consequently reducing the accuracy and reliability of leak detection using the NPW technique. The NPW technique may also be subject to a high rate of false leak detection where pressure fluctuations or anomalies unrelated to actual leaks can trigger alarms (Korlapati et al. 2022). This can result in unnecessary interventions and resource allocation. These challenges highlight the considerations and complexities associated with implementing the NPW technique for leak detection in pipelines. Addressing these challenges through advanced equipment, noise reduction techniques, improved data analysis algorithms, network-specific optimizations, and strategies to reduce false detections can enhance the effectiveness and reliability of the NPW technique.

5.5.2.4 Negative Pressure Waves Technique—A Case Study

Figure 5.15 represents a water pipeline network used in a study conducted in Texas United States by Fathi et al. (2023). The network comprises multiple interconnected components involved in the water distribution system with more than 5 Km pipeline. There are 15 water sources shown in green arrows in the network responsible for pumping water into the system. Each source has an accumulation tank, and the water pumping process is initiated when the specified tank limit is reached. The injection rate from each source is adjusted based on the total capacity of the network to maintain the desired fluid level in the accumulation tank. The system operates in a periodic manner, where water pumping and adjustment of injection rates occur repeatedly. All the water injected in the network will be disposed through single injection wells shown as red arrow (point 1). The system also includes 4 pressure transducers installed in points 19, 20, 21, and 24. Data from the network was collected over a period of three months and includes information on pressure, flow rates, and pump status. The resolution of the recorded data is 1 minute for flow rates, 1 second for pressures, and pump status is represented using binary values (1 for on and 0 for off) for each source.

Figure 5.16 displays the pressure profiles at 20 different locations within the network over time. It shows how the pressure changes at each location as water flows through the pipeline. Figure 5.17 also shows the flow rates at 15 injection sites within the network. These flow rates indicate the amount of water being pumped into the system from each source. The derivative of the pressure is plotted in Fig. 5.19, which represents the rate of change of pressure over time. This information can provide insights into pressure fluctuations and potential anomalies in the system. To accurately pinpoint anomalies that may indicate a leak, a dynamic control limit line is

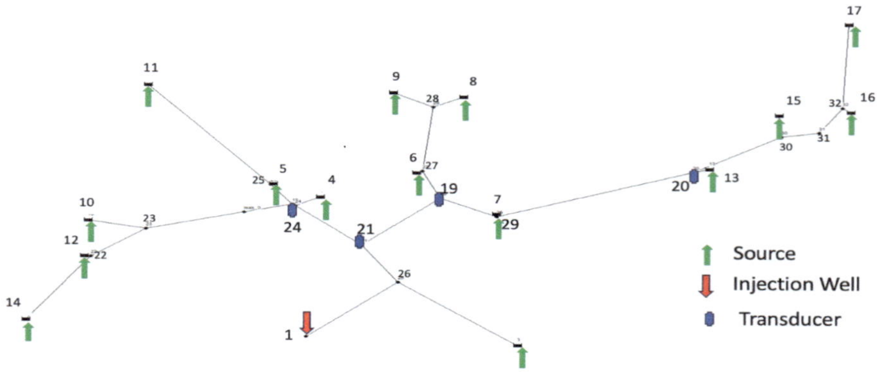

Fig. 5.15 Water transport network south Texas modified (Fathi et al. 2023)

employed. This control line sets upper and lower thresholds based on a specified number of standard deviations from the mean of pressure derivatives. When the data significantly deviates from this moving window, it signals an event. However, due to the fluctuating nature of pressure and flow rate caused by pump scheduling, the mean determining these control limit lines cannot remain static. To address these variations, a rolling window approach is implemented. This method involves calculating the mean and standard deviation over a window of the preceding X number of data points for each timestep. This approach captures the local behavior of the pressure data, allowing to adapt the dynamic control limit lines to the specific characteristics of each location. By considering dynamic changes in the mean, the accuracy of the control limit lines can be enhanced. This decision on the number of proceeding data points to obtain the dynamic mean is influenced by the resolution of flow rate and pressure measurements, enabling the control limit lines to be customized for the unique characteristics of each location. While the rolling window approach effectively handles the challenge of a non-constant mean, it's essential to note that initially setting the control limit line at three times the standard deviation was based on a widely used threshold in the statistical process control (Fig. 5.18).

Figure 5.19 includes a control limit line that sets the upper and lower boundaries for pressure values. Any pressure events that exceed these limits or fall below the lower limit are considered potential leak events. These events are then recorded and compared with flow events to make sure the fluctuations in pressure do not relate to pump events. Following are the steps 1 to 5 required to be completed for leak detection using NPW using a workflow presented by Fathi et al. (2023).

Step 1: Identification and Organization of Flow and Pressure Events.

Flow rate and pressure events exceeding the upper and lower boundary limits of the control limit line are identified. These events are organized separately for flow and pressure. For pressure events, only values below the lower boundary limit are considered, indicating potential pressure drops from leaks. Pump events, affecting both upper and lower limits of the control limit line, are also considered. Event tables

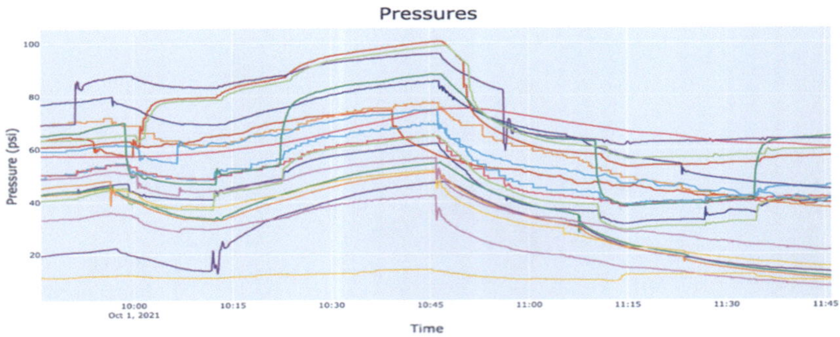

Fig. 5.16 Water transport network pressure measurements modified (Fathi et al. 2023)

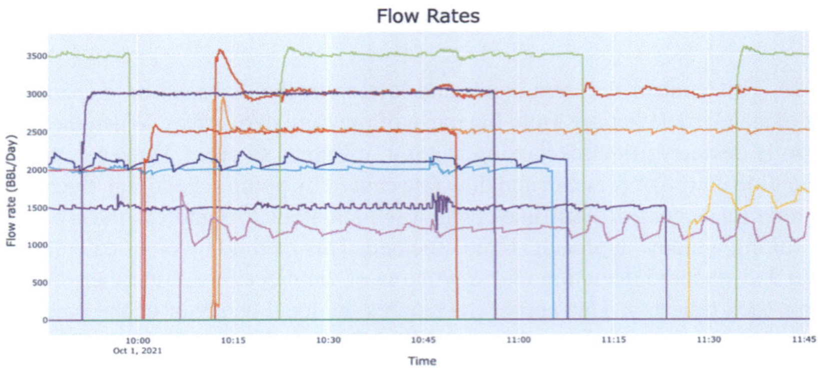

Fig. 5.17 Water transport network flow rate measurements modified (Fathi et al. 2023)

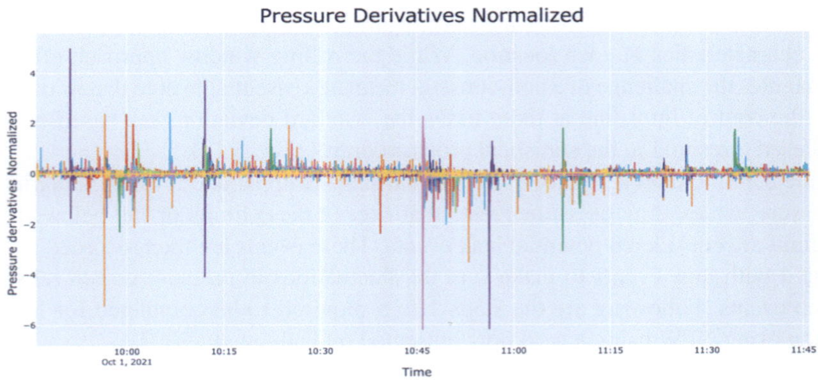

Fig. 5.18 Normalized pressure derivatives modified (Fathi et al. 2023)

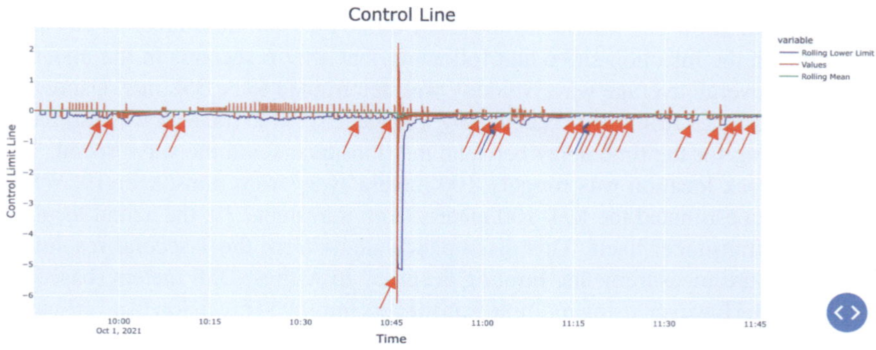

Fig. 5.19 Water transport network sample pressure derivative and control limit line presentation modified (Fathi et al. 2023)

are created, including relevant information like time stamps, actual values, derivative values, rolling means, rolling standard deviations, and control limit line bounds.

Step 2: Combining Flow and Pressure Events.

Flow events from all 15 injection sites are combined into a consolidated flow event log. The associated time stamps are extracted. Similarly, pressure events from the 20 pressure transducers are integrated into a unified pressure log.

Step 3: Filtering of Pressure Events.

Pressure events that coincide with flow events or occur within 60 seconds after a flow event are removed from the pressure log. This filtering step eliminates pressure events associated with pump activities, allowing the focus on potential leak events.

Step 4: Sorting and Grouping of Pressure Events.

The remaining pressure events are sorted by magnitude. Pressure events occurring within a 20-second time span are grouped together. Each group represents a sequence of events that may indicate a leak in the network.

Step 5: Cross-Validation Procedure.

Remaining events undergo a cross-validation procedure to ensure their validity. Each event is tracked to verify its consistent appearance in nearby measurement sites. The arrival time of the event is cross-checked based on the relative distance between the transducers. If all tests are satisfied, the event is considered a leak event and proceeds to further investigation.

This step-by-step workflow provides a systematic approach to detect and investigate potential leaks in a pipeline network, combining information from flow and pressure events while considering the characteristics of pump activities. By filtering out non-leak-related events and applying cross-validation procedures, the workflow aims to identify and prioritize potential leak events for further investigation.

The results show one leak identified at the inline pressure transducer 19 and after 3 seconds in the injection site 6 and following that after 6 seconds in the injection site 7. The overall average wave velocity was determined to be 358 m/s. Using this value, the distance from the first detecting transducer (19) to the leak was estimated by accounting for the time delay between the transducers and the wave speed. The calculated leak location was roughly 100 meters away from transducer 19. While the field data estimated the leak 100 meters from transducer 19, the actual location was at the transducer itself. This discrepancy stems from the 1-second resolution of the pressure measurements, limiting accuracy to within ~358 meters (based on wave speed). However, using multiple transducers improved precision to a 100-meter radius, which is considered highly accurate given the equipment constraints.

The NPW workflow tackles difficulties related to noisy data, rapid pressure decay, and the necessity for robust detection of leak events. The key advantages of the workflow include its effectiveness in detecting and localizing leaks in complex network configurations and minimizing false positives. The workflow is adaptable to different fluid systems, opening potential applications in various industries. However, the current focus is on single leak detection, and future research can explore extending it to handle multiple leaks. With further advancements, NPW holds promise for practical implementation in various industries.

5.5.2.5 Industry Standards and API Requirements

The requirements for an effective Leak Detection System (LDS) can be summarized as being sensitive to small changes, being reliable and accurate and having a robust performance. The LDS must be capable of detecting small leaks and detecting them quickly. It should minimize fluid loss by promptly identifying even minor leaks in the pipeline. The user's trust in the LDS is the most important point to consider. The system should accurately report real alarms while avoiding false alarms. False alarms can disrupt operations and lead to unnecessary interventions and downtime. Accurate detection and precise information about the leak's magnitude and position enables the operators to promptly react and effectively implement the mitigation strategies. The LDS should demonstrate robust performance even under non-ideal circumstances. For instance, in the event of a transducer failure, the system should promptly detect the failure and continue monitoring, even though this might result in reduced sensitivity. The ability to withstand such challenges ensures continuous monitoring and minimizes the risk of undetected leaks.

In addition to these general requirements, there are specific API standards and regulations related to leak detection in the United States including API 1130, 1149, and American 49 CFR 195.

API 1130 addresses computational pipeline monitoring for liquids (Maderia et al. 2024). It provides guidelines and requirements for implementing effective leak detection systems in liquid pipelines. API 1149 addresses the variability uncertainties within pipelines and their influence on the effectiveness of leak detection. It aids in comprehending and effectively managing uncertainties linked to leak detection procedures. American 49 CFR 195 governs the transportation of hazardous liquids through pipelines in the United States. It includes provisions and guidelines for leak detection and safety measures. Compliance with these API standards and regulatory requirements ensures that the LDS meets industry best practices and operates in accordance with legal and safety guidelines. These standards collectively can enhance the effectiveness and reliability of leak detection systems.

5.5.3 CO_2 Leak Detection in Subsurface Geological Storage

Subsurface geological carbon storage is a proven technology that has shown successful technical and economical implementation around the world. However, to ensure the safe and sustainable storage of CO_2, robust leak detection techniques required to monitor CO_2 storage in long term. Different techniques such as geophysical, geochemical, and remote sensing approaches have been practiced for CO_2 leak detection in subsurface geological storage.

Various techniques, both at the surface and downhole, have been suggested for the detection of CO_2 leakage from geologic CO_2storage reservoirs. Surface methodologies include geophysical approaches such as Seismic Monitoring Electrical Resistivity Tomography and Gravity (Yang et al. 2019). The geophysical surface monitoring strategy involves the deployment of multiple survey lines at the surface that measures changes in the seismic velocity of subsurface materials. These changes in velocity results in adjustments in seismic travel times due to the presence of CO_2. In cases where significant amount of CO_2presented the elastic properties of the subsurface can also be affected that also influences the measured seismic response at the surface (Yang et al. 2013). The seismic surveys help to detect CO_2 plume distribution and potential CO_2 leakage from caprock when time lapsed 4D seismic is used. In this case the 3D survey will be deployed at different times, and comparison with the baseline will be used to interpret the CO_2 movements. This technique has large coverage with good resolution, however, it's extremely expensive and is not sensitive to small leaks.

Electrical Resistivity Tomography is another geophysical technique that can detect alterations in bulk electrical resistivity attributed to CO_2 leakage. This detection results from the combined impact of resistive CO_2 and conductive species concentrations arising from CO_2 dissolution. The CO_2dissolved in brine will change the brine resistivity that can be captured by ERT technique. The ERT can provide either two-dimensional or three-dimensional resistivity or contrast of electrical resistivity distributions. The disadvantage of ERT is the shallow depth applicability of the technique up to 200 meters that limits its application for deep storage monitoring (Sattar et al. 2020).

The gravity method is a geophysical imaging technique which is sensitive to mass distribution (Walker et al. 2023). It relies on observed density changes over time due to CO_2 displacement. The effectiveness of the gravity method depends on the magnitude of the density variations. In the event of a supercritical CO_2 leak from the reservoir, CO_2 rises due to buoyancy, as it reaches shallow depths due to decrease in pressure it changes from the supercritical CO_2 to the gas CO_2 which results in significant changes in the density that can be detected and recorded at the surface.

Geochemical methods can also be used to detect CO_2 leakage to the surface such as soil-gas monitoring where CO_2 concentrations and isotopic signatures are measured near the surface (Themann, et al. 2009).

Remote sensing such as satellite and aerial monitoring using advanced spectrometers, hyperspectral imaging, or drones and aircrafts can also be used to detect CO_2 concentrations. However, these techniques are more effective in CO_2 leakage from surface facilities as they have very large coverage make it hard for small leak detections.

The downhole approaches include downhole sampling (of water or soil) downhole chemistry monitoring downhole pressure monitoring and combination of downhole pressure and chemistry monitoring (Buscheck et al. 2019).

Downhole soil sampling utilizes gas chromatography to measure various soil-gas components of soil samples taken from different depths, including methane (CH_4), oxygen (O_2), nitrogen (N_2), volatile organic compounds (VOCs), and CO_2 (Zhao et al. 2017). Downhole water sampling involves collecting water samples from specific well depths to analyze water composition, enabling the assessment of CO_2 leakage impact on water quality. Parameters such as pH, dissolved oxygen, concentrations of specific ions, $CO_2$and total dissolved solids (TDS) are considered in this process (Kharaka et al. 2008). In downhole chemistry monitoring, a set of geochemical indicator parameters such as dissolved inorganic carbon (DIC), dissolved CO_2or total dissolved solids is identified based on groundwater geochemistry characterization (Yang et al. 2015). Specialized sensors or probes are used for real-time or periodic measurements within the well, enabling continuous monitoring to track changes over time.

Downhole Pressure Monitoring involves measuring changes in subsurface pressure using high-precision downhole gauges. Detection of pressure changes caused by CO_2 leakage is achieved by comparing predicted pressure profiles with measured pressure profiles. Monitoring typically occurs in a well adjacent to the CO_2 injection well. The depth of leakage along the wellbore is a critical factor influencing CO_2, total dissolved solids (TDS), and pressure plumes. Supercritical CO_2 migrating to depths below 800 m undergoes a phase change to gaseous CO_2, resulting in a significant pressure rise and spreading of TDS and pressure plumes (Trainor-Guitton et al. 2016; Thomas et al. 2019).

Downhole water and soil sampling provide detailed insights into groundwater and subsurface gases, respectively. In contrast, chemistry monitoring and pressure monitoring focus on geochemical indicators and subsurface pressure changes, respectively. Downhole pressure monitoring is more reliable due to its high precision in detecting CO_2 leakage that offers reliable data on subsurface pressure changes. Surface techniques, like Electrical Resistivity Tomography, offer cost-effective monitoring with moderate accuracy but require complex interpretation.

Downhole methods, while more accurate and readily available, are more expensive and the choice depends on budget, desired precision, and the urgency of CO_2 monitoring needs.

5.6 Surveillance Techniques for CO_2 Pipelines

Pipeline surveillance and leak detection are two closely related but distinct concepts in pipeline safety management. Leak detection as discussed earlier focuses specifically on identifying and locating leaks within a pipeline system. In contrast, pipeline surveillance includes a broader range of activities planned for maintaining the overall integrity and security of the pipeline. It includes continuous monitoring, routine inspections, and protective measures against external activities and threats against the facility. This can be done by using aerial surveillance, ground patrols or drone inspections. Aerial surveillance can be done using aircrafts equipped with high resolution cameras or infrared sensors to monitor large areas. This can also be done using drone inspections as well. The visual inspection by inspectors driving along the pipeline is also an option where the pipeline is limited and accessible for ground patrols.

Video cameras and surveillance systems such as closed-circuit television (CCTV) can also be used for continuous monitoring of the pipeline to prevent theft and sabotage.

5.7 Conclusion

The safe and efficient transportation of CO_2 is a critical component of carbon capture and storage (CCS) systems. This ensures that captured carbon is delivered to storage sites without compromising environmental or operational safety. This chapter has explored technologies and strategies that support CO_2 transport monitoring, leak detection, and surveillance.

In this chapter different methodologies for CO_2 transport, including pipelines and shipping, each with distinct economic and logistical considerations, are discussed. Pipeline networks offer cost-effective solutions for large-scale transport but require rigorous safety protocols. Shipping is more flexible and used for remote storage sites, however, it faces challenges in pressurization and temporary storage. Both methods demand robust leak detection systems to mitigate risks.

The chapter highlighted leak detection technologies, ranging from internal monitoring systems to external techniques. Case studies, such as application of the Negative Pressure Waves technique for real-time leak detection and localization showed gcreat promises to prevent leaks from transportation lines. Surveillance systems emerged as another critical layer of protection, integrating remote sensing, drones, and AI-driven analytics to monitor pipeline integrity and detect anomalies. These

Fig. 5.20 Images designed by Freepik

systems not only enhance safety but also align with regulatory frameworks that require continuous monitoring of CO_2 transport infrastructure. Looking ahead, the scalability of CCS will depend on advancements in smart monitoring technologies, and standardized regulations to address gaps in the current systems (Fig. 5.20).

References

A. Al Kindi, V.E. Sebastian, H. Al Hashemi, Non-intrusive corrosion monitoring technology, in *SmartPIMS and MatPIMS. AMPP Annual Conference + Expo*, (Nashville, TN, United States, 2025, April)

D. Appah, V. Aimikhe, W. Okologume, Assessment of gas leak detection techniques in natural gas infrastructure: a review, in *SPE Nigeria Annual International Conference and Exhibition*, (Lagos, Nigeria, 2021). https://doi.org/10.2118/208236-MS

G. Brown, S. Fasham, P. Tomlinson, R. Crook, *A case study on the implementation of an acoustic automatic leak detection sonar (ALDS) in GoM and its application to current and future Brazilian fields* (OTC Brasil, Rio de Janeiro, Brazil, 2015). https://doi.org/10.4043/26270-MS

S. Bond, Ship design for CO_2 transportation. In Global Gas Development Science Day, London, England (2024)

J. Burger, J. Nöhl, J. Seiler, P. Gabrielli, P. Oeuvray, V. Becattini, A. Reyes-Lúa, L. Riboldi, G. Sansavini, A. Bardow, Environmental impacts of carbon capture, transport, and storage supply chains: Status and the way forward. Int. J. Greenhouse Gas Control, **132**, 104039 (2024), ISSN 1750-5836. https://doi.org/10.1016/j.ijggc.2023.104039

T.A. Buscheck, K. Mansoor, X. Yang, H.M. Wainwright, S.A. Carroll, Downhole pressure and chemical monitoring for CO_2 and brine leak detection in aquifers above a CO_2 storage reservoir. Int. J. Greenh. Gas Con. **91**, 102812 (2019). https://doi.org/10.1016/j.ijggc.2019.102812

D.A. Camerini, J.P. von der Weid, C.S. Camerini, C.E. Maia, Proceedings of the 2004 international pipeline conference. 2004 international pipeline conference, volumes 1, 2, and 3, in *Leak detector pig for oil pipelines*, (Calgary, Alberta, Canada, ASME, 2004), pp. 1065–1068. https://doi.org/10.1115/IPC2004-0095

C. Chairunissa, D.K. Amanu, G. Astari, E. Indrayana, SPE/IATMI Asia Pacific Oil & Gas Conference and Exhibition, Virtual, in *Kedung Keris full well stream pipeline fiber optic leak detection system*, (2021). https://doi.org/10.2118/205776-MS

P.L.C. Clarkson. Liquid CO_2 carriage by sea: an introduction. (2022). https://www.clarksons.com/home/news-and-insights/2022/liquid-co2-carriage-by-sea-an-introduction/

B. Eisler, G.A. Lanan, Fiber optic leak detection systems for subsea pipelines, in *Offshore Technology Conference*, (Houston, TX, United States, 2012). https://doi.org/10.4043/23070-MS

W. El Sherbeny, I. Nuic, G. Hasan, A. Abdesslam, T. Hassane, Magnetic flux leakage (MFL) technology provides the industry's most precise pipe integrity and corrosion evaluation, accurately characterizing casing and tubing strength, in *Technology Overview and Case History. SPE North Africa Technical Conference and Exhibition*, (Cairo, Egypt, 2015). https://doi.org/10.2118/175871-MS

F. Engel, A. Kather, Improvements on the liquefaction of a pipeline CO2 stream for ship transport. Int. J. Greenh. Gas Con. **72**, 214–221 (2018)

Fahy, R. (2023). Quantification & reduction of methane (GHG) emissions through aerial intelligence. ADIPEC, Abu Dhabi, UAE. https://doi.org/10.2118/216027-MS

E. Fathi, M.F. Adenan, N. Moryan, F. Belyadi, H. Belyadi, High-precision single-leak detection and localization in single-phase liquid pipelines using the negative pressure wave technique: an application in a real-field case study. SPE J. (2023). https://doi.org/10.2118/217455-PA

E. Fathi, F. Belyadi, M.F. Adenan, C. Pacheco, Application of fiber optics for completion design optimization: a methodological approach and key findings. Fuels **5**(1), 33–52 (2024b)

E. Fathi, G.B. Liwuitekong Dongho, B. Heidari, T.R. Carr, F. Belyadi, I. Bilgesu, Optimizing deep geothermal drilling for energy sustainability in the Appalachian Basin. Sustainability **16**(18), 8053 (2024). https://doi.org/10.3390/su16188053

Global Centre for Maritime Decarbonization. (2024) Landmark study on offloading onboard captured carbon dioxide. https://www.gcformd.org/landmark-study-on-offloading-onboard-captured-carbon-dioxide/

T. Grant, A. Guinan, C. Shih, S. Lin, D. Vikara, D. Morgan, D. Remson, Comparative analysis of transport and storage options from a CO_2 source perspective. Int. J. Greenh. Gas Con. **72**, 175–191 (2018). https://doi.org/10.1016/j.ijggc.2018.06.012

W.Y. Hong, A techno-economic review on carbon capture, utilisation and storage systems for achieving a net-zero CO_2 emissions future. Carbon Capture Sci. Technol. **3**, 100044 (2022). https://doi.org/10.1016/j.ccst.2022.100044

M. Hussain, T. Zhang, M. Chaudhry, I. Jamil, S. Kausar, I. Hussain, Review of prediction of stress corrosion cracking in gas pipelines using machine learning. Machines **12**(1), 42 (2024). https://doi.org/10.3390/machines12010042

IEAGHG, The status and challenges of CO_2 shipping infrastructures (2020). https://ieaghg-publications.s3.eu-north-1.amazonaws.com/Technical+Reports/2020-10+The+Status+and+Challenges+of+CO2+Shipping+Infrastructures.pdf

Y.K. Kharaka, J.J. Thordsen, E. Kakouros, G. Ambats, W.N. Herkelrath, S.R. Beers, J.T. Birkholzer, J.A. Apps, N.F. Spycher, L. Zheng, R.C. Trautz, H.W. Rauch, K.S. Gullickson, Changes in the chemistry of shallow groundwater related to the 2008 injection of CO_2 at the ZERT field site, Bozeman, Montana. Environ. Earth Sci. **60**, 273–284 (2010). https://doi.org/10.1007/s12665-009-0401-1

N.V.S. Korlapati, F. Khan, Q. Noor, S. Mirza, S. Vaddiraju, Review and analysis of pipeline leak detection methods. J. Pipeline Sci. Eng. **2**(4), 100074 (2022). https://doi.org/10.1016/j.jpse.2022.100074

E. Larson, C. Greig, J. Jenkins, E. Mayfield, A. Pascale, C. Zhang, J. Drossman, R. Williams, S. Pacala, R. Socolow, E. Baik, R. Birdsey, R. Duke, R. Jones, B. Haley, E. Leslie, K. Paustian, A. Swan, *October 29* (Potential pathways, infrastructure, and impacts. Princeton University, Net-zero America, 2021) https://netzeroamerica.princeton.edu/

I. Madeira, H.R.B. Orlande, M.J. Colaço, Leak detection of pipelines. In Handbook of Pipeline Engineering (pp. 833–855). Springer (2024). https://doi.org/10.1007/978-3-031-33328-6_20

D. Morgan, A. Sheriff, M. Mark-Moser, G. Liu, T. Grant, G. Creason, D. Vikara, L. Cunha, Modeling the cost of onshore CO_2 pipeline transport and onshore CO_2 saline storage, in *SPE/AAPG/SEG CCUS Conference*, (TX, United States, Houston, 2024)

K. Orchard, CO_2 transport: the role of CO_2 shipping, in *UK CCS Research Centre Workshop*, (2019)

M. Ozaki, T. Ohsumi, CCS from multiple sources to offshore storage site complex via ship transport. Energy Procedia **4**, 2992–2999 (2011). https://doi.org/10.1016/j.egypro.2011.02.157

P.W. Parfomak, Carbon dioxide pipelines: safety issues (CRS report IN11944). Congressional Research Service (2022). https://crsreports.congress.gov/product/pdf/IN/IN11944

S. Parker, Riviera, in *New Wave of LCO2 Ships Emerges, but Ports Remain a Bottleneck*, (2024) https://www.rivieramm.com/news-content-hub/news-content-hub/new-wave-of-lco2-ships-emerges-but-ports-remain-a-bottleneck-80312

J. Reynolds, A. Kam, PSIG Annual Meeting, in *An evaluation of negative pressure wave leak detection: challenges, limitations and use cases*, (London, United Kingdom, 2019)

M.A. Sattar, M.M. Garcia, R. Banasiak, L.M. Portela, L. Babout, Electrical resistance tomography for control applications: quantitative study of the gas-liquid distribution inside a cyclone. Sensors **20**, 6069 (2020). https://doi.org/10.3390/s20216069

R.A. Schultz, S. Williams-Stroud, B. Horváth, J. Wickens, H. Bernhardt, W. Cao, P. Capuano, T.A. Dewers, R.A. Goswick, Q. Lei, M. McClure, U. Prasad, B.A. Schwartz, H. Yu, S. Voegeli, Q. Zhao, Underground energy-related product storage and sequestration: Site characterization, risk analysis and monitoring. Geological Society, London, Special Publications, **528**(1), 37–59 (2023). https://doi.org/10.1144/SP528-2022-66

L. Shawn, C. Yue, D. Ryan, S. Shahidi, Evaluation of Internal Leak Detection Techniques. Paper presented at the PSIG Annual Meeting, New Orleans, Louisiana (2015)

SIGTTO, *SIGTTO Information Paper No.15:A Listing of Design Guidelines for Liquefied Gas Terminals (Referencing Ports and Jetties)* (Witherby & Co. Ltd., Livingston, UK, 1997)

H. Spidle, Smart methane emission detection system development. Oil & Natural Gas 2020 Integrated Review (2020)

Splash247.com, *Shipping of carbon dioxide readies to go mainstream* (Splash Extra, 2023) https://splash247.com/shipping-of-carbon-dioxide-readies-to-go-mainstream/

S. Themann, H.M. Schmidt, D. Esser, Measurement, monitoring, and verification of CO_2 storage: An Integrated Approach, in *SPE International Conference on CO_2 Capture, Storage, and Utilization*, (San Diego, CA, United States, 2009). https://doi.org/10.2118/129127-MS

C. Thiberville, Y. Wang, P. Waltrich, W. Williams, S.I. Kam, Modeling of smart pigging for pipeline leak detection. SPE Prod. Oper. **35**, 610–627 (2020). https://doi.org/10.2118/198648-PA

A. B. Thomas, M. Kayyum, Y. Xianjin, M. W. Haruko, A. Susan, Carroll, Downhole pressure and chemical monitoring for CO_2 and brine leak detection in aquifers above a CO_2 storage reservoir. Int. J. Greenhouse Gas Control, **91**, 102812 (2019), ISSN 1750-5836. https://doi.org/10.1016/j.ijggc.2019.102812. https://www.sciencedirect.com/science/article/pii/S1750583619301677

S. Topham, S. Bazzanella, S. Luhr, L. Zhao, A. Otto, D. Stolten, Carbon dioxide, in *Ullmann's Encyclopedia of Industrial Chemistry*, (Wiley-VCH Verlag GmbH & Co. KGaA, 2014), p. 43

W. Trainor-Guitton, K. Mansoor, Y. Sun, S. Carroll, Merits of pressure and geochemical data as indicators of CO2/brine leakage into a heterogeneous, sedimentary aquifer, Int. J. Greenhouse Gas Control, **52**, 237–249 (2016), ISSN 1750-5836. https://doi.org/10.1016/j.ijggc.2016.07.002

B. Wetenhall, H. Aghanjani, H. Chalmers, S.D. Benson, M.-C. Ferrari, J. Li, J.M. Race, P. Singh, J. Davison, Impact of CO_2 impurity on CO_2 compression, liquefaction, and transportation. Energy Procedia **63**, 2764–2778 (2017)

K. Walker, A.P. Garcia, J. Nunn, G. Lyman, SPE Energy Transition Symposium, in *CO_2 leak detection and conformance verification using borehole gravity*, (Houston, TX, United States, 2023). https://doi.org/10.2118/215734-MS

M.H. Weik, *Fiber Optics Standard Dictionary*, 3rd edn. (Springer, 1997). https://doi.org/10.1007/978-1-4615-6149-4

X. Yang, T. Buscheck, K. Mansoor, Z. Wang, K. Gao, L. Huang, D. Appriou, S. Carroll, Assessment of geophysical monitoring methods for detection of brine and CO_2 leakage in drinking water aquifers. Int. J. Greenhouse Gas Control **90**, 102803 (2019). https://doi.org/10.1016/j.ijggc.2019.102803

X. Yang, R.N. Lassen, K.H. Jensen, M.C. Looms, Monitoring CO_2 migration in a shallow sand aquifer using 3D crosshole electrical resistivity tomography. Int. J. Greenhouse Gas Control **42**, 534–544 (2015). https://doi.org/10.1016/j.ijggc.2015.09.005

C. Yang, K. Romanak, S. Hovorka, R.M. Holt, J. Lindner, R. Trevino, Near-surface monitoring of large-volume CO_2 injection at Cranfield: Early field test of SECARB phase III. SPE J. **18**(3), 486–494 (2013). https://doi.org/10.2118/163075-PA

X. Zhao, H. Deng, W. Wang, et al., Impact of naturally leaking carbon dioxide on soil properties and ecosystems in the Qinghai-Tibet plateau. Sci. Rep. **7**, 3001 (2017). https://doi.org/10.1038/s41598-017-02500-x

Zero Emissions Platform & Carbon Capture & Storage Association. (2024). Achieving a European market for CO_2 transport by ship

Chapter 6
Geological CO_2 Storage: Mechanisms and Applications

6.1 Chapter Overview and Objectives

6.1.1 Overview

In this chapter we will explore different aspects of underground geological storage of carbon dioxide and considerations regarding the best practices that enable efficient and permeant storage. We will start by introduction to the volume of CO_2 storage needed on a global scale and its significance in reducing greenhouse gas emissions. We will then investigate the concept of supercritical CO_2 and its properties. Next, we will discuss the different rock types that are suitable for CO_2 storage, focusing on their capacity and trapping mechanisms. Additionally, we will explore different types of storage sites, where we highlight their unique characteristics and suitability for CO_2 storage. To illustrate these concepts, we will examine a real-world example of the Illinois Basin–Decatur Project (IBDP) injection site and discuss the practical application of geological storage techniques.

6.1.2 Learning Objectives

Upon completion of this chapter, you will be able to:

1. Explain the importance of geological storage for long-term CO_2 sequestration,
2. Identify the volume of CO_2 storage needed and to optimize the injection ad storage strategies,
3. Explain the concept of supercritical CO_2 and its role in efficient storage,
4. Recognize different rock types suitable for CO_2 storage and understand their capacity and trapping mechanisms,

E. Fathi and F. Belyadi, *Subsurface CO_2 Sequestration*,
https://doi.org/10.1007/978-3-032-28296-5_6

5. Analyze various storage site options, such as depleted oil and gas fields, saline aquifers, and coal seams, and evaluate their suitability for CO_2 storage based on geological and engineering considerations,
6. Evaluate the potential of geological storage as a viable and effective option for CO_2 mitigation, considering its technical, and environmental aspects,
7. Analyze real-world examples like the Illinois Basin–Decatur Project for insights into operational aspects and challenges of geological storage.

6.2 Introduction to Geological Storage of CO_2

As previously discussed, industries such as power generation, steel, cement, and petrochemicals emit massive volumes of CO_2, that requires effective capture and storage solutions. We have also explored various CO_2 capture technologies, along with their capabilities for compression and transportation either for storage or utilization. Given the scale of CO_2 emissions, geological storage has become one of the most viable long-term solutions. This method involves injecting compressed CO_2 in a supercritical state into deep underground formations to securely be stored for thousands of years. Suitable geological storage sites include deep saline aquifers, depleted oil and gas reservoirs, igneous rocks (e.g., basalts), shale and coalbed methane reservoirs, and active oil and gas reservoirs, where CO_2 is used for Enhanced Oil Recovery (EOR) and Enhanced Gas Recovery (EGR) while simultaneously being stored. Each of these storage options must be carefully selected and monitored to ensure long-term storage of carbon dioxide.

6.3 The Rationale for Geological Storage

Geological storage, also referred to as underground or subsurface storage, is widely considered as an effective method for securely storing captured carbon dioxide (CO_2). The primary goal of geological storage is to ensure long-term storage of CO_2 from the atmosphere, typically for periods exceeding 10,000 years, this time frame is a critical requirement for meaningful greenhouse gas reduction. The viability of geological storage is supported by its proven ability to store massive quantities of hydrocarbons over geologic time frame without significant leakage. This characteristic of geological storage sites makes subsurface formations a highly favorable option for CO_2 storage. Furthermore, the drilling, completion, and injection technologies essential for carbon storage in CCUS projects are already well-established within the oil and gas industry. The existing infrastructure further enhances the economic feasibility of deploying CCUS in geological storage sites.

6.3.1 Advantages of Geological Storage

There are several key reasons why geological storage is favored for achieving this long-term carbon sequestration objective. Firstly, geological formations, such as saline aquifers and depleted oil and gas reservoirs, offer vast storage capacity that can accommodate significant amounts of CO_2 emissions (Wei et al. 2013; Johnsson 2011). These formations have been naturally trapping fluids, including hydrocarbons and water, for millions of years, indicating their potential for long-term storage of CO_2.

Secondly, geological storage provides multiple layers of natural barriers that help contain the stored CO_2. These barriers include impermeable cap rocks or seals that prevent upward migration of CO_2, baffles and intermediate thin layers, as well as the trapping mechanisms within the geological formations themselves, such as structural, residual, solubility, and mineral trapping (Li et al. 2023). These trapping mechanisms work together to immobilize the CO_2 and minimize the risk of leakage outside of storage site. Furthermore, geological storage offers the advantage of permanence. Once the CO_2 is injected into a suitable geological formation and securely trapped, it undergoes a gradual process of mineralization, where it reacts with the surrounding rock formations and becomes permanently stored as mineral carbonates. This mineralization process further enhances the long-term stability and permanence of the stored CO_2. This process can be accelerated in specific conditions such as injecting CO_2 in igneous formations such as Basalt. Basalt is rich in calcium (Ca), magnesium (Mg), and iron (Fe) silicates, which easily react with dissolved CO_2 to form stable carbonate minerals (Snæbjörnsdóttir et al. 2014). The high reactivity of these minerals significantly speeds up the carbonation process compared to other rock types like sandstone or shale.

Another important aspect of geological storage is its compatibility with existing infrastructure and practices in the oil and gas industry. Depleted oil and gas reservoirs, for example, can be repurposed for CO_2 storage, leveraging the knowledge and expertise gained from decades of hydrocarbon reservoir development and production. This compatibility allows for cost-effective implementation and facilitates the utilization of existing infrastructure, such as wells and pipelines, reducing the need for extensive new construction. Geological storage offers a scientifically and technically viable solution for keeping captured CO_2 out of the atmosphere for extended periods, in line with the long-term goals of climate change mitigation. Underground storage capacity of geological formations along with multiple trapping mechanisms, and ensuring compatibility with existing infrastructure, geological storage presents a practical pathway toward achieving a sustainable and low-carbon future (Goodman et al. 2011).

6.4 Estimating CO_2 Storage Volumes and Capacity Assessment

Accurate estimation of CO_2 storage capacity is a fundamental requirement for successful CCUS project implementation, as it determines the suitability of geologic formations to accommodate transferred CO_2 volumes from emission sources (Bachu et al. 2007). This assessment primarily depends on three critical parameters: pore space availability derived from formation properties (including total/effective porosity, thickness, and areal extent); injectivity potential governing injection feasibility and efficiency (can be obtained using injection step rate test); and long-term caprock integrity ensuring permanent containment. Injectivity, defined as the ability to inject CO_2 at optimal rates and pressures, is influenced by different interdependent factors including fluid properties (CO_2 density, viscosity), rock characteristics (porosity, permeability), rock and fluid interactions (relative permeability, capillary pressure), geomechanical constraints (fracture pressure, stress regimes), and well completion design (open-hole vs cased/perforated, well orientation, stimulation treatments).

These properties are evaluated through static analysis conducted pre-injection using geological data to estimate theoretical capacity, and dynamic analysis performed during injection operations incorporating real-time monitoring and reservoir simulation to validate and refine storage estimates. The integration of these methods enables comprehensive characterization of storage formations and quantification of critical operational considerations including injection rate optimization, pressure management, and long-term containment assurance.

6.4.1 Storage Rock Properties

The success of carbon storage in subsurface formation highly depends on rock properties of the storage and caprock formations. The main rock properties that need to be quantified include porosity, permeability, mineralogy, clay content, geomechanical properties such as Young's modules and Poisson ratio, baffles and natural fractures. Additional important parameters that come into effect when rock and fluid interactions are considered include interfacial tension, wettability, and capillary pressure. These parameters each play a significant role in defining the storage capacity in terms of both volume and rate of injection as well as containments characteristics in terms of trapping mechanism and its efficiency to contain CO_2 for long-term storage. In the following section we briefly introduce these parameters and techniques used to quantify these parameters.

Total porosity (ϕ_t) represents the total void space fraction in a rock, while effective porosity (ϕ_e) defines the interconnected pores accessible to fluid flow. A formation rock could have significant total porosity, however, if these pore spaces are not interconnected, they will not be participating in the CO_2 storage. On the other hand, the effective porosity represents the fraction of pore spaces that is interconnected and accessible to fluid flow. A great example of rock with extremely high total porosity but limited effective porosity is carbonate rocks like limestones with isolated vugs.

Figure 6.1 presents scanning electron microscope (SEM) images of two well-known sandstones Berea and Keuper. These high-resolution SEM images reveal complex details of the rocks' pore networks at the microscale, including pore shapes (roundness and sphericity), pore-throat configurations, surface textures, and mineral cements. On the left, the Berea sandstone appears well-sorted with approximately 22% porosity, while the Keuper sandstone exhibits better-rounded grains and slightly higher porosity (~24%). Both samples display strong pore connectivity and a pore size distribution ranging from 8 to 16 μm (Scheurer et al. 2022). In contrast, Fig. 6.2 shows an outcrop of high-porosity travertine (limestone). While the outcrop reveals large, visible pores that suggest significant storage potential, most of these pores are isolated vugs rather than interconnected pore network. Consequently, they don't effectively contribute to porosity or CO_2 flow. By examining rocks across scales, from outcrop observations to micrometer/nanometer-scale SEM imaging, we can comprehensively characterize pore systems in both sandstones and carbonates which is critical for assessing CO_2 storage viability of formations.

Absolute permeability (k) quantifies a rock's ability to transmit fluids which directly controls CO_2 injectivity and plume migration. While absolute permeability describes single-phase flow, relative permeability becomes critical in CO_2 storage scenarios where multiple fluids (CO_2, brine, and/or hydrocarbons) are present. The relative permeability to each phase governs how these fluids move and interact within the pore space. The relative permeability to each phase has direct relationship with the saturation of that phase in the reservoir. As CO_2 is injected into the reservoir and CO_2 saturation increase the water saturation goes down (water drainage) and consequently relative permeability to water decreases, the water later might reinvade the pore increasing saturation of water (imbibition) that results in CO_2 capillary/residual

Fig. 6.1 Scanning electron microscope (SEM) image of Berea sandstone in left and Keuper in right Scheurer et al. 2022

Fig. 6.2 High porosity limestone (Travertine) from Wyoming USA James St. John

trapping. This behavior will help us to understand the dynamics of CO_2 plume at early time and late time of injection and post injection.

Wettability refers to the affinity of a rock surface to be in contact with one fluid over another. This property significantly influences CO_2 capillary trapping in reservoirs by controlling the distribution of CO_2 and brine within pore spaces. In water-wet systems, the rock surface is coated with water, forcing CO_2 to occupy the central portion of pores and pore throats. This configuration enables water to override and trap CO_2 as residual saturation within the pore network.

Capillary pressure, defined as the pressure difference across fluid interfaces in the pore network, determines which fluid occupies different sized pores and establishes the minimum pressure required for CO_2 to enter and displace water. The interfacial tension between CO_2 and brine further governs these interactions by regulating the strength of the fluid-fluid interface. Generally, these properties control the effectiveness of both structural and capillary trapping mechanisms that are fundamental for secure long-term CO_2 storage.

These parameters are typically quantified through a combination of well logging, core analysis, thin-section microscopy, and outcrop studies. Well logging (e.g., gamma, density, resistivity, neutron, NMR, sonic logs, lithology, PE, and geomechanical logs) provides in situ estimates of porosity, fluid saturation, permeability, geomechanical properties and mineralogy of the formation. Core measurements (routine/core analysis, mercury injection capillary pressure (MICP), triaxial testing, and relative permeability tests) deliver direct data on pore structure and flow behavior. Thin-section, X-ray Diffraction (XRD), X-ray Fluorescence (XRF), and outcrop studies reveal mineralogy, pore networks, and diagenetic features that influence storage capacity and seal integrity.

When it comes to carbon dioxide storage, the selection of an ideal reservoir rock is a fundamental component. The injected carbon dioxide moves within the reservoir rock in an upward direction due to its buoyancy. To prevent its escape to the surface, a seal rock acts as a barrier, confining the carbon dioxide within the reservoir. Additionally, any lateral movement is constrained by a trap, ensuring the containment of the carbon dioxide within the designated storage area.

The main question to answer here is what type of rock is considered the best to serve as a reservoir for carbon dioxide storage?

The answer lies in the characteristics of the rock itself. Ideally, the reservoir rock should hold properties that facilitate efficient carbon dioxide storage and minimize the risk of leakage. These properties include high effective porosity, which provides required space for storing the carbon dioxide, and high permeability, allowing for the movement and distribution of the injected carbon dioxide within the rock. Several rock types exhibit these desired properties, including porous sandstone, limestone, and dolomite formations. These rocks often possess interconnected pore networks and sufficient permeability to accommodate the injected carbon dioxide. Additionally, the presence of natural traps, such as anticlines, sealing fault lines or structural formations, can further enhance the containment of the carbon dioxide within the reservoir. However, it is important to note that the suitability of a rock type for carbon dioxide storage also depends on the specific geological conditions and the characteristics of the carbon dioxide source. Site-specific considerations, such as the presence of suitable seal rocks and the proximity to emission sources, are essential factors to evaluate when determining the best type of rock for carbon dioxide storage. Recently, depleted unconventional shale oil and gas reservoirs are also targeted for CO_2 storage. As mentioned earlier the captured CO_2 is compressed and injected subsurface for storage in super critical stage. Followings we investigate the properties of supercritical CO_2 and its importance in securely storing CO_2.

6.4.2 *Properties of Supercritical CO_2*

The phase behavior of pure CO_2 can be represented by a PVT (Pressure-Volume-Temperature) diagram, as illustrated in Fig. 6.3. This figure highlights key features that define CO_2's phase transitions. The triple point represents the unique combination of pressure and temperature where CO_2 coexists in equilibrium as a solid, liquid, and gas simultaneously. For CO_2, the triple point occurs at a pressure of 74.95 psi and a temperature of −56.6 °C (−69.9 °F). Another important point in PVT diagram is the critical point at which CO_2 transitions into a supercritical fluid which is a distinct state presenting hybrid liquid/gas property. Above the critical temperature, CO_2 cannot be liquefied regardless of pressure, and above the critical pressure, liquid and gas phases cannot co-exist (become indistinguishable). For CO_2, the critical point is reached at a pressure of 1084.56 psi and a temperature of 31.1 °C (88 °F). Above both critical values, CO_2 behaves as a supercritical fluid, combining liquid-like density (0.5–0.8 g/cm^3) with gas-like diffusivity and viscosity. This enhances its solvent capacity and mobility which is critical for geological storage and industrial applications.

The vapor pressure line marks the equilibrium boundary between liquid and gaseous CO_2. Below the critical temperature and above this line, CO_2 remains liquid

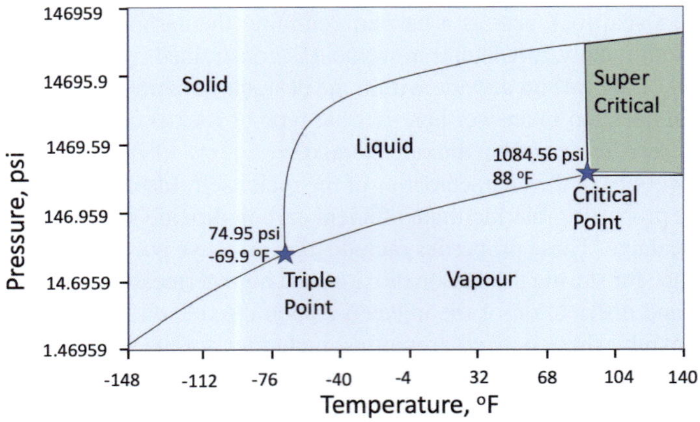

Fig. 6.3 Carbon dioxide phase diagram

and under this line represents the gas phase. The PVT diagram of **pure** CO_2 illustrates the transitions between solid, liquid, and gaseous states, the critical point, vapor pressure line, and the unique behavior of CO_2 as a supercritical fluid above its critical temperature and pressure.

6.4.3 Supercritical CO_2 Density and Behavior

The density vs temperature diagram of CO_2 is also used to describe the behavior of CO_2 at different temperature and pressure based on changes in its density. Figure 6.4 illustrates the behavior of carbon dioxide under different temperature and pressure conditions. The diagram consists of several distinct regions representing the phases of CO_2. At low temperatures and pressures, below the dashed blue line (two phase region) CO_2 exists as a gas. In this region, the density of CO_2 decreases with increasing temperature while maintaining a relatively low density. At higher pressures and lower temperatures, CO_2 transitions into a liquid state above the dashed blue line (two phase region). In this region, the density of CO_2 increases with increasing pressure and decreasing temperature, as CO_2 molecules come closer together. The boundary between the gas and liquid regions is known as the two-phase region (contained with blue dash line). In this region, both gas and liquid phases coexist in equilibrium, and the density remains constant along the phase boundary. As temperature and pressure increase further, CO_2 enters the supercritical region. Above approximately 31.3 °C (88.3 F) and roughly 7.4 MPa (1084.56 psi), CO_2 is in a supercritical state. In this state, CO_2 exhibits properties of both a gas and a liquid, with densities as low as 100 kg/m^3 at 7 MPa. As pressure increases, CO_2 in the supercritical region can have a liquid-like density of approximately 500–800 kg/m^3. The supercritical boundaries in the diagram represent the points where CO_2 transitions from a gas to a

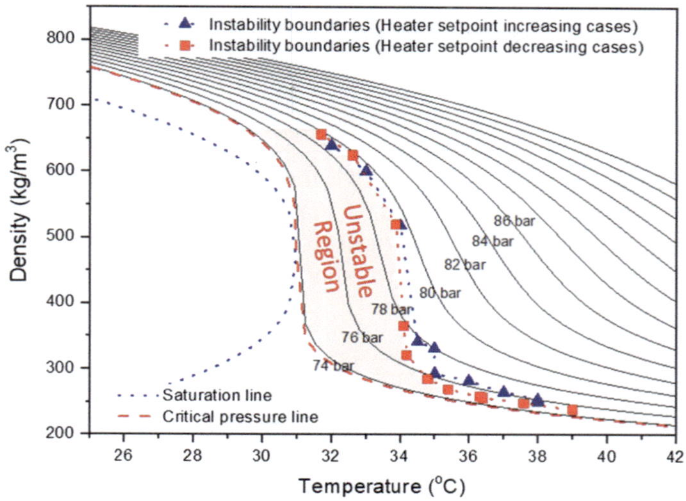

Fig. 6.4 Density vs temperature diagram of CO_2 Min et al. (2018)

supercritical fluid or from a supercritical fluid to a liquid. These boundaries indicate the specific temperature and pressure conditions at which these transitions occur.

The storage volumes required for storing CO_2 emissions are immense, particularly when considering the emissions from large coal-fired power plants. For instance, a single power station can release approximately 1 million tons of CO_2 annually. It's worth noting that there are currently over 9000 coal-fired power plants worldwide (IEA) that indicates the magnitude of CO_2 emissions that need to be captured and stored only from coal-fired power plants. The advantage of storing CO_2 in underground geological formations is as depth of storage increase the hydrostatic pressure will increase which means we will deal with more compressed CO_2 with higher density that require less storage volume compared with surface volume of CO_2. To put this into perspective, if we consider the density of CO_2 as around 2 kilograms per cubic meter (2 kg/m^3) at surface conditions, and if we need to store approximately 5000 million cubic meters of CO_2 emissions, by simply compressing the CO_2 emissions to supercritical CO_2, which has a higher density, around 500 kg per cubic meter (500 kg/m^3) at surface conditions, the storage volume required will decrease to around 20 million cubic meters of supercritical CO_2. This means supercritical CO_2 occupies ~250 times less volume than gaseous CO_2 at surface conditions. Followings are the details of the calculation:

1. Mass (m) of the 5000 million cubic meters of CO_2 emissions with density of 2 kg/m^3 is
   $m = \text{Volume}_{@\text{surface}} \times \rho_{@\text{surface}} = 5 \times 10^9\ \text{m}^3 \times 2\ \text{kg/m}^3 = 10^{10}\ \text{kg (10 million metric tons)}$
2. Volume (V) of the 10 million metric tons of CO_2 at supercritical condition with density (r) of (500 kg/m^3) is

$$\text{Volume}_{@\text{supercritical}} = m/\rho_{@\text{supercritical}} = \left(10^{10}\ \text{kg}\right)/\left(500\ \text{kg/m}^3\right) = 2 \times 10^7\ \text{m}^3\ \left(20\ \text{million m}^3\right).$$

This shows 99.6% decrease from 5000 million $m^3$of CO_2 at surface to 20 million m^3in pore space required to store sCO_2.

Now assuming the underground geological storage formation at depth of 8000 ft. with 128 °F temperature (2440 m and 53 °C) the required storage volume shrinks to ~13.9 million m^3 following similar calculations. This highlights the importance of supercritical CO_2 behavior as a function of reservoir pressure and temperature in estimating the capacity of storage sites.

6.5 Trapping Mechanisms for CO_2 Storage

As discussed earlier captured CO_2 from emitters (sources) are transported to the wellhead via pipeline and injected into the wellbore to be permanently stored in underground geological formations. This process relies on multiple trapping mechanisms to ensure long-term storage of CO_2. Understanding these mechanisms is significantly important for the successful implementation of CCUS projects.

CO_2 trapping mechanisms can be divided in two class of primary and secondary trapping mechanisms. Primary trapping mechanisms are the initial geological features that act as barriers to limit the lateral flow of CO_2. These mechanisms are significant in confining CO_2 within a designated storage reservoir. They include "Structural traps" and "Stratigraphic traps". Secondary trapping mechanisms are additional processes that further decrease the mobility of CO_2 and enhance its confinement within a storage reservoir. These mechanisms can be created during fluid flow and complement the primary trapping mechanisms. They include "Capillary trapping", "Dissolution trapping," and "Mineral trapping."

These primary and secondary trapping mechanisms work together to confine CO_2 within the reservoir and decrease its mobility, ensuring the long-term storage and containment of CO_2 in geological formations (Hosseininoosheri et al. 2018).

6.5.1 Overview of Primary Trapping

Primary CO_2 trapping mechanisms include structural and stratigraphic traps. Structural traps are formed by geological structures that deform the rock layers, creating impermeable barriers to fluid flow. The natural physical barriers within the reservoir such as anticlines or faulted compartments restrict the lateral movement of CO_2. Over time, structural trapping plays a significant part in accumulating and retaining CO_2 within the reservoir. On the other hand, stratigraphic traps are formed by variations in the arrangement and properties of the rock layers, such as depositional pinch-outs, diagenetic traps and Reef or carbonate buildup traps.

6.5.1.1 Anticline Structural Trapping

Anticlines, which are arch-like folds in geological formations, provide highly effective structural traps for CO_2 storage, they are form of tight folds like Synclines (Fig. 6.5), however, convex upward. In these traps, reservoir rocks are folded into a dome shape, creating a natural container capped by impermeable layers (e.g., shale or salt). The CO_2 injected into the reservoir migrates upward due to buoyancy but is confined beneath the caprock. The vertical distance between the highest point of the anticline "Crest" and the lowest elevation of the anticline "Spill" defines the closure and storage capacity of anticline. These are the most favorable structural trapping mechanism due to high integrity to confine the CO_2 and easier monitoring of CO_2 after injection. The choice of injection point in anticline varies based on the anticline characteristics and application. The most common technique would be drilling the injection well at the sides of the fold "Flank" and away from the crest as possible. This ensures better distribution of the CO_2 and slower accumulation of CO_2 beneath the cap rock, therefore reducing the risk of CO_2 leakage. Figure 6.6 illustrates an anticline structure being utilized for CO_2 storage, where injection well is drilled in the flank of an anticline leading to CO_2 upward movement due to buoyancy and being confined by the anticline seal. However, drilling of injection well at the Crest is also an option for small anticlines with high filling rate. Injection in Crest is much easier and faster, however, significantly increases the risk of caprock fracturing and CO_2 leakage and requires robust monitoring and managed pressure build up.

Fig. 6.5 Large syncline in Hancock Maryland (E. Fathi Photo)

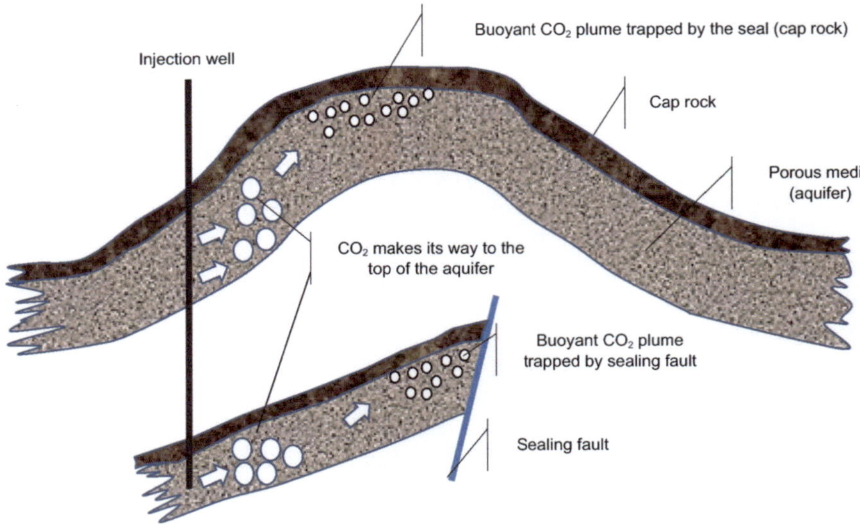

Fig. 6.6 Image from Ajayi et al. (2019)

6.5.1.2 Sealed Faults and Fault-Related Structures

Faulted formations serve as structural traps where fault displacement creates isolated formation capable of storing CO_2. Faults can be either conductive (allowing fluid flow) or sealing (impermeable). Impermeable faults act as barriers, confining CO_2 within the reservoir. A sealed fault refers to a fault that does not have significant permeability, preventing the upward migration and escape of CO_2 along the fault planes. When considering faults as reservoir traps for CO_2 storage, it is essential to assess the fault's sealing capacity. This involves evaluating the fault's sealing properties, such as the presence of clay minerals or mineralized zones, which can act as barriers to fluid flow. Proper characterization and understanding of the fault's sealing capabilities are essential to ensure the long-term containment of the injected CO_2. Furthermore, it is important to consider the maximum allowable pressure within the reservoir to avoid reactivating the fault (Gao et al. 2016). Faults have a certain threshold of pressure beyond which they may experience slip or movement. By monitoring and controlling the pressure levels within the CO_2 storage reservoir the risk of fault reactivation can be minimized. This ensures the structural integrity and containment of the stored CO_2. Considering both the sealing capacity of the fault and the maximum allowable pressure helps to ensure the effectiveness and safety of fault-based trapping mechanisms in CO_2 storage.

In some cases, faulting contrasts impermeable layers (e.g., shale) against porous reservoir rocks (e.g., sandstone) that further enhances the trapping mechanism. However, faulted traps carry higher leakage risks, particularly if injection pressures reactivate faults, converting them into conductive pathways. Thus, detailed geological and geomechanical assessments are critical to ensure long-term storage integrity in such formations. Figure 6.6 also shows a sealing fault confining the injected CO_2 in the aquifer. This fault forms a vertical barrier that intersects the upward movement of the CO_2 plume. The fault effectively traps the CO_2 within a specific area by preventing lateral movement.

6.5.1.3 Stratigraphic Trapping Mechanism

Stratigraphic traps are formed by variations in the arrangement and properties of the rock layers, which is different than the structural traps due to formation deformation. Stratigraphic traps rely on change in rock properties such as porosity and permeability or deposition pattern to confine CO_2 within the storage formation. Stratigraphic traps include depositional pinch-outs, diagenetic traps and Reef or carbonate buildup traps (Dolson 2018). Depositional pinch out is the most important stratigraphic trap while others stratigraphic traps require careful characterization to ensure the integrity of the confinements due to highly heterogeneous nature of the processes. Depositional pinch out occurs when a permeable reservoir rock like sandstone gradually thins and discontinuous (pinches out) surrounded by less permeable or impermeable rock layers. These formation variations from permeable reservoir to impermeable barrier prevent the lateral flow of fluids, traps injected CO_2 within the permeable reservoir boundaries.

Post depositional chemical reactions such (diagenesis) alters rock properties such as porosity and permeability through different scaling and cementation process. These cementation and scaling will lead to reduced permeability and porosity developing natural barrier to follow providing confinements to trap CO_2. Reef or carbonate build up traps where highly porous reef limestone surrounded with impermeable tight mudstone can also act as CO_2 storage sites (Ma et al. 2024). Figure 6.7 shows the outcrop of the Zuni Sandstone (light-colored unit) overlain by the darker Dakota Formation sandstone in New Mexico, USA. The Zuni Sandstone is an eolian (wind-deposited) Jurassic unit that marks the southern pinch-out of coeval fluvial and lacustrine deposits in the region.

6.5.1.4 Salt Domes as both Structural and Stratigraphic Trapping Mechanism

Salt domes are subsurface structures formed by the upward movement of buoyant salt deposits. Salt domes are primarily classified as structural traps; however, they can also form stratigraphic trapping. In forming structural trapping salt moves upward and deforms overlying sedimentary layers and creates folded and faulted dome that works as structural closures. The impermeable nature of salt makes these domes highly effective cap rocks to confine CO_2 in storage site. When CO_2 is injected into adjacent reservoir rocks it moves upward due to buoyancy until it hits the salt dome, which acts as a natural caprock and prevents vertical movement of CO_2. However, despite their sealing efficiency, salt domes face high risks including potential salt

Fig. 6.7 The outcrop of Zuni sandstone (light-colored unit) overlain by the darker Dakota formation sandstone in New Mexico, USA. *Image* by Kent G. Budge 2020

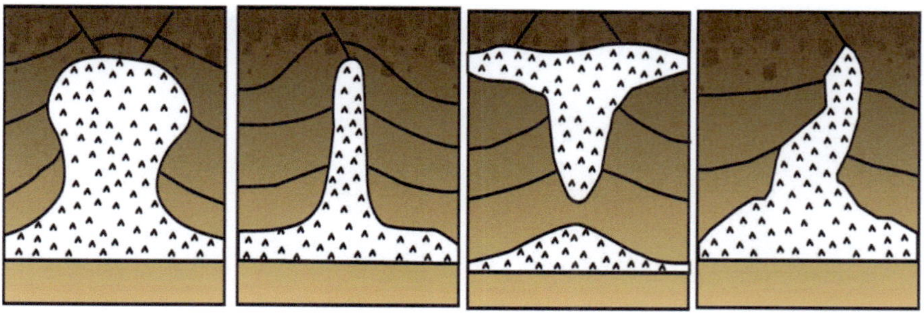

Fig. 6.8 Stratigraphic and salt dome trapping mechanism Zamani and Kenz (2024)

dissolution as it contacts with CO_2 saturated brine and mechanical integrity issues as pressure builds up.

Salt domes can also create stratigraphic traps when reservoir rocks (e.g., sandstones) pinch out against the salt body, forming lateral seals. In this case the salt body is not imposing deformation but imposing variation in rock properties. This dual trapping mechanism combining structural and stratigraphic elements enhances storage security. However, site selection requires careful evaluation of salt stability, reservoir geometry, and potential fluid-salt interactions to mitigate long-term risks.

As mentioned earlier, in CO_2 storage projects, salt domes serve as seals rather than storage reservoirs (Zamani and Kenz 2024). The salt itself is almost never targeted for injection due to its low permeability, mechanical instability (e.g., creep-induced wellbore deformation), and dissolution risks. Instead, CO_2 is injected into porous reservoir rocks adjacent to the salt dome, such as neighboring sandstones or carbonates, where the salt provides lateral and top seals. Figure 6.8 shows 4 different salt formations that can act as either structural or stratigraphic traps for CO_2.

6.5.2 *Overview of Secondary Trapping*

Secondary trapping mechanisms enhance CO_2 storage security by gradually immobilizing CO_2 through physical and chemical interactions between injected CO_2 and reservoir rock and fluids. These time-dependent interactions operate along with primary structural traps to ensure long-term storage of CO_2. The key secondary trapping processes include capillary trapping, Dissolution trapping, and Mineral trapping.

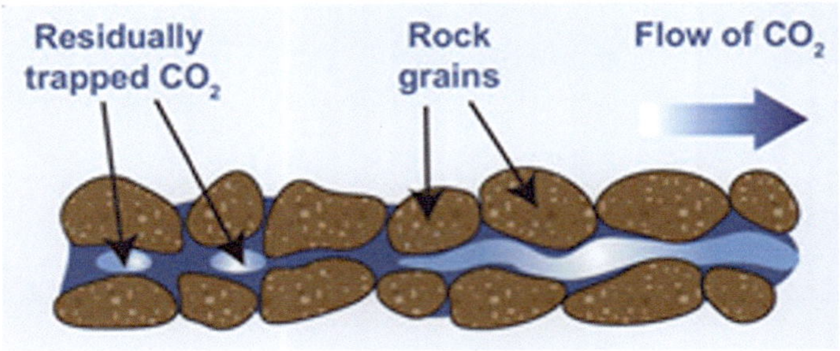

Fig. 6.9 Capillary trapping figure from NETL, US DOE (https://www.netl.doe.gov/carbon-management/carbon-storage/faqs/carbon-storage-faqs)

6.5.2.1 Capillary Trapping

One important secondary trapping mechanism in CO_2 storage is capillary (residual gas phase) trapping. Capillary trapping occurs when the capillary forces between the CO_2 and the pore spaces in the reservoir rock prevent the CO_2 from flowing freely, therefore, the CO_2 is physically trapped within the pore spaces of the reservoir rock (Bourg et al. 2015). The capillary pressure and surface tension forces specially working to immobilize CO_2 within smaller pores and narrow pore throats, and effectively trapping CO_2 as residually trapped CO_2 as shown in Fig. 6.9. After the injection of CO_2, the CO_2 plume begins to migrate upward and spread laterally under the influence of buoyancy forces and occupies pores previously filled with formation water (formation water drainage). As the plume rises, water starts to infiltrate back to the pore spaces previously occupied by CO_2 at the base of the plume, displacing a portion of the CO_2 and starting the water imbibition However, the displacement of all the CO_2 by water is limited due to capillary forces that results in significant volumes of CO_2 being trapped as residual CO_2 saturation Generally, 20–50% of CO_2 will be trapped per unit volume of pore space and abandoned in place as the plume thins and spreads. This trapped CO_2 remains permanently stored in the reservoir and increase the residual or irreduacable CO_2 saturation. This process will result in relative permeability hysteresis that needs to be considered when simulation CO_2 saturation dynamics during and after injection. This trapping mechanism contributes to the long-term storage of CO_2 as a residual phase, even after initial migration (Zapata et al. 2020).

6.5.2.2 Dissolution Trapping

Dissolution trapping is another important CO_2 trapping mechanism where injected CO_2 dissolves into the formation water (brine or aqueous phase) present in the reservoir rock. The dissolved CO_2 remains confined within the reservoir due to its reduced mobility. As CO_2 dissolves, it forms carbonic acid (H_2CO_3) through

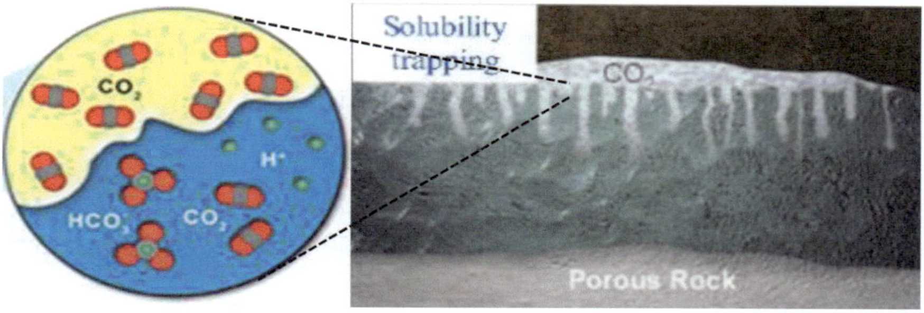

Fig. 6.10 Left Linear CO_2 molecules at top reacts with hydrogen molecules generating carbonic acid at the CO_2 brine interface a (modified from NETL, US DOE, https://netl.doe.gov/carbon-management/carbon-storage/faqs/carbon-storage-faqs). Right solubility trapping and downward movement of higher density CO_2-saturated brine (modified from Cao et al. 2020)

diffusion at the CO_2-brine interface, these geochemical reactions further immobilize the CO_2. The dissolution of CO_2 into brine increases the fluid density as well causing the CO_2-saturated brine to sink downward within the reservoir. This density-driven movement promotes broader dispersion of CO_2, counteracting buoyancy-driven rise and reducing leakage risks (Chen et al. 2023). Over time, dissolved CO_2 may react with minerals, leading to precipitation and permanent mineralization. By mitigating vertical CO_2 migration, dissolution trapping improves storage efficiency and complements other trapping mechanisms, such as structural and capillary trapping. Figure 6.10 shows the solubility trapping forming carbonic acid (H_2CO_3) at the CO_2-brine interface and downward movement of higher density CO_2-saturated brine.

Figure 6.11 shows the schematic of CO_2 solubility trapping in a saline aquifer where the injected supercritical CO_2 gets in contact with saline aquifer. CO_2 dissolved in brine generate CO_2-rich brine which moves horizontally and vertically down due to higher density (density driven). Supercritical CO_2 moves upward due to buoyancy effect till it hits the caprock and extends laterally. As pressure builds up and more CO_2 dissolved in brine the density driven convection will lead to downward movement and developing of viscous fingering of CO_2-rich brine in formation water (Chen et al. 2023).

6.5.2.3 Mineral Trapping

Mineral trapping is the most permanent CO_2 storage mechanism that help to securely store CO_2 under the ground. In this process, the dissolved CO_2 chemically reacts with the surrounding rocks, forming new minerals like calcite or magnesite (Khandoozi et al. 2023). Super critical CO_2 dissolution in brine as discussed earlier forms acidic solution (carbonic acid) that moves downward due to higher density and can react with rock minerals and form calcite ($CaCO_3$) precipitation. This process immobilizes the CO_2 in solid form. The rate of mineral trapping can vary depending on the

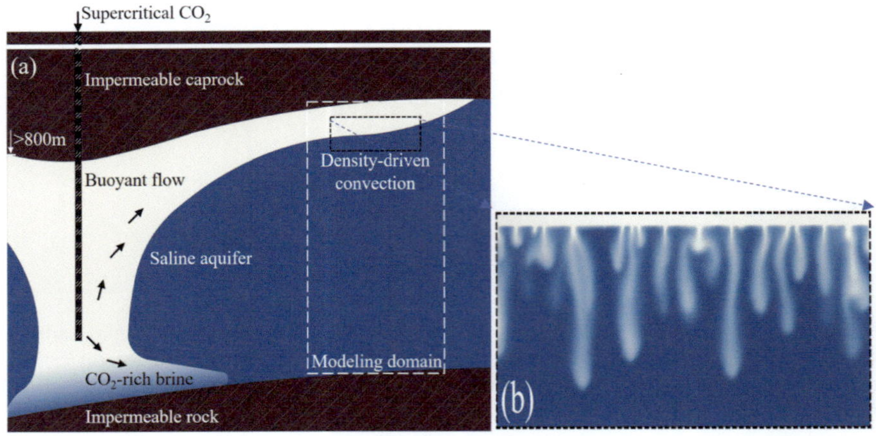

Fig. 6.11 CO_2 solubility trapping in saline aquifers: **a)** Sequestration setup, **b)** density-driven convection with fingering, modified from Chen et al. (2023)

reactivity of the rocks, temperature, pressure, brine chemistry, and CO_2 pressure. Reactive rocks such as ultramafic rocks (olivine, Serpentine) and Basalts can undergo mineral trapping relatively quickly. One of the most successful mineral trapping CO_2 storage projects is the Orca Plant in Iceland where the injected CO_2 dissolved in water reacts with Basalt and reach more than 95% mineralization in less than 2 years.

Figure 6.12 illustrates CO_2 mineralization where dissolved CO_2 (as carbonic acid) reacts with magnesium-rich rock minerals (e.g., olivine or serpentine) to form solid magnesite ($MgCO_3$). This process immobilizes CO_2 ensuring long-term stability. These secondary trapping mechanisms, including residual gas phase trapping, dissolution trapping, and mineral trapping, are critical parts in ensuring the long-term containment and storage of CO_2 in the reservoir (McPherson 2009). Understanding and accounting for these mechanisms are essential in the assessment and design of effective CO_2 storage projects.

6.5.2.4 Other Gas Trapping Mechanism—Hydrodynamic Traps

Hydrodynamic trapping occurs when the continuous movement of groundwater within a sedimentary basin can influence the behavior of CO_2 storage by restricting its movement (Bachu et al. 1994; Larkin et al. 2010). If the flow of groundwater is sufficiently strong like charging aquifer, it can cause the CO_2-water contact to deviate from the horizontal plane due to hydrodynamic shear stress. This deviation from horizontal alignment can affect the distribution and accumulation of CO_2 within the reservoir. In certain situations, CO_2 may accumulate without the presence of a complete trap closure. This means that while the CO_2 may not be fully confined within a sealed reservoir, its accumulation is still possible due to the dynamic flow of fluids. Understanding these hydrodynamic processes is necessary for assessing

Fig. 6.12 CO_2 trapping in form of mineralization and formation of Magnesite (Mg CO3) modified from NETL, US DOE. https://netl.doe.gov/carbon-management/carbon-storage/faqs/carbon-storage-faqs). Right Mineral trapping (modified from Cao et al. 2020)

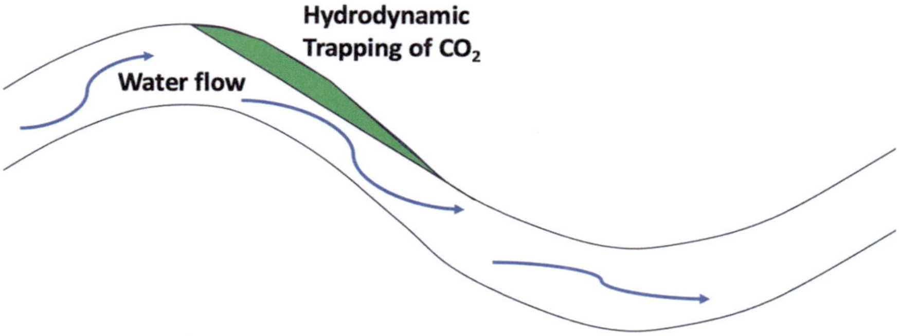

Fig. 6.13 Hydrodynamic trapping of CO_2 plume in dynamic aquifer

the effectiveness and long-term stability of CO_2 storage in hydrodynamic traps. Hydrodynamics trapping relies on fluid dynamics and is a time-dependent process that develops after injection thus classified as other secondary trapping mechanism. Figure 6.13 shows the schematic of continuous flux of groundwater that resulted in hydraulic trapping of CO_2 plume within the storage site.

6.5.3 Trapping Mechanism Efficiency and Evolution Over Time

Structural trapping mechanism is responsible to confine 60–80% of injected CO_2 depending on reservoir porosity and permeability. It's a dominant trapping mechanism that acts during the whole life of the storage site from day one to over 1000 years. Capillary trapping is the second most important trapping mechanism that can retain 15–30% of injected CO_2. The effectiveness of this process highly relates to rock properties and heterogeneity of the formation. This process can start from weeks to months during and after injection. Dissolution trapping can also take anywhere between 5 and 15% of injected CO_2 as dissolved CO_2. Finally, the slowest trapping mechanism is the mineralization that is negligible in short term but has great capacity to store CO_2 permanently in long terms. However, in right conditions in terms of presence of reactive minerals and temperature and pressure this process is shown successful even in less than two years after injection. This process is valid for CO_2 storage in sedimentary formations, however, CO_2 sequestration in igneous rocks like Basalts follow different efficiency and evolution with time, due to their high reaction rate that leads to rapid mineralization and mineral trapping of CO_2. The CarbFix project demonstrated this process by injecting CO_2 dissolved in water. This method eliminates buoyancy, preventing CO_2 migration toward the surface. As a result, structural and capillary trapping mechanisms which are critical in conventional storage become negligible (Snæbjörnsdóttir et al. 2016). Snæbjörnsdóttir et al. 2017 showed that solubility trapping dominates initially, while mineral trapping becomes the primary mechanism within two years. This rapid mineralization, driven by chemical reactions between CO_2 and basalt, must be explicitly accounted for in geological storage design (Kim et al. 2023).

Different reservoir simulators can be used to quantify each CO_2 trapping mechanisms. Fathi and Belyadi used CMG-GEM compositional reservoir simulator to quantify each trapping mechanisms. The structural trapping mechanism can be quantified monitoring CO_2 saturation plume during (AoR) and after injection (PISC). The capillary trapping mechanism can be modeled using hysteresis in relative permeability curve. Drainage (CO_2 entering the pores) and imbibition (water reentering preoccupied pores by CO_2) cycles will result in increasing residual CO_2 saturation. The dissolution and mineralization trappings can also be quantified using aqueous (sCO_2 and Brine) and aqueous mineral reactions (Thibeau et al. 2007).

6.6 Underground Reservoirs for CO_2 Sequestration

Geological sequestration of carbon dioxide involves injecting captured CO_2 from emitters into deep subsurface formations, where CO_2 hold supercritical characteristics and can be securely stored for thousands of years. These storage sites must provide adequate porosity and permeability to accommodate and transmit CO_2, along with a thick caprock and reliable trapping mechanisms to contain CO_2 for significantly long time.

Various types of geological formations are suitable for CO_2 storage, each offering unique characteristics with advantages and limitations. These include deep saline aquifers, depleted oil and gas reservoirs, basalt formations, unmineable coal seams, and shale formations. Among these, saline aquifers offer the largest global storage potential due to their vast distribution and high pore volume. Depleted hydrocarbon reservoirs are also attractive, given their well-characterized geology and existing infrastructure and opportunity for enhanced oil recovery to compensate part of the cost of the CO_2 sequestration. Basalts have also shown great success in storing CO_2 permanently through mineralization in projects such as Orca. Other storage sites are in conceptualization process and has not been practiced in field scales.

The selection of a storage reservoir depends on both technical and non-technical factors. Technical considerations include reservoir quality, depth, pressure, temperature, rock type, reservoir fluid properties and proximity to CO_2 sources and infrastructure. Non-technical factors involve compliance with federal, state, and local regulations, as well as public acceptance.

Understanding of the different reservoir types available for storage and their advantages and limitations is critical for designing effective CCS projects and ensuring safe, long-term CO_2 containment. The Carbon Storage Atlas—Fifth Edition (Atlas V) (2015), published by the U.S. Department of Energy (DOE), evaluates the potential for geologic CO_2 storage across different underground formations. It provides quantitative bulk estimates of storage capacity in oil and gas reservoirs, unmineable coal seams, and saline formations, using volumetric calculations.

6.6.1 Saline Formations

Saline formations are sedimentary rocks saturated with formation water (Brine) with high porosity and permeability, and good confinement. These geological structures are found extensively across onshore and offshore sedimentary basins. Figure 6.14 illustrates extensive saline formation distributions in North America providing great potential for developing significant CO_2 storage sites. These formations suitable for CO_2 storage are characterized by formation waters with high salinity levels exceeding 50,000 mg/L total dissolved solids (TDS). Saline formations hold great promise as CO_2 storage sites due to their significant thickness and extensive spatial coverage, often exceeding that of oil and natural gas reservoirs or coal seams. Due to wide spread of saline formations, it is highly probable to find ideal saline formation sites for CO_2 storage near CO_2 sources. This significantly reduces the need for extensive pipeline transportation.

To ensure the effectiveness of CO_2 storage, saline formations must demonstrate the ability to trap both CO_2 and pressurized brine, preventing any leakage into underground sources of drinking water (USDW) or the atmosphere (Dempsey et al. 2014). Trapping mechanisms can involve structural or stratigraphic like those observed in hydrocarbon trapping. Alternatively, a combination of primary and secondary trapping mechanisms like capillary, dissolution, and mineralization trapping along with an extended flow path can offer efficient trapping.

It is critical to quantify the integrity of the saline formations (reservoir) and their caprock to determine if the reservoir can handle the intended injection volumes by accommodating the required pressure changes. This can be done by performing minifrac or diagnostic fracture injection test in both reservoir and caprock providing maximum operating pressure that guarantees the integrity of the reservoir and caprock. Furthermore, it's important to examine existing faults to identify

Fig. 6.14 Assessed and non-assessed saline formation distributions in North America (NETL Carbon Storage Atlas fifth edition)

any limitations on the anticipated pressure buildup to minimize potential of fault reactivation or any induced seismicity.

Deep aquifers are also underground formations that contain water with high salt content sometimes considered as CO_2 storage site. These aquifers are considered ideal for CO_2 storage due to their large storage capacities and widespread availability. The high salinity of the water makes them less desirable for other uses, making them suitable for long-term CO_2 storage (Kelemen et al. 2019). The Citronelle Project stored CO_2 in the Paluxy Formation in Alabama is an example of a deep saline aquifer that has been studied for CO_2 storage potential. Its geological characteristics and location make it a promising candidate for storing large volumes of CO_2.

With all the similarities between saline formations and deep aquifers they cannot be treated the same when it comes to storing CO_2. Aquifers need more detailed site characterization in terms of their heterogeneity and connectivity. It is not uncommon for deep aquifers to have hydraulic communication with other adjacent permeable formations, underground flows or seawater intrusions posing risk of CO_2 leakage, pressure changes, and brine displacement risk.

6.6.2 CO_2 Storage in Hydrocarbon Reservoirs

Oil and natural gas fields, whether currently in use or not, present geological features that make them well-suited for CO_2 storage. In this context, a "field" represents an area covering either a single reservoir or multiple reservoirs linked to the same geological structural feature or stratigraphy. A "reservoir" is an underground formation with good porosity and permeability, confined by impermeable rock, and marked by a single natural pressure system. The impermeable rocks that overlay oil and natural gas reservoirs have successfully confined the upward movement of hydrocarbons over millions of years as cap rock. These same geologic conditions that trap oil and gas can also serve as effective traps for CO_2 storage. Moreover, due to extensive research and exploration efforts in oil and gas fields there are plenty of production history, well log data, and other relevant information for these fields exist that can help to quantify the storage site as well as reducing the cost of storage site characterization. Additionally, the presence of substantial infrastructure within these fields presents opportunities for repurposing and utilizing existing facilities for CO_2 storage. An added advantage of injecting CO_2 into mature oil fields is the potential for enhanced oil recovery (EOR). This process, known as CO_2 enhanced oil recovery, can result in the production of additional oil. Thus, the utilization of oil and gas fields for CO_2 storage not only allows for secure storage of CO_2 but also offers the possibility of extracting additional resources from these fields. The geologic conditions, extensive data availability, and existing infrastructure associated with oil and natural gas fields make them highly favorable candidates for CO_2 storage. These conditions are offering both secure storage options and potential benefits such as CO_2 enhanced oil recovery (You et al. 2021).

6.6.2.1 CO_2 Storage in Depleted Oil Reservoirs

Depleted oil reservoirs are the optimal candidates for CO_2 storage due to their proven ability to retain hydrocarbons over geologic timescales (millions of years) (Wei et al. 2023; Awan and Kirmani 2025). Extensive historical production and pressure data during primary production, waterflooding, and possibly tertiary recovery (e.g., Enhanced Oil Recovery) of these reservoirs are recorded, such as EOR field at Cranfield, MS (Hosseini et al. 2013). This data reveals long-term saturation dynamics and pressure behavior of the reservoir that can be used for forecasting CO_2 migration, trapping mechanisms, and long-term containment during carbon storage projects. Additionally, the effect of change in geomechanical properties of the reservoir during multiple stages of production and injection (CO_2 flooding, or water alternate gas flooding) enables quantification of hysteresis effects on reservoir properties that can be used to further quantify the structural integrity of the storage site.

While storage capacity in depleted reservoirs is typically smaller than in saline formations, their well-characterized geology and existing infrastructure (e.g., wells, pipelines, seismic data) significantly reduce deployment risks and costs, making them highly attractive for early-stage carbon capture and storage (CCS) projects. However, to ensure secure long-term CO_2 storage in depleted oil reservoirs, robust monitoring protocols including pressure surveillance, seismic imaging, and leakage detection must be implemented. Figure 6.15 shows the distribution of oil reservoirs in North America as possible targets for CO_2 storage.

Fig. 6.15 Oil reservoirs distributions in North America (NETL Carbon Storage Atlas fifth edition)

6.6.2.2 CO_2 -Enhanced Oil Recovery (EOR)

Conventional oil production includes three stages of primary, secondary, and tertiary recovery. Primary recovery relies on the reservoir's natural energy including pressure depletion, gas cap expansion, aquifer support, or gravity drainage. This phase typically recovers only 10–25% of the original oil in place (OOIP) due to declining pressure. To extend production, secondary recovery techniques like waterflooding are employed, where injected water sweeps additional oil to the producers, while maintains reservoir pressure. This will increase the recovery from 25 to 40% of OOIP. Therefore, over half of their original oil remains in the reservoir that require additional advanced methods, "tertiary recovery."

Among tertiary techniques, CO_2 enhanced oil recovery (CO_2EOR) has been widely used for decades. Injected CO_2 dissolves in oil and reduces oil viscosity, improves mobility, and repressurizes the reservoir resulting in enhances oil recovery, while a portion of CO_2 becomes permanently trapped underground through structural, residual, or solubility mechanisms (Lei et al. 2016; Faruque Hasan et al. 2014; Hill et al. 2013). This dual benefit through increased oil production and simultaneous CO_2 storage makes CO_2-EOR an attractive hydrocarbon extraction method. While continuous CO_2 flooding is standard for large-scale projects, alternatives like water-alternating-gas (WAG) injection can optimize sweep efficiency by combining water's mobility control with CO_2's displacement capabilities (Wang et al. 2023), this also helps in enhancing the CO_2 capillary trapping. The choice of technique depends on reservoir characteristics and economic factors. CO_2 flooding offers the highest storage potential but risks early breakthrough in heterogeneous formations. WAG injection balances oil recovery and storage by mitigating CO_2's high mobility while improving vertical sweep. WAG also help to improve the capillary trapping of CO_2. Figure 6.16 depicts the steps of the process where first CO_2 is captured from industrial sources or direct air capture facilities. The captured CO_2 is typically purified and separated from other gases. The captured CO_2 is then transported to the injection site (wellhead) and injected deep into an oil reservoir. Once injected, the CO_2 forms a plume within the reservoir, spreading and migrating through the rock formation. The CO_2 plume is represented by a cloud-like shape expanding within the reservoir. The injected CO_2 mixes with the existing oil in the reservoir, creating a CO_2 miscible oil zone. This interaction improves the oil's mobility and facilitates its recovery. As the CO_2 gets dissolved in oil, plume of water is injected that pushes the oil and dissolved CO_2 ahead, resulting in an extra recoverable oil zone. This zone represents the additional oil that can be extracted due to the WAG injection. Production wells are used to extract the oil and dissolved CO_2 mixture from the reservoir. The produced fluid, containing both oil and dissolved CO_2, is brought to the surface. At the surface, the produced fluid undergoes separation processes. The CO_2 is separated from the oil and recycled to be reused for injection in the injection well. The separated CO_2 is combined with the captured CO_2 from the capture facility and reinjected back into the oil reservoir to continue the EOR process.

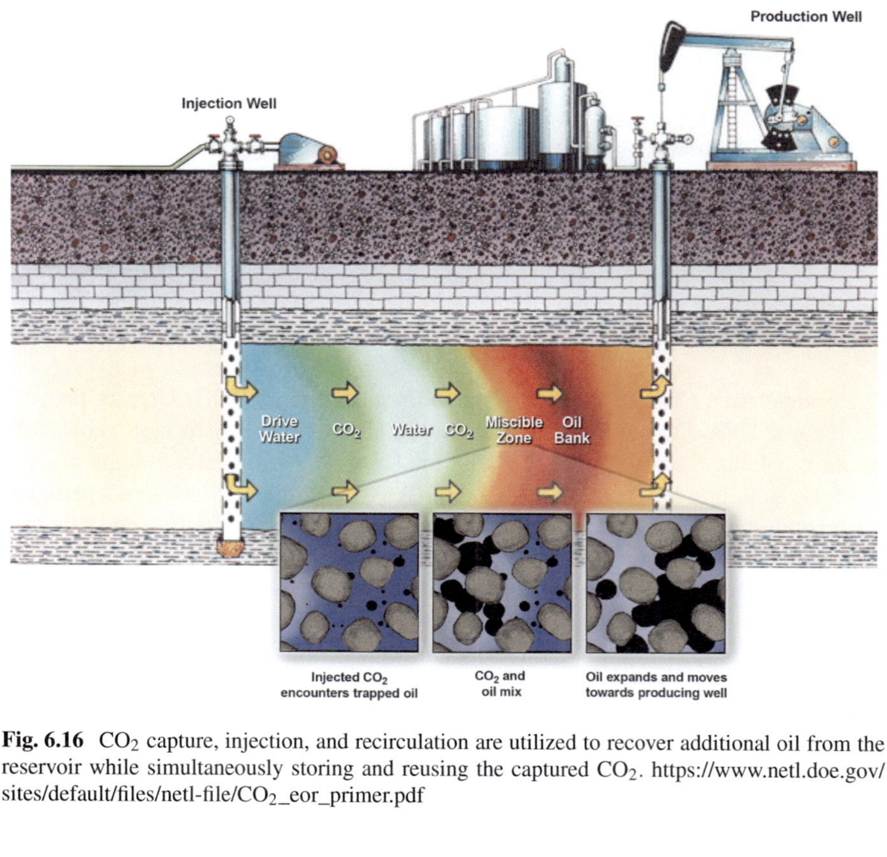

Fig. 6.16 CO_2 capture, injection, and recirculation are utilized to recover additional oil from the reservoir while simultaneously storing and reusing the captured CO_2. https://www.netl.doe.gov/sites/default/files/netl-file/CO2_eor_primer.pdf

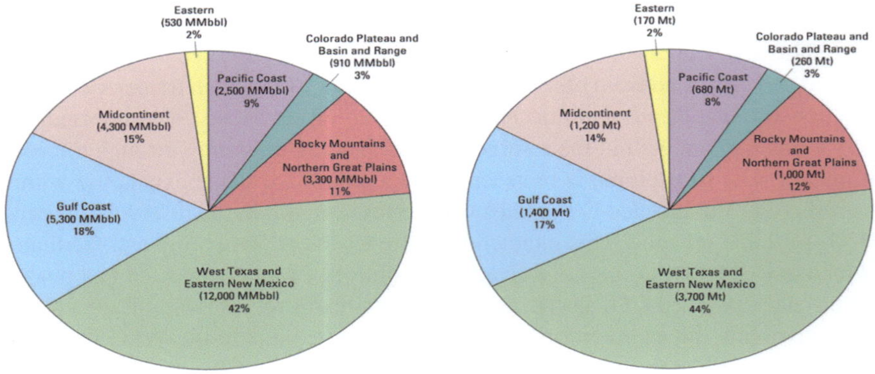

Fig. 6.17 (Left) Regional average estimates of enhanced oil recovery (CO_2-EOR) in million BBLs process and (right) accompanying CO_2 masses in million tons of carbon dioxide that could be stored using CO_2-EOR, Warwick et al. (2022)

In 2020, the U.S. Geological Survey (USGS) performed a probabilistic study to quantify the potential of CO_2 enhanced oil recovery in conventional onshore oil reservoirs in terms of enhanced oil recovery and CO_2 storage across the United States in USGS Circular and Data Release (Warwick et al. 2022). The study estimated somewhere between 25 and 32 billion barrels of oil recovery with an average of 29 billion barrels from over 3500 oil reservoirs. This enhanced oil recovery also supported CO_2 storage capacity of 7.4–9.5 billion metric tons of CO_2 with average of 8.4 billion metric tons of potential storage. Study also highlighted that the Permian basin in West Texas/eastern New Mexico and Gulf coast to dominate both in terms of enhanced oil recovery and CO_2 storage as shown in Fig. 6.17.

The Scurry Area Canyon Reef Operators Committee (SACROC)

The Scurry Area Canyon Reef Operators Committee (SACROC) Unit in Texas is the longest USA CO_2 EOR project operating since 1972. Figure 6.18 shows key locations of the Southwest Regional Partnership's Phase II carbon sequestration projects, including a detailed cross-section (A-A') of the SACROC area's geology. The cross-section shows the Wolfcamp limestone (top layer), Wolfcamp shale and sandstone, Canyon Anticline oil reservoir (Canyon limestone with Cisco limestone cap) and Upper and Lower Strawn formations (base layers). Long injection and production data from SACROC provides unique data base to study different aspects of CO_2 EOR project in this field (Hosseininoosheri et al. 2018; Núñez-López and Moskal 2019; Hill et al. 2013; Fox 2009). These studies include effectiveness of CO_2 storage in EOR operations in terms of enhanced oil recovery and CO_2 storage capacities, potential risks and safety measures, and best practices for monitoring and verification. Advanced 3D modeling from well logs and seismic data combined with field observations indicated that the around 60% of injected CO_2 is stored

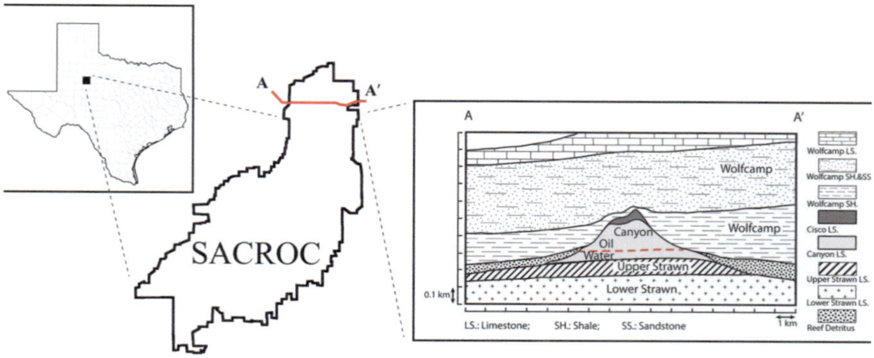

Fig. 6.18 Locations of major Southwest Regional Partnership on carbon sequestration (SWP) Phase II pilot projects. https://www.netl.doe.gov/coal/carbon-storage/atlas/swp/phase-II/sacroc

as dissolved gas in residual oil saturation and about 30% has been stored due to capillary trapping and remaining through other trapping mechanisms (Wei Jia, Brian McPherson, SACROC CMG model, 2018-01-10, https://edx.netl.doe.gov/dataset/sacroc-cmg-model, https://doi.org/10.2172/2529468).

6.6.2.3 CO_2 Storage in Depleted Natural Gas Reservoirs

Like depleted oil reservoir, depleted natural gas reservoirs also pose a great potential for large-scale CO_2 storage with similar advantages regarding detailed formation characteristics and presence of infrastructure (wells, pipelines, and monitoring systems) (Hamza et al. 2021; Guan et al. 2024). This enables cost-effective repurposing of depleted natural gas reservoirs for CO_2 storage. In addition to that storage capacities of depleted natural gas reservoirs commonly exceed those of depleted oil reservoirs. For instance, Fig. 6.19 shows the distribution of gas reservoirs in North America that can be considered as possible CO_2 storage site. Despite these common advantages the storage mechanisms show some differences compared to depleted oil reservoirs. In depleted natural gas reservoirs, storage of CO_2 primarily relies on structural and capillary trapping, with solubility trapping occurring as CO_2 mixes with remaining formation brine. The absence of hydrocarbon miscibility simplifies fluid dynamics modeling of CO_2 storage in these reservoirs; however still careful pressure management is required to avoid caprock fracturing during injection.

6.6.2.4 CO_2 Storage in Coal Seams

Coal seams present a great opportunity for carbon dioxide (CO_2) storage due to their strong affinity for adsorbing CO_2 much higher than that for methane. These characteristic forms the basis of a process known as Enhanced Coalbed Methane Recovery

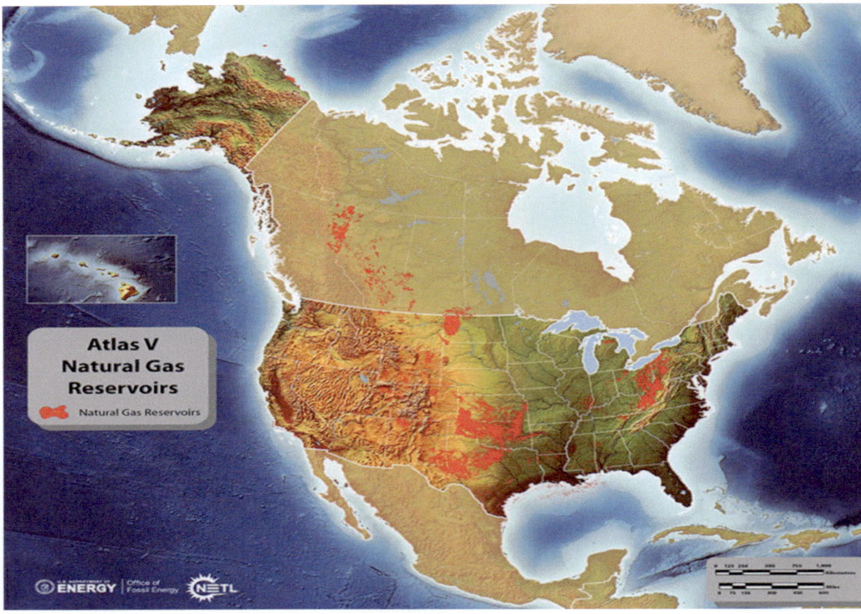

Fig. 6.19 Natural gas reservoirs distributions in North America (NETL Carbon Storage Atlas fifth edition)

(ECBM), in which CO_2 injection displaces methane from the coal, increasing methane production while also sequestering CO_2 (Koperna and Riestenberg 2009). This dual benefit of energy recovery and emission reduction has attracted significant research interest, particularly in regions with large coal reserves. Before implementing ECBM, it is critical to fully evaluate the coal seems mineralogy and characterize its physical and chemical properties. Site-specific conditions and local regulatory frameworks must also be considered when assessing the feasibility of CO_2 storage in coal seams. Unmineable coal seams, those that are too deep or too thin for economic extraction, are especially attractive candidates for storage projects.

Figure 6.20 shows the unmineable coal distribution in North America that presents great potential to be used as CO_2 storage sites.

As illustrated in Fig. 6.21, the ECBM process involves injecting CO_2 into a deep coal seam, where the CO_2 molecules are adsorbed onto the coal surface. This causes methane, which was previously adsorbed, to desorb and migrate toward a production well. The exchange of CO_2 and methane is driven by competitive adsorption, and the gas transport within the coal is governed by diffusion mechanisms. Near the injection well, counter-diffusion occurs, with CO_2 moving into the coal matrix and methane diffusing outward. Closer to the production well, both gases tend to co-diffuse due to convective forces generated by the pressure gradient, allowing excess CO_2 and the desorbed methane to migrate together toward the production point (Fathi and Akkutlu 2014; Godec et al. 2014).

Fig. 6.20 Unmineable coal distributions in North America (NETL Carbon Storage Atlas fifth edition)

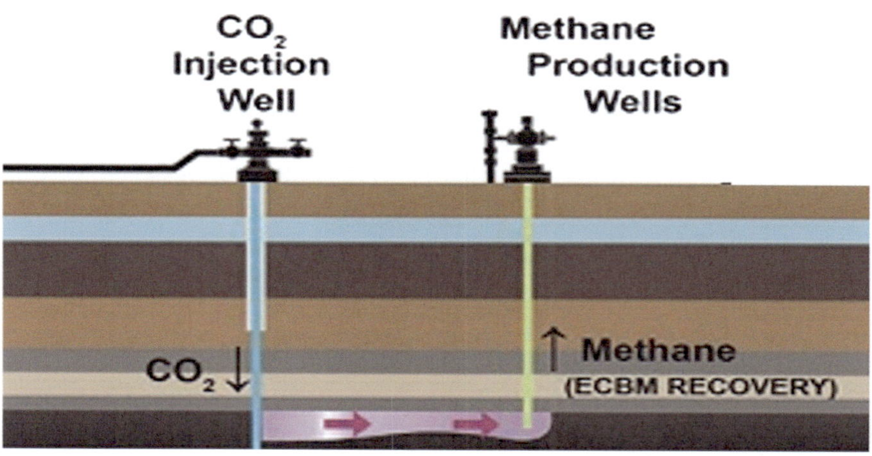

Fig. 6.21 CO_2 enhanced coalbed recovery modified from https://netl.doe.gov/carbon-management/carbon-storage/faqs/carbon-storage-faqs

However, injecting CO_2 into coal seams can cause the coal to swell, which may reduce permeability and hinder injectivity. This has been tested in the Pump Canyon project where the effectiveness of carbon dioxide (CO_2) storage in deep unmineable coal seams is tested (Koperna et al. 2009; Godec et al. 2014). As shown in Fig. 6.22 using the high-resolution pressure and temperature data from downhole memory gauges (DMG) show significant injectivity decline during CO_2 injection. The injection rate dropped sharply from 1.15 m^3/s (3,500 Mscfd) to 0.08 m^3/s (250 Mscfd)

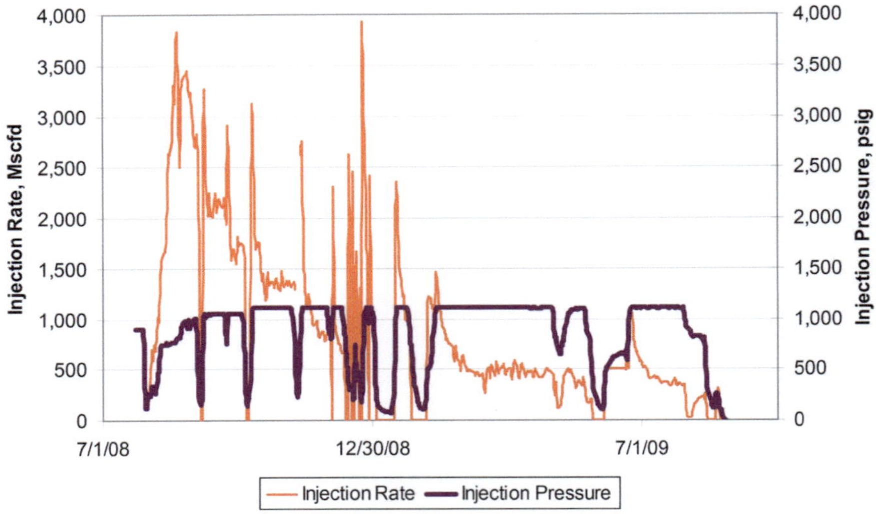

Fig. 6.22 Sample Injection rate and pressure measured during CO_2 injection in coal seams in the Pump Canyon project. NETL, US DOE https://www.netl.doe.gov/coal/carbon-storage/atlas/swp/phase-II/sanjuanecbm

over one year. This is due to multiple reasons including coal matrix swelling, and compaction due to CO_2 adsorption and injection pressure leading to permeability loss in coal seams. In many cases, especially at depths greater than 900 m (approximately 3000 feet), hydraulic fracturing may be necessary to maintain adequate flow. A successful CO_2 storage project in coal seams depends on a comprehensive understanding of the geological setting, appropriate engineering design, and robust monitoring strategies.

6.6.2.5 CO_2 Storage in Depleted Shale Gas Reservoirs

Shale formations offer a unique opportunity for subsurface CO_2 storage through pressure maintenance and physical adsorption of CO_2 very much like coalbed methane (Kang et al. 2011). Shale also exhibits higher CO_2 sorption capacity than methane (CH_4), particularly within their microporous organic-rich matrix (Busch et al. 2008). This property not only enables secure CO_2 sequestration but also suggests potential for CO_2 enhanced shale gas recovery (CO_2-ESGR), where injected CO_2 displaces adsorbed CH_4 due to its stronger affinity for organic materials. In case of CO_2 shale gas recovery, sweeping remaining gas trapped in natural fractures are the main enhance gas recovery as shale matrix is so tight that the matrix sweep efficiency is negligible. The CO_2 adsorption and methane desorption closer to injection well is also secondary recovery mechanism. The major considerations regarding CO_2-ESGR includes detailed natural fracture mapping, CO_2 rock and fluid interactions

Fig. 6.23 Distribution of Shale basins as potential CO_2 storage sites in North America (NETL Carbon Storage Atlas fifth edition)

and pressure managements. The injection pressure must not exceed the formation breakdown pressure to refrac the reservoir and should be les than fracture opening pressure to avoid early CO_2 breakthrough in the producer. The fracture opening pressure can be obtained using injection step rate test. Even though the CO_2-ESGR has not been commercially implemented depleted shale plays like the Barnett (Texas) and Marcellus (Northeastern U.S.) are prime candidates for CO_2 storage, due to existing infrastructure and proven capacity to retain gas over geologic timescales (Boswell 1996; Schepers et al. 2009). Figure 6.23 shows the distribution of shale gas reservoir in North America highlighting the great potential for CO_2 storage projects.

The process of CO_2-ESGR involves three coupled mechanisms operating across scales: (1) convective/dispersive flow of CO_2-CH_4 mixtures in fractures; (2) diffusive transport through the shale's secondary porosity (macropores/fractures); and (3) competitive adsorption in micropores, where CO_2 displaces CH_4. These interactions occur at disparate spatiotemporal scales from meter-scale hydraulic fractures to nanometer-scale pores creating a dynamic system where CO_2 storage and CH_4 recovery proceed simultaneously (see Fig. 6.23).

6.6.3 Storage in Basalt and Other Volcanic and Mafic Rocks

The chemical composition of certain volcanic rock types, particularly mafic rocks, like basalt which is rich in magnesium and iron provides a unique opportunity for CO_2 storage. These rocks have a high reactivity with CO_2 that potentially leads to the

Fig. 6.24 Basalt formation distributions in North America (NETL Carbon Storage Atlas fifth edition)

conversion of injected CO_2 solution into solid mineral forms (Rodosta et al. 2011). Basalt formations such as the Columbia River Basalts found in the Pacific Northwest are recognized as having significant potential for long-term CO_2 storage in form of mineralization. Figure 6.24 shows the distribution of Basalt formations across North America highlighting the great potential of CO_2 storage in basalt formations.

The capacity of Basalt formations to store CO_2 highly depends on their porosity and permeability such that required injectivity is achieved and enough surface contact becomes available to facilitate the reaction and mineralization. Therefore, much research has been focused on enhancing the mineralization reactions and optimizing CO_2 flow within basalt formations (Deng et al. 2025; Cao et al. 2024).

Even though in this case the main trapping mechanism is mineralization, however, as this process might takes few years to be completed the confinement ability of these formations, specifically their ability to prevent CO_2 leakage, still needs to be confirmed through further research and field studies.

The Orca plant is a successful CCUS project that stands out as a project integrated with use of renewable energy where direct air capture technology is used to capture CO_2 from the atmosphere. Captured CO_2 are then mixed with water and injected into volcanic basalt rock, where the injected CO_2-water mixture reacts with Basalt. This reaction forms calcium carbonate and converts to hydride of Sulfur (H_2S) within around four months (85% H_2S mineralized), Clark et al. 2020. Trials have shown that Icelandic basalt can effectively sequester CO_2 in solid rock within about two years, (Clark et al. 2020; Matter et al. 2016; Pogge von Strandmann et al. 2019). It has the capacity to capture and store up to 4000 tons of carbon annually. To make sure the process of capturing and sequestration does not generate extra CO_2 geothermal energy is used to provide energy for the whole process including CO_2 capture and storage.

6.7 Case Study: Illinois Basin–Decatur Project (IBDP) Site

The Illinois Basin–Decatur Project (IBDP), led by the Midwest Geological Sequestration Consortium (MGSC) and funded by the U.S. Department of Energy, demonstrated the feasibility of large-scale CO_2 storage in deep geological formations. One million metric tons of CO_2 from an Archer Daniels Midland Company (ADM) ethanol plant were captured and injected into the Mt. Simon Sandstone, a saline reservoir at 2100 m (6890 feet) depth, in three years' time span (Finley 2014).

The Mt. Simon Sandstone is a Cambrian-period formation spanning the Midwest, is a key storage formation within the Illinois Basin. At the IBDP site, the sandstone exceeds 457 m (1500 feet) thick with high porosity in its lower layers due to feldspar dissolution (Heath et al. 2013). The Eau Claire Formation which is a 212-m (695-foot) shale layer acts as an impermeable caprock over Mt. Simon sandstone confines the injected CO_2 (Lahann et al. 2014). CO_2 injection started in November 2011 where the CO_2 was initially injected into the Argenta formation before being stored in the Mt. Simon Sandstone reservoir. Due to its buoyancy and low viscosity, the injected CO_2 formed plumes that migrated upward displacing formation brine until encountering the Eau Claire Formation caprock. Reservoir heterogeneity, injection parameters, and thermal gradients influenced plume behavior, which was carefully monitored using multiple complementary techniques. Advanced monitoring techniques, including pressure measurements, microseismic surveys, and 3D seismic imaging is implemented to ensure secure CO_2 storage (Couëslan et al. 2014). Figure 6.25 illustrates the IBDP site layout including an injection well (CCS1), a monitoring well (VW1), and a CO_2 pipeline connecting to surface compressors.

Detailed well log analysis is performed to study CO_2 plume behavior in the reservoir (Fig. 6.26). Formation MicroImager (FMI™) logs identified numerous low-permeability baffles between 5371.5 and 7198 ft. depth that restricted vertical flow and promoted lateral spreading of injected CO_2, while natural fractures provided alternative migration pathways. These baffles were characterized using multiple methods: high-resolution FMI image logs (averaged over 5-ft and 10-ft intervals) provide detailed distribution of baffles and natural fractures, while combination of Gamma Ray values >90 and Effective Porosity <0.01 provided identified baffles with lower resolution (Adenan et al. 2024). Time-lapse Pulsed Neutron Logging between 2014 and 2019 showed contrasting plume behavior between wells that confined near the injection well (CCS#1) but vertically migrated at the monitoring well (VW#1) due to pressure buildup.

Microseismic monitoring (2012–2018) also complemented these findings, revealing cyclical migration patterns with clustered events near the injection point, indicating pressure-driven breakthroughs through baffles (Iyegbekedo et al. 2024).

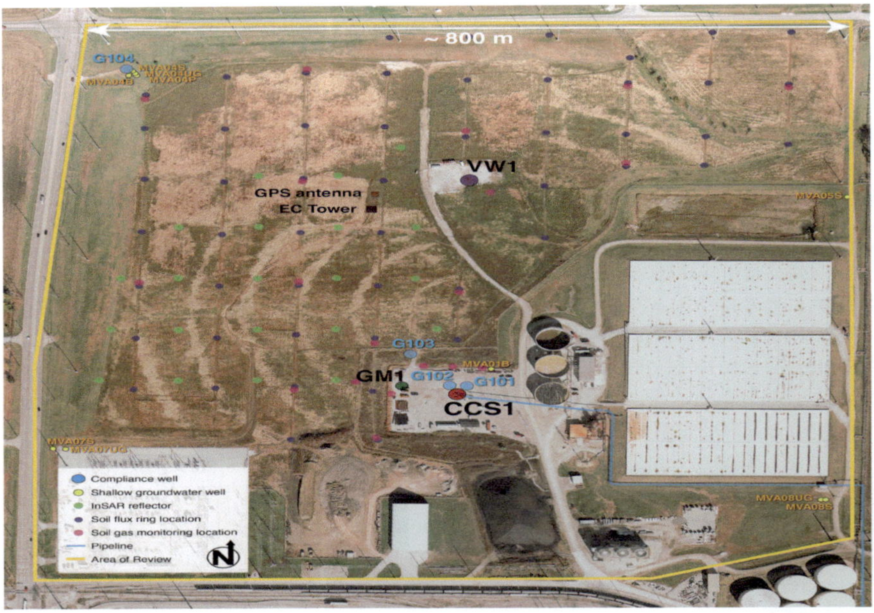

Fig. 6.25 IBDP site layout including an injection well (CCS1), a monitoring well (VW1), and a CO_2 pipeline connecting to surface compressors https://netl.doe.gov/coal/carbon-storage/atlas/mgsc/phase-III/ibdp

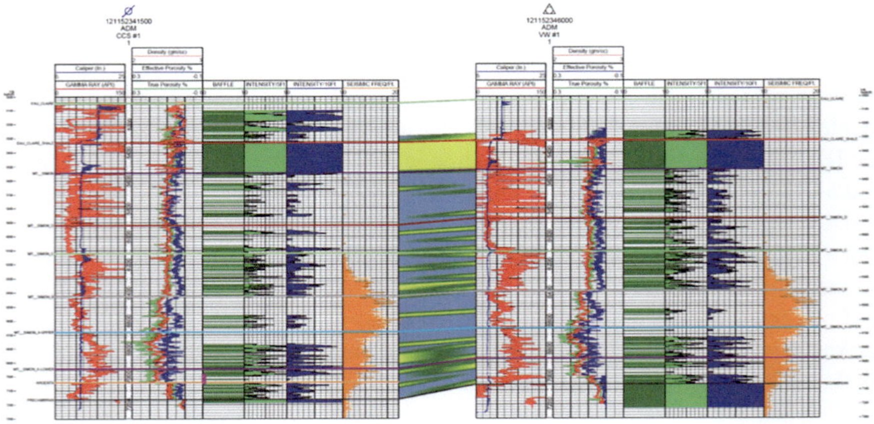

Fig. 6.26 IBDP well log analysis of both injection (CCS#1) and monitoring (VW#1) wells. From left well logs in first track includes caliper in blue and gamma ray in red, second track includes the density in red, total porosity in blue and true or effective porosity in green, third track includes Baffles obtained using the hard cut offs for porosity and gamma ray, the fourth and fifth tracks are fracture intensities per 5 ft and 10 ft obtained using FMI image logs, and sixth track is the seismic intensities. The same list of logs repeated for monitoring well (VW#1) as well

The integrated dataset, including gamma ray, porosity, and baffle intensity measurements (Fig. 6.26), demonstrated how reservoir heterogeneity controls CO_2 distribution. FMI logs confirmed baffles as leaky seals composed of cemented or shale-rich layers, explaining the observed alternating phases of lateral containment and vertical migration that characterize plume behavior in this storage system.

IBDP Well Pad Dashboard developed by Dr. Brian Panetta (http://ibdpdashboard.herokuapp.com/) presents an overview of the Illinois Basin–Decatur Project (IBDP) site, specifically focusing on two wells drilled in the area. The first component of the dashboard displays the microseismic events recorded during the injection and post injection, providing both a vertical and surface map view. This allows for the visualization of the distribution of these events in a three-dimensional space. Dashboard provides insights into the subsurface activities and potential reservoir behavior. Additionally, the dashboard includes a structural map of all the layers including the Precambrian formation at the bottom, and Eau Claire Shale as caprock and in between different Mt. Simon units that provides a geological perspective of the underlying rock formations. This visualization tool helps in understanding the subsurface characteristics and the potential pathways for fluid migration. Furthermore, the dashboard highlights the injection and reservoir details of the CO_2 storage project.

Figure 6.27 shows the majority of microseismic events highlighted as colored dots are concentrated within the Mt. Simon B and Argenta intervals, with limited activity extending into the Precambrian basement. These events clearly demonstrate buoyancy-driven CO_2 migration toward the top of Mt. Simon B. During the injection and monitoring period the plume appears to have reached the base of Mt. Simon C (top of Mt. Simon B) a tight zone composed of very fine sandstone that marks the vertical extent of CO_2 migration.

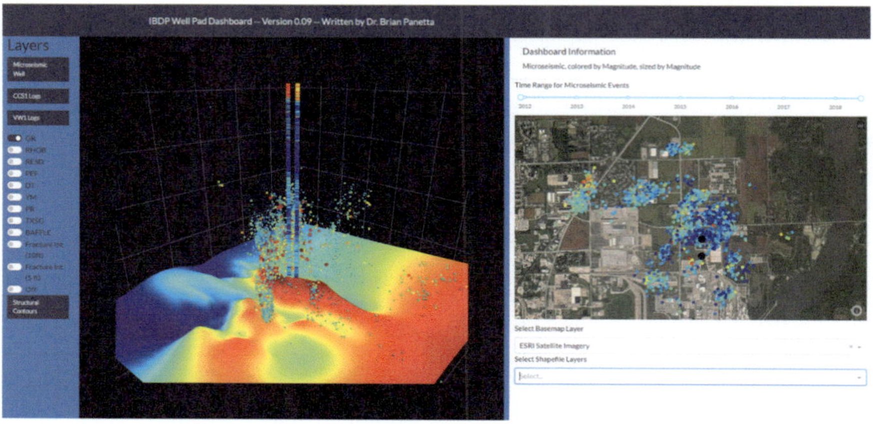

Fig. 6.27 3D visualization of the injection and monitoring wells in IBDP at left and surface map of seismic events collected during injection and post injection at right (dashboard by Dr. Brian Panetta) http://ibdpdashboard.herokuapp.com/

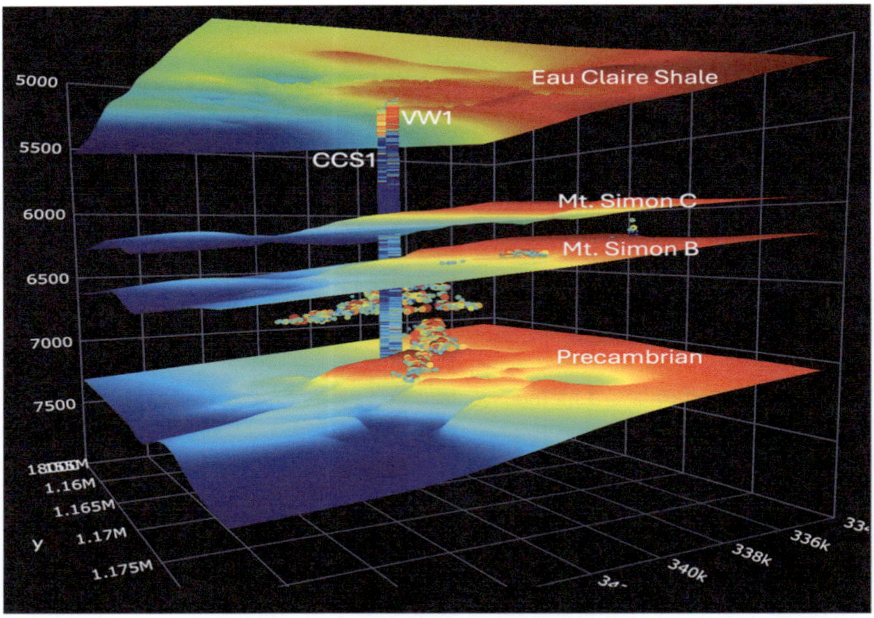

Fig. 6.28 Microseismic events because of CO_2 injection showing upward movement of CO_2 to the bottom of Mt. Simon B and extending laterally and breaking through with some distance from injection and moving upward to the base of Mt.Simon C (dashboard by Dr. Brian Panetta). http://ibdpdashboard.herokuapp.com/

The analysis of microseismic activity between the Precambrian basement and Mt. Simon B, shows general upward movement of CO_2 while it's also expanded laterally in different horizons because of presence of impermeable baffles (numerous baffles were detected through detailed FMI image logging). Baffles initially restrict vertical migration until sufficient pressure builds for breakthrough. The concentration of events near formation boundaries further supports the concept of alternating containment and migration phases that characterize CO_2 storage in heterogeneous reservoirs like the Mt. Simon Sandstone (see Fig. 6.28).

6.8 Conclusion

Geological storage of carbon dioxide stands as the most viable technology in global efforts to mitigate climate change. This technology offers sustainable solution for permanently sequestering billions of tons of CO_2 underground. This chapter has systematically examined the scientific principles, practical considerations, and real-world applications of geological CO_2 storage.

Fig. 6.29 images designed by Freepik

A key insight from this chapter is the critical importance of rock properties and trapping mechanisms in determining storage feasibility and long-term sustainability. Supercritical CO_2, with its unique phase behavior, enables efficient storage in geological formations, under combination of trapping mechanism such as structural, capillary, dissolution, and mineral trapping. The chapter investigated how different reservoir types from deep saline aquifers and depleted oil/gas fields to basalt formations offer distinct advantages in terms of capacity, injectivity, and mineralization of CO_2 underground. For instance, depleted hydrocarbon reservoirs benefit from existing infrastructure and well-characterized geology, while basaltic rocks provide the added advantage of rapid mineral carbonation.

The Illinois Basin–Decatur Project (IBDP) served as a success story, illustrating the practical challenges and feasibility of large-scale CO_2 injection in a saline formation. Looking ahead, scaling geological storage to meet global climate targets will require the following:

1. advanced capacity assessments to identify optimal storage sites worldwide,
2. developing infrastructure for CO_2 transportation network,
3. innovations in monitoring technologies to verify long-term containment, and,
4. policy frameworks that incentivize storage development while addressing Regulatory Liability and community concerns (see Fig. 6.29).

References

M.F. Adenan, E. Fathi, T.R. Carr, B. Panetta, Advancing subsurface analysis: Integrating computer vision and deep learning for the near real-time interpretation of borehole image logs in the Illinois Basin-Decatur project. Interpretation **12**(2), SB19–SB27 (2024)

T. Ajayi, J.S. Gomes, A. Bera, A review of CO_2 storage in geological formations emphasizing modeling, monitoring and capacity estimation approaches. Pet. Sci. **16**, 1028–1063 (2019). https://doi.org/10.1007/s12182-019-0340-8

M.M.A. Awan, F.U.D. Kirmani, Evaluating the potential of depleted oil reservoirs for CO_2 sequestration through simulation modeling. Pet. Res. **10**(2), 425–431 (2025)

S. Bachu, D. Bonijoly, J. Bradshaw, R. Burruss, S. Holloway, N.P. Christensen, CO_2 storage capacity estimation: methodology and gaps. Int. J. Greenh. Gas Con. **1**(4), 430–443 (2007). https://doi.org/10.1016/S1750-5836(07)00086-2
S. Bachu, W.D. Gunter, E.H. Perkins, Aquifer disposal of CO_2: hydrodynamic and mineral trapping. Energy Convers. Manag. **35**(4), 269–279 (1994)
R. Boswell, Play UDs: Upper Devonian black shales, In J. Roen and B. Walker, eds., The atlas of major Appalachian gas plays: West Virginia Geological and Economic Survey, **25**, 93–99 (1996)
I.C. Bourg, L.E. Beckingham, D.J. DePaolo, The nanoscale basis of CO_2 trapping for geologic storage. Environ. Sci. Technol. **49**(17), 10265–10284 (2015). https://doi.org/10.1021/acs.est.5b03003
A. Busch, S. Alles, Y. Gensterblum, D. Prinz, D.N. Dewhurst, M.D. Raven, H. Stanjek, B.M. Krooss, Carbon dioxide storage potential of shales. Int. J. Greenh. Gas Control. **23** 2008297–308 (2008). 1750-5836. https://doi.org/10.1016/j.ijggc.2008.03.003
C. Cao, H. Liu, Z. Hou, F. Mehmood, J. Liao, W. Feng, A Review of CO_2 Storage in View of Safety and Cost-Effectiveness. Energies, **13**, 600 (2020). https://doi.org/10.3390/en13030600
X. Cao, Q. Li, L. Xu, Y. Tan, A review of in situ carbon mineralization in basalt. J. Rock Mech. Geotech. Eng. **16**(4), 1467–1485 (2024)
Y. Chen, S. Chen, D. Li, X. Jiang, Density-driven convection for CO_2 solubility trapping in saline aquifers: Modeling and influencing factors. Geotechnics **3**(1), 70–103 (2023)
D.E. Clark, E.H. Oelkers, I. Gunnarsson, B. Sigfússon, S.Ó. Snæbjörnsdóttir, E.S. Aradóttir, S.R. Gíslason, CarbFix2: CO_2 and H_2S mineralization during 3.5 years of continuous injection into basaltic rocks at more than 250°C. Geochim. Cosmochim. Acta **279**, 45–66 (2020)
M.L. Couëslan, V. Smith, G. El-Kaseeh, J. Gilbert, N. Preece, L. Zhang, J. Gulati, Development and implementation of a seismic characterization and CO_2 monitoring program for the Illinois Basin–Decatur project. Greenh. Gases: Sci. Technol. **4**(5), 626–644 (2014). https://doi.org/10.1002/ghg.1452
D. Dempsey, S. Kelkar, R. Pawar, Passive injection: a strategy for mitigating reservoir pressurization, induced seismicity and brine migration in geologic CO_2 storage. Int. J. Greenh. Gas Con. **28**, 96–113 (2014). https://doi.org/10.1016/j.ijggc.2014.06.002
H. Deng, C. Yu, Q. Jiang, X. Shi, X. Zhou, X. Chen, Accelerated CO_2 mineralization in basalt via reaction process control: mineralization effect, mineral evolution, and reservoir property implications. Gas Sci. Eng. **140**, 205648 (2025)
J. Dolson, Z. He, B.W. Horn, Search and Discovery, in *Advances and perspectives on stratigraphic trap exploration: making the subtle trap obvious*, (2018), p. 60054
E. Fathi, I.Y. Akkutlu, Multi-component gas transport and adsorption effects during CO_2 injection and enhanced shale gas recovery. Int. J. Coal Geol. **123**, 52–61 (2014)
R.J. Finley, An overview of the Illinois Basin-Decatur project. Greenh. Gases: Sci. Technol. **4**(5), 571–579 (2014). https://doi.org/10.1002/ghg.1423
C.E. Fox, *CO_2 EOR Carbon Balance at SACROC [PowerPoint Slides]* (Kinder Morgan CO2 Company, Houston, TX, United States, 2009)
Q. Gao, Y. Cheng, E. Fathi, S. Ameri, Analysis of stress-field variations expected on subsurface faults and discontinuities in the vicinity of hydraulic fracturing. SPE Reserv. Eval. Eng. **19**(1), 54–69 (2016)
M. Godec, G. Koperna, J. Gale, CO_2-ECBM: a review of its status and global potential. Energy Procedia **63**, 5858–5869 (2014). https://doi.org/10.1016/j.egypro.2014.11.619
A. Goodman, A. Hakala, G. Bromhal, D. Deel, T. Rodosta, S. Frailey, M. Small, D. Allen, V. Romanov, J. Fazio, N. Huerta, D. McIntyre, B. Kutchko, G. Guthrie, U.S. DOE methodology for the development of geologic storage potential for carbon dioxide at the national and regional scale. Int. J. Greenh. Gas Con. **5**(4), 952–965 (2011). https://doi.org/10.1016/j.ijggc.2011.03.010
D. Guan, P. Gao, Z. Jiang, Q. Fan, Q. Li, Y. Zhou, L. Zhang, J. Zhao, Y. Song, L. Yang, Spatial evolution of CO_2 storage in depleted natural gas hydrate reservoirs and its synergistic efficiency analysis. Appl. Energy **376**(Part A), 124247 (2024)

A. Hamza, I.A. Hussein, M.J. Al-Marri, M. Mahmoud, R. Shawabkeh, S. Aparicio, CO_2 enhanced gas recovery and sequestration in depleted gas reservoirs: a review. J. Pet. Sci. Eng. **196**, 107685 (2021)

Heath, J. E., Bauer, S. J., Broome, S. T., Dewers, T. A., & Rodriguez, M. A. (2013). Petrologic and petrophysical evaluation of the Dallas Center structure, Iowa, for compressed air energy storage in the mount Simon sandstone (SAND2013-7819). Sandia National Laboratories

M.M. Hasan, F. Faruque, F. Boukouvala, E.L. First, C.A. Floudas, Nationwide, regional, and statewide CO_2 capture, utilization, and sequestration supply chain network optimization. Ind. Eng. Chem. Res. **53**(18), 7489–7506 (2014). https://doi.org/10.1021/ie402931c

S.A. Hosseini, H. Lashgari, J.W. Choi, J.-P. Nicot, J. Lu, S.D. Hovorka, Static and dynamic reservoir modeling for geological CO_2 sequestration at Cranfield, Mississippi, U.S.A., Int. J. Greenhouse Gas Control, **18**, 449–462, ISSN 1750-5836 (2013). https://doi.org/10.1016/j.ijggc.2012.11.009

P.P. Hosseininoosheri, S.A. Hosseini, V. Nuñez-López, L.W. Lake, Impact of field development strategies on CO_2 trapping mechanisms in a CO_2-EOR field: A case study in the Permian Basin (SACROC unit). Int. J. Greenh. Gas Con. **72**, 92–104 (2018). https://doi.org/10.1016/j.ijggc.2018.03.002

B. Hill, S. Hovorka, S. Melzer, Geologic carbon storage through enhanced oil recovery. Energy Procedia **37**, 6808–6830 (2013). https://doi.org/10.1016/j.egypro.2013.06.614

I. Iyegbekedo, E. Fathi, T.R. Carr, F. Belyadi, Machine learning-driven quantification of CO_2 plume dynamics at Illinois Basin Decatur project sites using microseismic data. Energies **17**(17), 4421 (2024)

F. Johnsson, Perspectives on CO_2 capture and storage. Greenh. Gases: Sci. Technol. **1**(2), 119–133 (2011). https://doi.org/10.1002/ghg.9

S.M. Kang, E. Fathi, R.J. Ambrose, I.Y. Akkutlu, R.F. Sigal, CO_2 storage capacity of organic-rich shales. SPE J. **16**(4), 842–855 (2011)

P. Kelemen, S.M. Benson, H. Pilorgé, P. Psarras, J. Wilcox, An overview of the status and challenges of CO_2 storage in minerals and geological formations. Frontiers in Climate **1**, 9 (2019). https://doi.org/10.3389/fclim.2019.00009

S. Khandoozi, R. Hazlett, M. Fustic, A critical review of CO_2 mineral trapping in sedimentary reservoirs—from theory to application: pertinent parameters, acceleration methods and evaluation workflow. Earth Sci. Rev. **244**, 104515 (2023)

K. Kim, D. Kim, Y. Na, Y. Song, J. Wang, A review of carbon mineralization mechanism during geological CO_2 storage. Heliyon **9**(12), e23135 (2023). https://doi.org/10.1016/j.heliyon.2023.e23135

G. Koperna, A. Oudinot, G. McColpin, J. Heath, N. Liu, A. Wells, G. Young, 2009 International Coalbes & Shale Gas Symposium, in *CO_2-ECBM/storage activities at the San Juan Basin's Pump Canyon test site*, (2009)

G.J. Koperna, D. Riestenberg, SPE International Conference on CO_2 Capture, Storage, and Utilization, in *Carbon dioxide enhanced coalbed methane and storage: Is there promise*, (San Diego, CA, United States, 2009). https://doi.org/10.2118/126627-MS

R. Lahann, J. Rupp, C. Medina, An evaluation of the seal capacity and CO_2 retention properties of the Eau Claire formation (Cambrian). Environ. Geosci. **21**(2), 83–106 (2014). https://doi.org/10.1306/eg.04281313010

R.G. Larkin, SPE Improved Oil Recovery Symposium, in *Hydrodynamic trapping of CO_2 geosequestered in saline aquifers*, (Tulsa, OK, United States, 2010). https://doi.org/10.2118/128205-MS

H. Lei, S. Yang, L. Zu, Z. Wang, Y. Li, Oil recovery performance and CO_2 storage potential of CO_2 water-alternating-gas injection after continuous CO_2 injection in a multilayer formation. Energy Fuel **30**(11), 8922–8931 (2016). https://doi.org/10.1021/acs.energyfuels.6b01307

N. Li, W. Feng, J. Yu, F. Chen, Q. Zhang, S. Zhu, Y. Hu, Y. Li, Recent advances in geological storage: Trapping mechanisms, storage sites, projects, and application of machine learning. Energy Fuel **37**(14), 10087–10111 (2023). https://doi.org/10.1021/acs.energyfuels.3c01433

Y. Ma, S. Yi, M. Wang, Biomimetic mineralization for carbon capture and sequestration. Carbon Capture Sci. Technol. **13**, 100257 (2024)

J.M. Matter, M. Stute, S.Ó. Snæbjörnsdóttir, E.H. Oelkers, S.R. Gíslason, E.S. Aradóttir, B. Sigfusson, I. Gunnarsson, H. Sigurdardóttir, E. Gunnlaugsson, G. Axelsson, H.A. Alfredsson, D. Wolff-Boenisch, K. Mesfin, D. Fernandez de la Reguera, J. Taya, J. Hall, K. Dideriksen, W.S. Broecker, Rapid carbon mineralization for permanent disposal of anthropogenic carbon dioxide emissions. Science **352**(6291), 1312–1314 (2016)

B.J. McPherson, Looking ahead: Research agenda for the study of carbon sequestration. Geophys. Monogr. Ser. **183**, 305–316 (2009). https://doi.org/10.1029/2008GM000792

I.H. Min, S.-G. Kang, C. Huh, Instability Analysis of Supercritical CO_2 during Transportation and Injection in Carbon Capture and Storage Systems. Energies, **11**, 2040 (2018). https://doi.org/10.3390/en11082040

V. Núñez-López, E. Moskal, Potential of CO_2-EOR for near-term decarbonization. Front. Clim. **1**, 5 (2019). https://doi.org/10.3389/fclim.2019.00005

P.A.E. Pogge von Strandmann, K.W. Burton, S.O. Snaebjornsdottir, B. Sigfusson, E.S. Aradottir, I. Gunnarsson, H.A. Alfredsson, K.G. Mesfin, E.H. Oelkers, S.R. Gislason, Rapid CO_2 mineralisation into calcite at the CarbFix storage site quantified using calcium isotopes. Nat. Commun. **10**(1), 1983 (2019)

T. Rodosta, J. Litynski, S. Plasynski, L. Spangler, et al., US Department of Energy's regional carbon sequestration partnership initiative: update on validation and development phases. Energy Procedia **4**, 3457–3464 (2011). https://doi.org/10.1016/j.egypro.2011.02.271

K.C. Schepers, B. Nuttall, A.Y. Oudinot, G. Reinaldo, Reservoir Modeling and Simulation of the Devonian Gas Shale of Eastern Kentucky for Enhanced Gas Recovery and CO_2 Storage. Paper presented at the SPE International Conference on CO_2 Capture, Storage, and Utilization, San Diego, California, USA, November 2009. https://doi.org/10.2118/126620-MS

C. Scheurer, R.E. Hincapie, E. Neubauer, A. Metz, D. Ness, Sorption of nanomaterials to sandstone rock. Nano **12**, 200 (2022). https://doi.org/10.3390/nano12020200

S.Ó. Snæbjörnsdóttir, F. Wiese, T. Fridriksson, H. Ármansson, G.M. Einarsson, S.R. Gislason, CO_2 storage potential of basaltic rocks in Iceland and the oceanic ridges. Energy Procedia **63**, 4585–4600 (2014)

S.Ó. Snæbjörnsdóttir, E.H. Oelkers, K. Mesfin, E.S. Aradóttir, K. Dideriksen, I. Gunnarsson, E. Gunnlaugsson, J.M. Matter, M. Stute, S.R. Gislason, The chemistry and saturation states of subsurface fluids during the in-situ mineralisation of CO_2 and H_2S at the CarbFix site in SW-Iceland. Int. J. Greenhouse Gas Control **58**, 87–102 (2017). https://doi.org/10.1016/j.ijggc.2017.01.007

S. Snæbjörnsdóttir, S.R. Gislason, CO_2 storage potential of basaltic rocks offshore Iceland. Energy Procedia **86**, 371–380 (2016). https://doi.org/10.1016/j.egypro.2016.01.038

S. Thibeau, L. X. Nghiem, H. Ohkuma, A Modeling Study of the Role of Selected Minerals in Enhancing CO_2 Mineralization During CO_2 Aquifer Storage. Paper presented at the SPE Annual Technical Conference and Exhibition, Anaheim, California, U.S.A. (November 2007). https://doi.org/10.2118/109739-MS

H. Wang, Z. Kou, Z. Ji, S. Wang, Y. Li, Z. Jiao, M. Johnson, J.F. McLaughlin, Investigation of enhanced CO_2 storage in deep saline aquifers by WAG and brine extraction in the Minnelusa sandstone, Wyoming. Energy **265**, 126379 (2023). https://doi.org/10.1016/j.energy.2022.126379

P.D. Warwick, E.D. Attanasi, M.S. Blondes, S.T. Brennan, M.L. Buursink, S.M. Cahan, C.A. Doolan, P.A. Freeman, C.Ö. Karacan, C.D. Lohr, M.D. Merrill, R.A. Olea, J.L. Shelton, E.R. Slucher, B.A. Varela, National assessment of carbon dioxide enhanced oil recovery and associated carbon dioxide retention resources—Results (U.S. Geological Survey circular 1489) (2022). https://doi.org/10.3133/cir1489

B. Wei, B. Wang, X. Li, M. Aishan, Y. Ju, CO_2 storage in depleted oil and gas reservoirs: A review. Advances in Geo-Energy Research, **9**(2), 76–93 (2023). https://doi.org/10.46690/ager.2023.08.02

N. Wei, X. Li, Y. Wang, R.T. Dahowski, C.L. Davidson, G.S. Bromhal, A preliminary sub-basin scale evaluation framework of site suitability for onshore aquifer-based CO_2 storage in China. Int. J. Greenhouse Gas Control **12**, 231–246 (2013). https://doi.org/10.1016/j.ijggc.2012.10.012

J. You, W. Ampomah, A. Morgan, Q. Sun, X. Huang, A comprehensive techno-eco-assessment of CO_2 enhanced oil recovery projects using a machine-learning assisted workflow. Int. J. Greenhouse Gas Control **111**, 103480 (2021). https://doi.org/10.1016/j.ijggc.2021.103480

O.A.M. Zamani, D. Knez, Well integrity in salt cavern hydrogen storage. Energies **17**(14), 3586 (2024). https://doi.org/10.3390/en17143586

Y. Zapata, M.R. Kristensen, N. Huerta, C. Brown, C.S. Kabir, Z. Reza, CO_2 geological storage: Critical insights on plume dynamics and storage efficiency during long-term injection and post-injection periods. J. Nat. Gas Sci. Eng. **83**, 103542 (2020). https://doi.org/10.1016/j.jngse.2020.103542

Chapter 7
Class VI Well Permits and CCS Project Lifecycles

7.1 Overview and Objectives

7.1.1 Overview

In this chapter, we will explore Class VI well permits for carbon storage, focusing on critical aspects of carbon management. We begin with the importance of carbon management in mitigating greenhouse gas emissions and addressing climate change. Next, we delve into Regional Carbon Sequestration Partnerships (RCSP) and their participation in advancing carbon sequestration technologies. By the end of the chapter, you'll grasp the regulatory and technical aspects of obtaining a Class VI well permit for carbon storage, emphasizing the importance of best practices for success.

7.1.2 Learning Objectives

Upon completion of this module, you will be able to:

1. Applying the concept of carbon management and its significance in addressing climate change challenges,
2. Learning the regulatory requirements and considerations involved in obtaining a Class VI well permit for carbon storage projects, and,
3. Evaluate the Regional Carbon Sequestration Partnerships (RCSP) and their part in advancing carbon sequestration technologies.

E. Fathi and F. Belyadi, *Subsurface CO_2 Sequestration*,
https://doi.org/10.1007/978-3-032-28296-5_7

7.2 Introduction to Carbon Storage and Class VI Well Permits

As discussed in previous chapters, geological storage of carbon dioxide (CO_2) remains the most efficient and scalable technology for mitigating climate change within a practical timeframe. This well-established and scientifically proven approach offers a permanent solution to managing the vast volumes of CO_2 emitted by major industrial sectors such as power generation, steel, cement, and chemical manufacturing.

Underground formations particularly deep saline and depleted oil and gas reservoirs have demonstrated the capacity to securely store CO_2 for thousands of years, much like how they have retained hydrocarbons and formation water for millions of years. The oil and gas industry's extensive experience in drilling, completing, and injecting fluids into these subsurface formations provides a strong operational foundation and valuable technical analogy for supercritical CO_2 injection for storage in these reservoirs.

In contrast to underground geological storage, other carbon management approaches such as CO_2 utilization or carbon-negative technologies either lack the ability to provide permanent storage or are still in early stages of development and testing, requiring further advancements before full-scale deployment is feasible. For these reasons, underground CO_2 storage is widely recognized as the most viable and immediate solution for long-term carbon management.

In the United States, one of the primary regulatory frameworks governing carbon storage is the Class VI well permitting process, established by the *Environmental Protection Agency (EPA)* under the *Underground Injection Control (UIC)* Program.

Introduced in 2010, the Class VI well regulations are specifically designed to ensure the safe and secure geologic sequestration of CO_2. They provide comprehensive guidelines covering well drilling, completion, operation, monitoring, and closure to protect underground sources of drinking water and ensure long-term containment of injected CO_2 targeting storage formations capable of storing CO_2 for over 1000 years.

Due to the unique properties of supercritical CO_2 such as low density, high mobility, buoyancy, and corrosiveness the design and regulatory requirements for Class VI wells differ from those of Class II wells, which are typically used for enhanced oil recovery (EOR). Considering the differences in regulatory framework for oil and gas well in compared to CO_2 injection well, still oil and gas industry operators are uniquely positioned to lead the deployment of large-scale CO_2 sequestration projects due to their technical expertise and operational capabilities. However, they must fully address the rigorous requirements of Class VI wells. These include detailed site characterization, area of review and modeling (CO_2 saturation and pressure plume dynamics), well construction standards, injection depth constraints, monitoring and verification protocols, financial assurance, and closure and post-injection site care. Each of these elements will be discussed in detail throughout this chapter. While the Class VI permitting process is complex and rigorous, it is designed to ensure the long-term success and environmental integrity of CO_2 storage projects. This regulatory pathway represents a promising and evolving framework that supports the achievement of global climate goals, such as those outlined in the Paris Agreement. This approach will enable the continued operation of carbon-intensive industries in a sustainable and responsible manner.

7.3 Class VI Well Permitting and Underground Injection Control Workflow

The Class VI well permitting process for underground injection of CO_2 was established by the U.S. Environmental Protection Agency (EPA) in December 2010. Its purpose is to regulate CO_2 injection activities and ensure that CO_2 is stored safely and permanently in appropriate geological formations. Project developers planning to implement underground CO_2 sequestration projects in the United States must obtain a Class VI well permit from the EPA or from designated entities that has been granted primary enforcement authority, "Primacy" (Carbon Capture Coalition, 2024). It is important to note that each well that will be drilled for Carbon injection must receive its own individual permit.

The Class VI Primacy process, that gives states the authority to issue Class VI well permit, involves a structured, multi-level review, evaluation, and approval pathway spanning the state, regional, and federal (EPA headquarters levels). The process typically begins with the state where the state identifies key elements of the Underground Injection Control (UIC) program, develop draft rules, and plan public engagement activities. After detailed state-level review, the draft application and supporting documentation are forwarded to the EPA for further evaluation and feedback. Any comments and recommendations from the EPA must then be addressed, and a revised, final version of the application is prepared and submitted for state approval. Once approved at the state level, the regional EPA office reviews the final state submission, issues a memorandum of understanding (MOU), and forwards the complete application package to EPA headquarters. At the federal level, the EPA conducts a comprehensive review and publishes the draft and final application documents, along with EPA comments and state responses, and finalizes the incorporation of UIC-related regulations into the Code of Federal Regulations (CFR). Currently four states of North Dakota (ND), Wyoming (WY), West Virginia (WV), and Luisiana (LA) have the class VI permit Primacy and states of Texas (TX) is very close to obtain the Primacy with MOU issued by EPA in April 2025.

The class VI permit major focus is on effective containment of CO_2, predictability of the CO_2 dynamics, and verifiability of the whole process under EPA and UIC. However, in addition to the Class VI permit other permits are also required depending on the nature of the CCS project. These other permits can be classified in four groups of Environmental permits, CO_2 transport permits, Pore space rights and ownership, and long-term liability. For environmental permits the National Environmental Policy

Act (NEPA) is federal agency that evaluates the environmental impact of the CCS project. The environmental permits also include the air permit for capture facilities under clean air act (CAC) and water discharge permits under clean water act (CWA) both governed by EPA. CO_2 transport permits as discussed in Chap. 5 is governed under PHMSA. The pore space ownership and long-term liability are two critical discussions which are new and under development and have been treated differently in each state, we will discuss these in more detail in Chap. 8.

The process of obtaining Class VI well permits starts with submitting a Class VI well application for each individual well planned within the Geologic Sequestration Data Tool (GSDT). The application must follow the EPA templates that can be found at https://www.epa.gov. These include detailed site screening, selection, and characterization; AoR and PISC studies to clearly identify CO_2 pressure and saturation dynamics during and after injection; detailed injection and monitoring well characterization to ensure well integrity and an appropriate well construction plan; and a monitoring, verification, and reporting program that justifies and validates the previous site characterization, AoR, PISC, and well integrity studies. The application must also include a detailed operational plan, including injection rate and volume constraints, as well as any site-specific risks, mitigation measures, and emergency response plans, together with the required financial assurance to ensure the successful completion of the entire project. After submission of the application in the GSDT, the review process begins with the EPA, which typically lasts for 30 days. If any required documentation is missing or incomplete, a Notice of Deficiency (NOD) will be issued requesting additional information to satisfy the application requirements. Once the required documents have been completed and submitted, the application will proceed to technical review to ensure the protection of USDWs and the long-term containment of the CO_2. Any questions or comments identified during the review are communicated to the project developers through Requests for Additional Information (RAIs), with a 30-day timeframe to respond or request an extension. After all technical concerns have been addressed, the EPA prepares a draft permit that specifies the well operating conditions. This draft permit serves as the regulatory framework for the Class VI project. Following this stage, the draft permit is released for public review and comment. Any public comments received must be addressed by the project developer before the process can proceed. After this stage, the EPA makes a final permit decision authorizing the project developer to begin the project. The overall duration of the process varies depending on the complexity of the project, the quality and completeness of the application, and the timeliness of responses to questions and concerns raised during the various stages of the review process (https://www.epa.gov/uic/class-vi-wells-used-geologic-sequestration-carbon-dioxide).

7.3.1 Class VI Well Permitting Procedure

The class VI well permitting procedure is divided into four key phases of:

I. pre-construction,
II. pre-injection,
III. injection, and
IV. post-injection site care activities.

The pre-construction phase focuses on site screening and characterization, Area of review (AoR) evaluation, well drilling and testing plan, operation and project plan, financial responsibility and emergency response plan, Class VI well permit submission and permit approval.

The pre-injection phase includes finalizing the site characterization and apply any corrective action required before well drilling, perform well drilling, perform well testing and groundwater monitoring, finalize the project plan and financial commitments, and apply for any permit modifications and regulatory approvals before injection. Injection phase is when CO_2 is actively injected and continuous data collection and monitoring were performed for both reservoir and groundwater. This phase also includes submitting any regular reports to compliance with regulatory requirements including any abnormality in pressure, implement the adaptive management and emergency and mitigation strategies, and apply for any modifications on the permit. Finally post-injection site care that includes the well plugging and abandonment including permanently sealing the well, remove surface equipment and restore the site, implement the post-injection monitoring procedure and emergency response plan, and finally site closure and Regulatory Liability transfer.

United states department of energy (DOE) also recognized similar structure for developing and supporting (funding) CCS projects, however, with more details. DOE has divided the pre-construction phase into two phases of prefeasibility study and storage complexity feasibility study. The pre-injection phase is also divided in two phases of site characterization and permitting and construction phases (US EPA, US DOE, Carbon Capture and Storage 2022).

Phase 1 of CCS development project starts with focusing on forming a team, compiling data, conducting Uncertainty and Contingency Plannings, and addressing social impact studies. Acquiring available seismic and well data and assessing scenarios are important steps of this phase. The CCS Pre-Feasibility study lasts 12–18 months.

Phase 2 which is the storage complex feasibility phase spans 18–24 months and involves data collection, geological analysis, and evaluating contractual and regulatory requirements. Subsurface modeling, Uncertainty and Contingency Planning, monitoring requirements, and societal considerations are carefully assessed in this phase. In this phase detailed attention is also required to the Underground Injection Control (UIC) Program, specifically the dedicated carbon storage (DCS) under Class VI for permitting.

In Phase 3, site characterization and permitting takes less than 3 years. It includes detailed site characterization, UIC Class VI or Bureau of Safety and Environmental Enforcement (BSEE) permit submission, feasibility studies for CO_2 sources, CO_2 pipeline front-end engineering design (FEED) studies, storage field development plans, and environmental assessments under the National Environmental Policy Act (NEPA). Societal considerations and impacts continue to be addressed.

Lastly, Phase 4 involves drilling and completing injection and monitoring wells. Risk and mitigation plans are implemented and permits or authorizations (such as EPA UIC Class VI or BSEE) are obtained for injection.

Throughout all phases, the BSEE is a key in ensuring safety and environmental compliance. These phases are guided by regulatory frameworks such as UIC, Class VI DCS, BSEE, and NEPA, who provide a structured approach to CCS projects. This process is designed to ensure thorough analysis, adherence to regulations, and consideration of societal and environmental factors.

The DOE published Best Practice Manuals (BPMs) to provide standardized guidelines for key CCS project components, including site selection, characterization, monitoring, modeling, Uncertainty and Contingency Planning, field operations, and public engagement (www.netl.doe.gov/carbon-management/carbon-storage/strategic-program-support/best-practices-manuals). They aim to ensure consistency and enhance the success of all CCS activities. For this purpose, the granular approach is employed to identify the qualified sites for Class VI well permitting applications from site screening, where the potential basins in the region in proximity of CO_2 emitters are targeted, to site selection, where areas within the basins are recognized, to site characterization where potential sites are studies in detail for site characterization to obtain the qualified sites for permitting.

Another important focus of the BPMs is the classification of geologic formation storage. This involves categorizing different types of geological formations based on their suitability and capacity for CO_2 storage. Understanding the characteristics and behavior of these formations is required for effective carbon sequestration. The BPMs also address terrestrial sequestration of CO_2, which involves exploring options for storing carbon in terrestrial ecosystems such as forests and agricultural lands (Henry Boer 2013). This approach recognizes the potential of natural carbon sinks in mitigating CO_2 emissions.

Monitoring, Verification, and Accounting (MVA) are essential components of geologic storage projects, and the BPMs provide guidance on establishing robust MVA practices (Litynsk et al. 2008; Larkin et al. 2019). This includes monitoring the movement and behavior of CO_2 underground, verifying the effectiveness of storage operations, and accurately accounting for CO_2 stored. Risk management and simulation are also covered in the BPMs that helps project teams to identify and mitigate potential risks associated with geologic storage. Simulation techniques allow for modeling and predicting CO_2 behavior in different scenarios are also emphasized specially their capabilities in aiding in Uncertainty and Contingency Planning and decision-making (Hnottavange-Telleen et al. 2011). Operations for geologic storage projects are addressed to ensure safe and efficient implementation. This includes guidelines on well construction, injection procedures, and maintaining the integrity of storage sites throughout the project lifespan.

BPMs recognized the importance of public engagement, thus they emphasize on public outreach and education for geologic storage projects. Effective communication, stakeholder involvement, and transparent information sharing are essential for building public trust and acceptance. The BPMs provide a comprehensive framework for carrying out essential activities in CCS projects, promoting standardized approaches, risk mitigation, effective monitoring, and public engagement. These best practices are obtained because of extensive studies in the United States at different locations with significant diversity in geological, economical, regulatory and social conditions, therefore, using these guidelines will help to execute CCS projects more efficiently and with greater confidence in their environmental and societal impacts.

7.4 Carbon Management for CCS Projects

Carbon Management is significant in facilitating the decarbonization of industries and offsetting emissions that are challenging to decarbonize. This involves not only reducing emissions but also actively managing and mitigating the environmental risks associated with carbon capture, utilization, and storage activities. To effectively manage carbon emissions, it is essential to have a thorough understanding of the project and site-specific factors involved. This includes identifying, managing, mitigating, and avoiding environmental risks that may arise throughout the CCUS project lifecycle. Due to the unique characteristics of each project and their location, specific project needs should be identified to address potential environmental impacts effectively. At each stage of the carbon management process, there are distinct potential benefits and risks that need to be carefully considered. It is important to note that different federal and state laws and regulations apply to each of these activities, adding another layer of complexity to the implementation of carbon management strategies.

7.4.1 Project Definition and Management

Project Definition and Management involve the initial planning and establishment of a comprehensive framework for managing a process of CO_2 capture, transportation, injection and monitoring geologic storage for a specific project. In this path project definition is the first step to ensure a clear understanding of the project's objectives and to develop a detailed plan for subsequent stages, including contingency measures. During this phase, a set of *technical and economic criteria* needs to be defined that serve as guidelines for evaluating and ranking potential site candidates identified during the later site development process. These criteria help in making informed decisions and selecting the most suitable sites for further investigation and development. Effective project definition and management help project teams to lay a solid foundation for successful project execution and minimize risks associated with geologic storage project.

Project definition and management workflow can include several essential components following DOE best practices manual and the project definition can be divided into (i) project boundaries and objectives, (ii) carbon handling and storage approach, (iii) performance metrics, (iv) resources, (v) timelines, and (vi) uncertainty and contingency planning.

Developing the project boundaries and objectives establishes the overall framework and sets the direction for subsequent analysis. Developing a comprehensive plan for managing carbon dioxide including capture, transportation, and storage considerations is essential for success of the project. This involves assessing available technologies and determining the most suitable approach for the project. Establishing specific criteria to assess the feasibility, effectiveness, and sustainability of different project options are required to ensure best decision-making process. These criteria may include factors such as cost-effectiveness, environmental impact, regulatory compliance, and technical feasibility.

Evaluating the required resources for the project including human capital, funding, equipment, and infrastructure is the next step in project definition and management. This assessment ensures that the necessary resources are available and allocated appropriately. Next, detailed timeline and schedule for project implementation considering key milestones, dependencies, and critical paths is required. This helps to ensure that the project progresses in a timely and efficient manner. Identifying and assessing potential risks and uncertainties associated with the project is of most importance to develop any contingency plan. This involves evaluating various factors, such as technical challenges, regulatory compliance, market conditions, and stakeholder engagement, to proactively manage and mitigate risks throughout the project lifecycle. This workflow provides a structured approach to analyze and address key aspects of the project, followings each step is discussed in more details.

7.4.1.1 Project Boundaries and Objectives

The project boundaries and objectives start with defining the primary purpose of the project such as capture of specific amount of CO_2 emissions from identified industry sources, transport the captured carbon dioxide, and storing in underground geological formations. The secondary goal of the project also needs to be clearly stated such as compliance with local, state, and federal regulations and possibly inclusion of research and development and public outreach. The technical project boundaries and objectives encompass identifying the CO_2 emission sources and transportation options, the entire site evaluation process, ranging from initial site screening to detailed site characterization, monitoring and verification while considering the increasing costs associated with higher levels of detail. The primary focus is on identifying and mitigating uncertainties that may arise during the project's progression, such as geological factors, community considerations, modeling requirements, or surface site characteristics and stablishing clear contingency plans. It is essential to conduct a project scoping exercise for each potential site to ensure thorough evaluation.

The project definition plan should be flexible and serve as a foundation for adjustments as the project advances and circumstances evolve. Neglecting to accurately identify the project boundaries and objectives can result in unexpected delays and cost overruns, potentially leading to project failure. The clear project boundaries and objectives avoid unplanned expansion of the project that might impose extra cost beyond what has been planned leading to delays and failure to achieve the objectives of the project. This also can help in running a consistent language between different groups involved in the life cycle of the project.

7.4.1.2 Carbon Handling and Storage Approach

In developing a carbon handling and storage approach, several key considerations related to CO_2 should be addressed including CO_2 emission sources, CO_2 emission chemical characteristics, CO_2 volumes, injection sites, back up capacities, pressure management, and project time schedule. Additional considerations like possibility of using carbon management hubs to access shared facilities is of great importance to help the economics of the project.

Project developers need to identify and determine the planned source or sources of CO_2 intended for injection. This could include power plants, industrial facilities, or other CO_2-emitting or capture sources.

The next step is to define the maximum and minimum volumes of CO_2 that will be injected over the project's lifetime. This range should consider factors like anticipated injection rates of CO_2, project duration, and capacity of CO_2 capture facilities. For the maximum volume of CO_2 to be injected developers need to assess the number of injection sites that may be required based on the CO_2 volume and distance to capture facilities. This involves evaluating geological formations and their capacity to store the CO_2 effectively. In Tri-State CCS Hub project as an example over 120 facilities with more than total of 37 million tons of CO_2 emission per year identified, where around 20 million ton of CO_2 emission per year showed support to the project.

It is extremely important to make sure the continuity of CO_2 injection; therefore, developers need to evaluate the potential need for backup CO_2 capacity to ensure uninterrupted injection. This consideration accounts for scenarios such as equipment maintenance, unforeseen disruptions, or changes in CO_2 supply. Pressure management also needs to be considered to achieve desired pressure of CO_2 throughout the systems to optimize injection efficiency and prevent operational issues such as leakage or formation damage, this includes ensuring the CO_2 state as supercritical CO_2, identifying maximum surface treating pressure to avoid reservoir, and caprock breakdown or posing any wellbore stability issues such as casing burst or collapse or cement failure.

Understanding the chemical properties of the potential CO_2 gas stream is extremely important to achieve the efficient and safe injection and sequestration project. This includes the quantification of impurities and contaminants in CO_2 stream, as they can influence injection strategies, casing and cement selection, equipment selection, and overall project design. Additionally, the availability of delivery system options, such as existing pipeline infrastructure, should be considered to facilitate the transportation of CO_2 from the source to the injection sites efficiently and cost-effectively.

The team also need to determine a time schedule for the planned years of operation for the CO_2 injection project. This timeline impacts various aspects, including monitoring, reporting, and long-term uncertainty and contingency planning. Therefore, precise timing is essential for success of the project.

The pre-injection phase includes finalizing the site characterization and apply any corrective action required before well drilling, while drilling, perform well testing and groundwater monitoring, finalize the project plan and financial commitments, and apply for any permit modifications and regulatory approvals before injection.

7.4.1.3 Performance Metrics

Evaluation of CCS projects require a rigorous criterion and clear go/no go decision points at each stage of the project from pre-construction to post-injection. These criterions need to be set to guarantee the technical feasibility, economic plausibility, regulatory compliance, and risk mitigation plans of the project at each stage.

Primary factors for go/no-go decisions at each stage from pre-construction, pre-injection, injection, and post-injection site care are different and need specific attention. The pre-construction phase evaluation and go/no-go decisions could include the followings:

1. CO_2 emission source assessment including the capture facility readiness, capture volume and purity evaluation,
2. Storage potential assessment including basin screening for capacity and confinement characteristics evaluation, and,
3. Compliance with regulatory frameworks and permitting requirements.

Go/no-go decision is based on the evaluation of all the criteria above if the analysis proved that there is a sustainable predefined volume of captured CO_2 with good quality for storage will be available and at least one prospect storage site in the basin is identified that is following all the regulatory frameworks then we will proceed with the project. However, if any one of the criterions failed to pass there will be a no-go decision. This decision could be due to not reaching the volume of required CO_2 for storage or having intermittent sources of CO_2.

The pre-injection phase that includes site characterization and drilling activities also require its own criteria for go/no-go decisions that could include:

1. Techno economic analysis of the project to get clear understanding of cost of storage per tons of CO_2 captured this includes CO_2 emission transport and storage characterization cost,
2. Uncertainty and contingency planning such as leakage potential, induced seismicity, and CO_2 plume migration analysis.

If the cost aligns with project budget and storage site meets safety requirements there will be a proceed to injection phase decision, otherwise, there will be no go decision and termination of the project. This could happen simply due to identification of fault line in or proximity of storage formation using new seismic data that has not been identified earlier in the screening analysis.

As an example of such decision to no-go is the case of the Citronelle Dome CCS Project in Alabama where seismic survey revealed the presence of nearby faults intersecting the injection site that impose significant risk of CO_2 leakage or induced seismicity leading to a reevaluation of the storage location and declining the decision to proceed with identified storage site. In this case the no-go decision initiated to pause the injection plan and after identifying the new site with similar injectivity and capacity, however, without any risk of faults the revised permit is submitted for approval and decision is made to proceed with new site for injection.

For injection phase the go/no-go decision criteria could be defined using performance metrics such as actual CO_2 injection vs planned CO_2, or pressure behavior of injection well, the actual pressure builds up rate vs planned pressure build up, etc. The economics measures such as operational expenses (OPEX) such as maintenance of energy usage versus the allocated budget can also be used as decision criteria. In this case the decision will be made to proceed with injection if injection monitoring does not show any unexpected events or signs of leakage and the cost of CO_2 injection per ton stays in range of reallocated budget, otherwise, corrective actions are required before decision is made to proceed with injection.

Finally, the post-injection and closure criteria determine if the monitoring confirmed the stable storage that meets the regulatory compliance, therefore, the company can close the site and transfer the liabilities, or uncertainty persist in terms of pressure anomalies that require further monitoring before any decision for closure can be made.

Overall, the key factors across all the phases includes the following:

1. **Technical**: CO_2 capture, transport, storage, monitoring and safety.
2. **Economic**: Actual cost per ton of CO_2, captured, transported, stored vs allocated fundings.
3. **Regulatory**: Compliance with all the capture, transport, store and monitoring permits, and regulations in different local, state and federal levels.
4. **Social**: Stakeholder acceptance such as primary operators and investors, regulatory and government agencies, scientific and technical advisory boards, local community and landowners.
5. **Risk**: Leakage, reputational, financial risks of the project and mitigation and contingency plans.

7.4.1.4 CCS Project Resources & Management Plan

Defining a Carbon Capture and Storage (CCS) project requires careful planning and allocation of the necessary resources including human, financial, technical, and natural resources across all phases of the project. This begins with forming cross-functional teams that bring together the appropriate expertise for each stage of evaluation and execution.

Human resources require skilled personnel throughout the project including geologists, reservoir, drilling, completion engineers, economists, and policy analysts, who together ensure that the technical, economic, and regulatory aspects of the project are addressed. Legal teams and environmental consultants may also be engaged at different stages to ensure compliance and mitigate risks.

Project managers and an advisory board are critical in overseeing the overall progress, resolving challenges, and making informed, actionable decisions. Establishing a clear project management hierarchy and robust communication network is essential to define roles, align goals, and ensure effective distribution of project data, findings, and decisions across the team.

Technical resources in different phases of the CCS project require specific technical resources and data.

- During the pre-development phase, access to seismic and geological data from public and private databases is required, this includes well logs, core samples, and geological maps to characterize the site.
- In the pre-injection phase, high-resolution 3D seismic surveys are needed to define storage formations and evaluate seal integrity.
- During the injection and post-injection phases, ongoing monitoring may involve downhole seismic surveys, fiber-optic sensors, and downhole pressure gauges to track CO_2 behavior and ensure containment.

Financial resources for securing adequate funding are critical to the success of a CCS project. A comprehensive financial analysis should be conducted to estimate costs associated with each phase of the project and to identify potential decision points where revisions or delays may occur.

It is important to allocate contingency funds to cover unforeseen circumstances, which are common in large-scale projects. Furthermore, collaborating and sharing data with government agencies can help reduce costs and avoid duplication of efforts, particularly when leveraging existing public data or infrastructure.

Access to natural resources and infrastructure is another key consideration. Land access is required for conducting seismic surveys, constructing transportation routes for CO_2, and monitoring the storage site. Utilizing existing infrastructure such as pipelines, wells, and monitoring networks where possible can save time and costs while minimizing environmental impacts. Reviewing and incorporating existing infrastructure maps early in the planning stage helps avoid redundant work and optimizes resource use.

7.4.1.5 Timelines

The project definition phase should include a comprehensive and realistic timeline that reflects the time required to complete each stage of the project. This timeline should be based on a thorough assessment of planned activities, available resources, and potential risks. It is important to recognize that some tasks may need to be repeated, and additional data collection, analysis, or modeling could be necessary as the project progresses, leading to adjustments in the schedule. Therefore, contingency time should be built into the timeline to accommodate the repetition of analyses or evaluations across multiple regions, areas, or sites within the project's scope.

A reasonable schedule for a typical carbon capture and storage (CCS) project may include the following milestones:

- **Year 1**: Completion of a prefeasibility study, which includes basin screening, site selection, and negotiation of source agreements.
- **Year 2**: Preparation and submission of permitting documents, including the U.S. EPA Class VI well application.
- **Year 3**: Securing EPA approvals and initiating drilling activities.
- **Year 4**: Completion and testing of the injection well, followed by start of CO_2 injection operations.
- **Post-injection Phase**: Long-term site care and monitoring, which may extend for several decades to ensure the integrity of storage and compliance with regulatory requirements.

This phased approach ensures that each stage is adequately planned, executed, and verified before moving on to the next, while providing flexibility to address unforeseen challenges without significantly impacting the overall project timeline.

7.4.1.6 Uncertainty and Contingency Planning and Mitigation in CCS Project Development

The final component of the project definition phase involves conducting a comprehensive uncertainty and contingency planning to identify potential risks and developing tailored mitigation plans. A robust risk management framework for CCS projects encompasses identifying, characterizing, ranking, and mitigating risks across all phases, with continuous updates informed by stakeholder engagement and ongoing monitoring.

The uncertainty and contingency planning process begins with identifying and defining risk criteria specific to both internal and external project contexts. The internal context includes factors such as project objectives and boundaries, timeline, budget, and regulatory compliance, while the external context covers political, environmental, and social considerations. For each category, acceptable thresholds must be established. For example, in CO_2 storage, a leakage rate of less than 0.1% over 1000 years may serve as a benchmark for evaluating project viability. In site selection, criteria such as geological stability, proximity to faults, and groundwater protection laws should guide decision-making.

Risk identification should cover all project phases and distinguish between technical and regulatory risks.

Pre-construction Phase

- Technical Risk: Such as inadequate basin storage capacity, well design plan changes, or geotechnical concerns.
- Regulatory Risk: For example, unclear pore space ownership. Currently, many U.S. states lack explicit legislation governing pore space ownership, creating legal uncertainty and potential disputes over rights and responsibilities associated with underground storage, delay in acquiring permits and approval from EPA, NEPA, or PHAMSA.

Pre-injection Phase

- Technical Risk: As an example, presence of faults that compromise containment integrity, as seen in the Citronelle, Alabama CCS project, where faults led to delays and necessitated permit resubmissions.
- Regulatory Risk: Presence of technical risk that leads to additional approvals and extended timelines due to revised applications.

Drilling Phase

- Technical Risk: For example, well integrity failures, such as stuck pipe or lost tools, which can result in significant delays and costs, or issues during drilling that slows down the process and eats up the budget such as bit loss, or slow rate of penetration (ROP). Site access issues, rig issues, poor core recovery, poor logging, and HSE incidents.
- Regulatory Risk: One of the most critical regulatory risks during the drilling phase of a CCS project arises from the need to protect underground sources of drinking water (USDWs), as required under federal and state environmental laws particularly the U.S. Environmental Protection Agency's (EPA) Underground Injection Control (UIC) Program.

Injection Phase

- Technical Risk: If reservoir pressures exceed anticipated levels, the risk of compromising wellbore integrity, reservoir storage capacity, or caprock seal integrity increases. While reducing the CO_2 injection rate can be an effective mitigation strategy to avoid integrity issues, it may also jeopardize the project's ability to achieve the targeted CO_2 storage volume.
- Regulatory Risk: Such as CO_2 plume migration through undetected baffles or natural fractures, potentially leading to non-compliance with monitoring requirements. Exceeding allowable injection pressure or rate due to negligence.

Post-injection Phase

- Technical Risk: Long-term CO_2 leakage, raising challenges in regulatory liability transfer and ongoing monitoring responsibilities.
- Regulatory Risk: Such as failure to demonstrate long-term containment or inability to meet site closure criteria.

Once risks are identified, they should be characterized and ranked based on two key indices, Likelihood (probability of occurrence), scored 1 (very unlikely) to 5 (very likely), and Impact (severity of consequences), scored 1 (minimal) to 5 (catastrophic). The risk score is calculated as.

$$\text{Risk Score} = \text{Likelihood} \times \text{Impact}$$

This results in a range from 1 (lowest risk) to 25 (highest risk). For example, encountering an unforeseen fault in the reservoir with Likelihood = 2/5 and Impact = 5/5 results in the Risk Score = 10. While drilling-related issues (e.g., stuck pipe) with Likelihood = 4/5 and Impact = 4/5 result in Risk Score = 16. In this case, drilling issues would be prioritized over reservoir faults due to the higher risk score.

Risk levels and scores should be reviewed, refined, and updated periodically as the project evolves, typically through annual reviews and lessons-learned documentation, since risks and priorities can change over time. For each identified risk, a mitigation plan is developed, and the probability and impact of the risk are re-evaluated after implementing the mitigation measures. For example, risks associated with HSE incidents may be identified, and mitigation actions such as HSE awareness training and incident prevention training can be implemented. However, even after these trainings are conducted, there is still a possibility that HSE incidents may occur. The difference is that the likelihood and potential impact of such incidents are significantly reduced compared to the situation before the mitigation measures were implemented.

Mitigation strategies should correspond to the level of risk such as low, medium, and high-risk activities. High risk mitigation implements strong measures, such as no-go decisions, avoidance, or redesign (e.g., relocating the storage site due to faults, as occurred in Citronelle). Medium risk mitigation applies control measures, such as real-time monitoring and automated shut-off valves. While low risk mitigation prepares contingency plans, such as community benefit agreements to address opposition.

Mitigation measures must be deployed effectively and their success tracked throughout the project's lifecycle. Importantly, uncertainty and contingency planning and mitigation strategies should be transparent and shared with stakeholders including regulators, communities, and investors through open-access data platforms and regular communication, building trust and support for the CCS project. Figure 7.1 illustrates uncertainty and contingency planning for a CCS project. These risks extend beyond regulatory considerations and include factors that may impede the project's advancement to a contingent storage resource and ultimately a commercially viable storage capacity. Some examples of potential project risks include:

- Failure of CO_2 source or pipeline development,
- Technical or economic unsuitability of the selected reservoir,
- Equipment malfunctions,
- Inadequate acquisition of pore space or surface rights,
- Significant public opposition,
- Changes in legal and regulatory frameworks.

Fig. 7.1 Risk analysis and assessment workflow diagram of CCS project DOE/NETL-2017/1846 https://www.netl.doe.gov/sites/default/files/2018-10/BPM_RiskAnalysisSimulation.pdf

The uncertainty and contingency planning, paired with comprehensive mitigation plans, ensures that the project's scope, staffing, funding, schedule, and performance metrics are aligned and sufficient to support successful implementation.

7.5 Regional Carbon Sequestration Partnerships (RCSP)

The U.S. Department of Energy (DOE) established seven Regional Carbon Sequestration Partnerships (RCSPs) across the U.S. and parts of Canada to assess and advance the most suitable technologies, infrastructure, and regulatory frameworks for carbon storage in diverse geological regions (Litynski et al. 2006). The RCSPs successfully conducted small-and large-scale CO_2 injection tests, safely storing over 12 million metric tons (MMT) of CO_2 in geologic formations. The RCSP builds the foundation for commercial scale deployment of underground geological CO_2 storage projects under Carbon Safe program umbrella. This section provides a brief overview of the RCSP program, offering context for the subsequent detailed discussion on CCS project development phases and their anticipated outcomes.

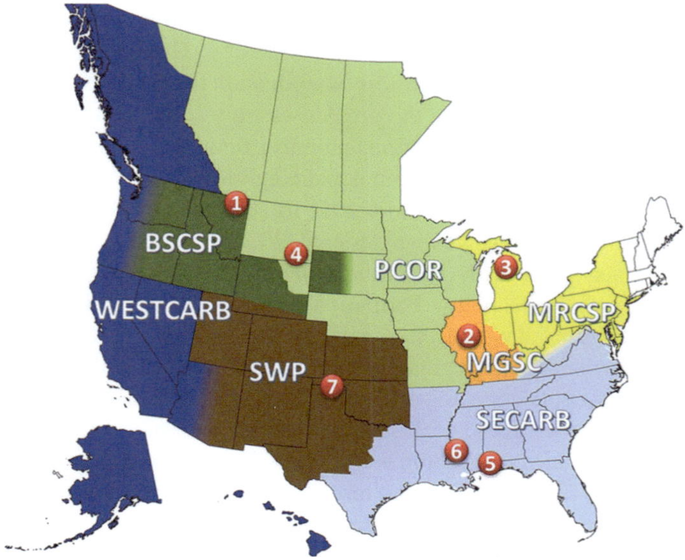

Fig. 7.2 Regional carbon sequestration partnerships carbon capture and storage US EPA, US DOE https://www.netl.doe.gov/sites/default/files/2018-10/BPM-SiteScreening.pdf

7.5.1 Overview and Goals

The seven Regional Carbon Sequestration Partnerships (RCSP) projects shown in Fig. 7.2 are geographically distributed across various regions of the United States. These include the Kevin Dome Project of the Big Sky Carbon Sequestration Partnership, Midwest Geological Sequestration Consortium–Illinois Basin–Decatur Project, Midwest Regional Carbon Sequestration Partnership–Michigan Basin Project, the Plains CO_2 Reduction Partnership–Bell Creek Field Project, Southeast Regional Carbon Sequestration Partnership–Citronelle Project, Southeast Regional Carbon Sequestration Partnership–Cranfield Project, and Southwest Carbon Sequestration Partnership–Farnsworth Unit Project.

7.5.2 The Kevin Dome Project of the Big Sky Carbon Sequestration Partnership

The Big Sky Carbon Sequestration Partnership (BSCSP) targeted the Kevin Dome, in Toole County, Montana as a potential model for secure, long-term carbon storage. The Kevin Dome is recognized as natural CO_2-rich geologic structure spanning 1942 km^2 (750 mi^2), trapping natural CO_2 for millions of years. The project focused on the Duperow Formation, a regionally extensive saline aquifer, to evaluate the

"CO_2 hub" concept, where BSCSP initially planned to extract native CO_2 from the dome and re-inject it down-dip into the same formation. This is due to three reasons first the natural CO_2 is purer and has less complication for presence of impurities in industrial CO_2 emissions. Second the project was to test the injectivity of the CO_2 and natural CO_2 was providing more economical option. This project also helped to better quantify and assess the potential and condition of natural CO_2 source. The project started the seismic surveys from 2011 to 2014, and in May 2014 started the drilling of first production well (Spangler et al. 2020).

The initial plan was to extract 1 million metric tons of naturally occurring CO_2 and reinjecting into the Duperow Formation's brine-filled flank. However, insufficient CO_2 production rates and brine salinity levels below EPA Class VI permit requirements (Technical and regulatory Risk) necessitated a shift in focus toward comprehensive data collection and analysis (Daley et al. 2017). The project successfully gathered extensive geological data including core samples from caprock and reservoir zones, well logs from cored wells and a 3D seismic survey of the transition zone between brine-filled and CO_2-rich reservoir areas. To study the behavior of CO_2 plume in the reservoir during injection a small-scale injection test performed injecting ~14,000 metric tons of reprocessed CO_2 in the reservoir. During the injection the integrity of the reservoir and caprock is monitored using advanced technologies like seismic, geochemical, and pressure monitoring.

Key findings highlighted the dome's structural stability and provided critical data on geochemical interactions between injected CO_2 and formation for forecasting injected CO_2 final state of storage (Daley et al. 2017).

7.5.3 The Midwest Geological Sequestration Consortium (Illinois Basin–Decatur Project)

The Illinois Basin–Decatur Project (IBDP), led by the Midwest Geological Sequestration Consortium (MGSC), demonstrated the first full-scale bioenergy carbon capture and storage (BECCS) (Greenberg 2016). From 2008 to 2015, the project followed complex regulatory pathway, where first a Class I Non-Hazardous permit is submitted to the Illinois EPA in January 2008 and finalized with a Class VI permit for CO_2 injection issued by US EPA in December 2014. It was an interesting approach as Class I Non-Hazardous permit issued in 2009 for well construction and initial operations. This process established a practice for CCS permitting while enabling the injection of 1 million metric tons of CO_2 from Archer Daniels Midland's (ADM) ethanol production into the Mt. Simon Sandstone between November 2011 and November 2014.

The IBDP project successfully demonstrated the entire CCUS value chain, from CO_2 capture at ADM's biofuel facility to secure storage in the Mt. Simon Sandstone at depths exceeding 7000 feet (Iyegbekedo et al. 2024). An advanced monitoring, verification, and accounting (MVA) program was deployed using innovative technologies ground sampling at (7000 ft), surface deformation tracking using radar, and laser tomography for atmospheric monitoring. These systems confirmed the reservoir's integrity, with no CO_2 leakage detected. Detailed discussion of CO_2 plume dynamics is discussed in earlier chapter.

7.5.4 *The Midwest Regional Carbon Sequestration Partnership (MRCSP)*

The Midwest Regional Carbon Sequestration Partnership (MRCSP) identified Core Energy's CO_2-enhanced oil recovery (EOR) fields as a promising CCS site with comparatively low regulatory risk. Located in the Niagaran Pinnacle Reef Trend, the site benefits from a long history of hydrocarbon production and active CO_2-EOR operations (Lewis et al. 2021). Geological characterization, well testing, and seismic studies revealed variability in reef geometry, lithology (dolomite vs. limestone), and salt plugging, which influenced storage potential. Dolomitized zones exhibited higher porosity, while limestone reservoirs were more heterogeneous. The overlying Salina B-salt served as an effective confining unit, supporting secure CO_2 containment and providing a framework for assessing storage in similar complex reservoirs.

This extensive reef system enabled MRCSP to test diverse carbon capture, utilization, and storage (CCUS) technologies across 10 fields with different geology, well configurations, and EOR stages (Fig. 7.3). The Michigan Basin Project aimed to inject one million metric tons of CO_2 while supporting EOR (Litynski et al. 2013). CO_2 sourced from Antrim Shale natural gas processing was compressed and delivered via pipeline to the injection sites. EOR operations expanded from a single field in the 1990s to 10 fields, allowing MRCSP to evaluate CO_2 storage across various operational scenarios. Between 2013 and 2019, 3.41 million metric tons of CO_2 were injected, with 1.53 million tons permanently stored and 1.06 million barrels of oil produced. Lifecycle analysis confirmed the operations were net carbon-negative, underscoring CO_2-EOR as a credible climate mitigation strategy.

7.5.5 *The Plains CO_2 Reduction Partnership–Bell Creek Field Project*

The Plains CO_2 Reduction Partnership (PCOR), led by the Energy and Environmental Research Center (EERC), is collaborating with Denbury Onshore LLC to study CO_2 storage at the Bell Creek oilfield in the Powder River Basin. Since 2010, characterization efforts have focused on the Muddy Formation, a sandstone reservoir at a depth of 4500 feet, to demonstrate safe, permanent CO_2 storage in conjunction with EOR operations. Ongoing studies have refined predictive simulations, enhancing

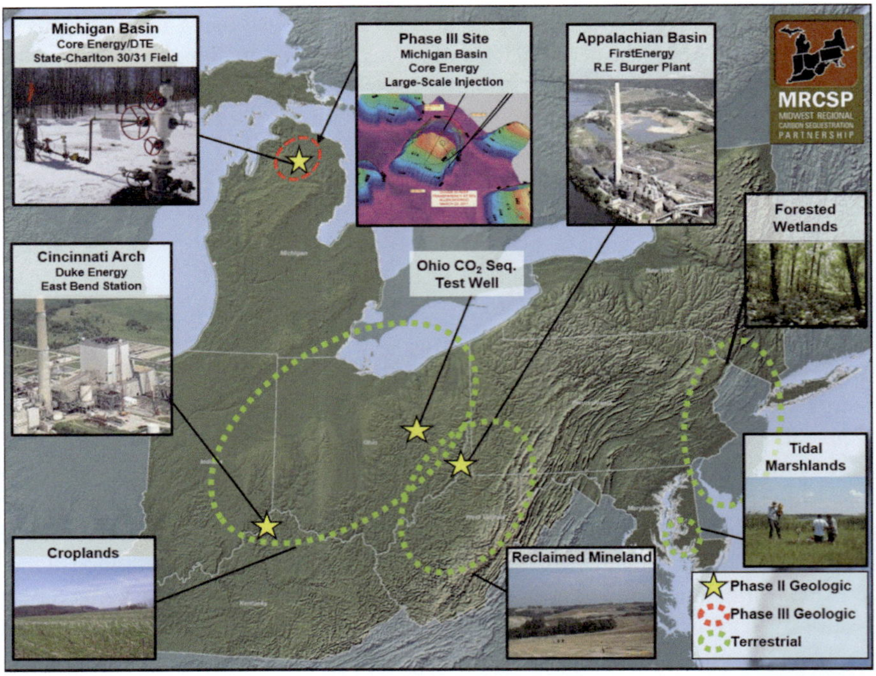

Fig. 7.3 The midwest regional carbon sequestration partnership (MRCSP) https://netl.doe.gov/coal/carbon-storage/atlas/mrcsp

confidence in tracking CO_2 movement and ensuring storage security (Hamling et al. 2013).

Approximately 50 million cubic feet per day of CO_2 is sourced from Wyoming's Lost Cabin and Shute Creek processing plants and transported via pipeline for injection into the Muddy Formation, where it enhances oil recovery through miscible flooding. Between May 2013 and July 2014, over one million metric tons of CO_2 were injected, with projections estimating an incremental oil recovery of 40–50 million barrels through CO_2-EOR (Rodosta et al. 2017).

The Bell Creek project demonstrates the feasibility of large-scale CO_2 storage within commercial EOR operations. This project delivers significant economic benefits by maximizing hydrocarbon recovery to incentivize the cost of CCS project.

7.5.6 *The Southeast Regional Carbon Sequestration Partnership (Citronelle Project)*

The Southeast Regional Carbon Sequestration Partnership (SECARB) Citronelle Project represents a significant demonstration of large-scale CO_2 storage in deep saline formations. This project presents a fully integrated system including CO_2 capture, transportation, subsurface injection, and advanced monitoring technologies to ensure safe and permanent storage. The Citronelle project is located near the Citronelle Dome in Alabama, where the project targets the Paluxy Formation as storage reservoir. The Paluxy formation is a deep saline reservoir at approximately 9400 feet depth, characterized by high porosity and permeability. The reservoir is securely sealed by the overlying Washita-Fredericksburg shale, which serves as an effective caprock to prevent upward migration of injected CO_2 (Fathi et al. 2023a, Esposito et al. 2008).

The Citronelle project benefits from its proximity to significant industrial CO_2 sources, especially the James M. Barry Electric Generating Plant (Plant Barry), a coal-fired power station in Bucks, Alabama. At Plant Barry, CO_2 is captured using advanced amine-based technologies, which efficiently remove CO_2 from flue gas streams. The captured CO_2 is then compressed and transported via a 12-mile CO_2 pipeline to the injection site at Citronelle.

Between 2012 and 2014, the project successfully injected over 114,000 metric tons of CO_2 into the saline formation situated above the producing zones of the Citronelle oilfield. This unique demonstration provided clear evidence that CO_2 storage in saline adquifers can coexist with ongoing hydrocarbon production operations, that indicated broader deployment of carbon storage in regions with active oil and gas fields. The Citronelle project contributed significantly to advancing the understanding of saline storage feasibility, system integration, and monitoring methods (Trautz et al. 2020).

7.5.7 *The Southeast Regional Carbon Sequestration Partnership (Cranfield Project)*

The Southeast Regional Carbon Sequestration Partnership (SECARB) Cranfield project located near Natchez, Mississippi, demonstrates large-scale CO_2 storage and enhanced oil recovery (EOR). Cranfield has provided critical insights into long-term carbon capture and storage (CCS) within saline formations while stimulating the abandoned Cranfield oilfield.

Since July 2008, Denbury Onshore, LLC, has operated commercial CO_2-EOR injections into the lower Tuscaloosa Formation which is a regionally extensive saline reservoir. The lower Tuscaloosa formation structural setting is a four-way closure above a salt dome which is typical of Gulf Coast hydrocarbon fields (Hovorka et al. 2013). There are faults within the formation, however, pressure testing of these faults did not indicate any leakage. For this project natural CO_2 is sourced from Jackson Dome which is one of the major natural CO_2 resources in the United States and transported using CO_2 pipeline to be injected at high volumes. By March 2015, the project had successfully monitored the injection and storage of over five million metric tons of CO_2. (Soltanian et al. 2016).

What makes the Cranfield project unique in compared to other CO_2-EOR project is in Cranfield the CO_2 injection is not started right after secondary water flooding. The Cranfield Oil reservoir has been shut in for decades prior to CO_2 injection, allowing pressures to stabilize near original hydrostatic conditions and fluids to re-equilibrate. This starting point for CO_2-EOR resembles the initial conditions expected in dedicated brine formation storage projects, offering cleaner data uncomplicated by previous recovery operations. The injection wells were also intentionally drilled below the oil-water contact in the water zone and at high injection rate to create experimental conditions like what is expected in deep brine reservoirs. This combination of representative geology and experimental design with initial reservoir conditions makes the Cranfield project particularly valuable for understanding CO_2 storage dynamics in both EOR and deep brine formations (Hovorka et al. 2011).

7.5.8 The Southwest Carbon Sequestration Partnership (Farnsworth Unit Project)

The Southwest Regional Partnership on Carbon Sequestration (SWP), in partnership with Chaparral Energy, developed the Farnsworth Unit (FWU) in Ochiltree County, Texas, as an integrated carbon capture and storage (CCS) and enhanced oil recovery (EOR) project. The FWU project demonstrates how anthropogenic CO_2 can be utilized to both enhance hydrocarbon recovery and achieve permanent geological storage at a commercial scale (Ampomah et al. 2016, Balch and McPherson 2016).

The project focuses on the Morrow B sandstone, a productive hydrocarbon-bearing formation that has historically yielded over 19 million barrels of oil and 44 billion cubic feet of natural gas. The estimated CO_2 storage capacity of the FWU is exceeding 25 million metric tons, which is an example for scalable CCS deployment in mature reservoirs. Geologically, the Morrow B sandstone lies on the northwestern shelf of the Anadarko Basin. This reservoir is effectively sealed by the overlying Thirteen Finger limestone and upper Morrow shale that provides robust containment for injected CO_2. The geological characteristics of the FWU make it an excellent reference for many midcontinent oil fields, illustrating the potential to extend the productive life of mature reservoirs through CO_2-EOR while simultaneously contributing to climate mitigation (Kutsienyo et al. 2023).

A distinctive feature of the FWU project is its exclusive use of anthropogenic CO_2 sourced from a fertilizer plant in Texas and an ethanol plant in Kansas. The CO_2 is transported to the site through existing pipeline infrastructure. By 2018, the project had securely stored more than 1 million metric tons of CO_2 in the subsurface that shows both the technical and economic viability of utilizing anthropogenic carbon in mature oil fields (Balch et al. 2025).

In addition to storage and production operations, the project implemented a comprehensive monitoring, verification, and accounting (MVA) program to ensure the integrity and effectiveness of CO_2 storage. Three wells drilled specifically for reservoir characterization also supported MVA activities. Advanced monitoring techniques including 3D seismic surveys, continuous microseismic monitoring, and tracer testing were deployed to track CO_2 plume migration, assess reservoir heterogeneity, and confirm caprock integrity. These efforts helped to enhance the understanding of reservoir dynamics under EOR and CCS conditions (Balch 2016).

7.5.9 *Evaluating CO_2 Transportation Economics in SWP's Farnsworth Unit (FWU) Project*

The economic feasibility of a carbon capture and storage (CCS) project is heavily influenced by CO_2 transportation costs particularly pipeline expenses from emission sources (e.g., power plants) to storage sites. Key variables such as terrain complexity, elevation changes, and the presence of environmentally or culturally sensitive areas can significantly impact infrastructure costs. A comparative analysis of two Southwest Regional Partnership (SWP) projects including Pump Canyon, New Mexico, and Farnsworth Unit (FWU), Texas illustrates how geographic and logistical factors affect project viability (McPherson and Balch 2024).

The Pump Canyon site faced substantial transportation difficulties due to its rough topography and critical wildlife habitats. These constraints would have made supplying anthropogenic CO_2 from an Arizona power plant extremely expensive. In contrast, the FWU project is supported from flat terrain, minimal environmental restrictions, and proximity to two industrial CO_2 sources (a fertilizer plant in Texas and an ethanol plant in Kansas). This strategic positioning of the FWU allowed for cost-effective pipeline deployment and operational flexibility that enabled the project to switch between CO_2 suppliers based on the market conditions.

Pipeline expenses vary dramatically depending on terrain. Data from Kinder Morgan highlights that constructing CO_2 pipelines in mountainous or wetland regions encounters significantly higher capital costs than in flat, and dry areas for instance, the table below summarizes approximate pipeline costs per inch-diameter per mile under different topographies and key challenges.

Terrain type	Capital cost ($/inch-diameter/mile)	Key challenges
Flat, dry	$50,000	Minimal permitting: Low labor/equipment needs
Mountainous	$85,000	Blasting rock and reinforcing slopes, erosion control, access logistics
Marsh/wetland	$100,000	Environmental mitigation; floating equipment
River crossings	$300,000	Horizontal drilling; regulatory compliance

(continued)

(continued)

Terrain type	Capital cost ($/inch-diameter/mile)	Key challenges
High-population zones	$100,000	Right-of-way acquisition; safety systems
Offshore (150–200 ft)	$700,000	Specialized vessels; corrosion protection

Given these cost discrepancies, Pump Canyon would have required either a nearby CO_2 source or other site-specific advantages to justify CCS development. In contrast, FWU's favorable geography and dual-source access reduced pipeline expenses, enhancing the project's economic sustainability (BEST PRACTICES: Site Screening, Site Selection, and Site Characterization for Geologic Storage Projects, 2017).

When selecting CCS sites, developers must weigh multiple factors:

1. Distance to CO_2 Sources: Longer pipelines increase capital and maintenance costs.
2. Terrain Difficulty: Rough landscapes elevate construction expenses and permitting complexities.
3. Land Use Constraints: Protected areas or cultural sites may restrict routing options.
4. Supply Redundancy: Proximity to multiple CO_2 sources (as with FWU) mitigates supply chain risks.

The FWU case demonstrates that optimizing transportation logistics is critical for CCS viability. Future projects should prioritize sites with manageable terrain, existing infrastructure, and flexible supply arrangements to minimize costs and ensure long-term feasibility. This approach not only improves economic returns but also supports scalable CCS deployment in alignment with climate goals.

7.5.10 Reliability of Anthropogenic CO_2 Supply in EOR: Farnsworth Unit Case Study

The Southwest Regional Partnership on Carbon Sequestration's case study of the Farnsworth Unit (FWU) provides clear understanding of the challenges of ensuring a reliable supply of anthropogenic CO_2 for carbon capture and storage (CCS) combined with enhanced oil recovery (EOR) operations. As discussed earlier, maintaining a consistent CO_2 supply remains one of the major technical risks faced by CCS projects. At the FWU, CO_2 is sourced from two industrial facilities including an ethanol plant in Kansas and a fertilizer plant in Texas where CO_2 is delivered to the injection site via dedicated pipeline infrastructure. This dual-source arrangement of CO_2 supply highlights the complexity of the life cycle economic analysis of the CCS projects (Morgan et al. 2021).

The ethanol plant's CO_2 emissions fluctuate in response to oil and gasoline market prices. When fuel prices rise, ethanol production becomes more economically competitive, leading to increased production and higher volumes of CO_2 byproduct. Conversely, during periods of low fuel prices, ethanol production and, consequently, CO_2 output decline. Federal biofuel subsidies introduce an additional layer of variability, as policy changes can abruptly affect ethanol production volumes. In contrast, the fertilizer plant's CO_2 emissions are primarily tied to natural gas prices, since natural gas is the main feedstock. When gas prices are low, fertilizer production is increased, generating more CO_2 byproduct. However, high gas prices can force production cuts, reducing CO_2 output. These market-sensitive production patterns create complex dynamics in CO_2 availability that CCS projects must be prepared to manage.

This variability poses significant design and operational challenges for CCS systems. Pipeline networks and compression facilities are typically optimized for steady flow rates; however, sizing infrastructure to handle peak supply leads to underutilization during periods of reduced production, therefore increasing the capital cost per ton of CO_2 delivered. More critically, temporary production stoppages or permanent facility closures can interrupt CO_2 deliveries, potentially forcing EOR operations to suspend injection or seek alternative, often more expensive, sources. Such interruptions jeopardize both oil production targets and CO_2 storage commitments, particularly for projects relying on carbon credits or 45Q tax incentives, which require verifiable sequestration volumes.

The FWU experience highlights several key risk mitigation strategies for CCS projects dependent on industrial CO_2 sources. Diversification of supply such as securing CO_2 from multiple independent sources, as implemented at FWU helps to minimize the impact of unforeseen disruptions. Additionally, using temporary storage sites, such as utilizing shallow subsurface formations as temporary CO_2 inventory, can provide an effective protection against supply interruptions (BEST PRACTICES: Site Screening, Site Selection, and Site Characterization for Geologic Storage Projects, 2017) (see Fig. 7.4).

Fig. 7.4 Images designed by Freepik

7.6 Conclusion

The successful deployment of carbon capture and storage (CCS) projects relies not only on technological capabilities but also on regulatory frameworks and project management. This chapter has provided a comprehensive examination of Class VI well permitting which is a critical regulatory mechanism established by the U.S. Environmental Protection Agency (EPA) to ensure the safe and permanent storage of CO_2 in underground formations. We explored the permitting workflow, carbon management strategies, and the role of Regional Carbon Sequestration Partnerships (RCSPs), to provide a multidimensional approach to advance CCS from concept to implementation.

Class VI permits mandate rigorous site characterization, monitoring, and reporting to mitigate risks such as leakage, groundwater contamination, and seismic activity. The permitting process includes multistep of application, technical review, public engagement, and compliance monitoring. Close collaboration between project developers, regulators, and communities are the key to successfully executing the requirements of Class VI well permits. Case studies like the Illinois Basin–Decatur Project and the Farnsworth Unit Project illustrate how RCSPs have demonstrated the feasibility of large-scale storage.

The chapter also highlighted the integrated nature of CCS project management, where defining scope, resource allocation, uncertainty and contingency planning, and timeline management are as critical as technical execution. Key lessons from RCSP initiatives reveal that early stakeholder engagement, transparent communication, and dynamic risk mitigation strategies are required for project success.

Looking ahead, the scalability of CCS will depend on

1. Facilitating the permitting processes without compromising safety, potentially through standardized guidelines,
2. Expanding RCSP—like collaborations to share data, reduce costs, and accelerate learning curves globally, and,
3. Integrating CCS project lifecycles with carbon markets and policy incentives (e.g., 45Q tax credits) to improve economic viability.

References

W. Ampomah, R.S. Balch, R.B. Grigg, R. Will, Z. Dai, M.D. White, Paper Presented at the SPE Improved Oil Recovery Conference, in *Farnsworth field CO_2-EOR project: Performance case history*, (Tulsa, Oklahoma, USA, 2016). https://doi.org/10.2118/179528-MS

R. Balch, R. Esser, N. Liu, *Monitoring CO_2 at an Enhanced Oil Recovery and Carbon Capture and Storage Project, Farnsworth Unit, Texas in CO_2 Summit II: Technologies and Opportunities* (Tri-State Generation & Transmission Association Inc. Frank Zhu, UOP/Honeywell Eds, ECI Symposium Series, Holly Krutka, 2016) https://dc.engconfintl.org/co2_summit2/30

R. Balch, B. McPherson, Integrating enhanced oil recovery and carbon capture and storage projects: a case study at Farnsworth field, Texas, in *Paper Presented at the SPE Western Regional Meeting*, (Anchorage, Alaska, USA, 2016). https://doi.org/10.2118/180408-MS

R. Balch, B. McPherson, M. Cather, A. Meyer, T. Xiao, W. Ampomah, R. Esser, M. Appold, M. White, N. Moodie, J. Heath, A. Luhman, G. El-kaseeh, K. McCormack, S. Lee, K. Hnottavange-Telleen, L. Huang, Z. Wu, R. Grigg, R. Will, *Southwest Regional Partnership on Carbon Sequestration: Phase III (Final Scientific/Technical Report)* (2025). https://doi.org/10.2172/2529468

H. Boer, Governing ecosystem carbon. Glob. Environ. Polit. **13**(4), 123–143 (2013). https://doi.org/10.1162/GLEP_a_00200

T.M. Daley, D. Vasco, J. Ajo-Franklin, L. Dobeck, L. Spangler, M. Leonti, Final Scientific/Technical Report for project "Geomechanical Monitoring for CO_2 Hub Storage: Production and Injection at Kevin Dome" (2017). https://doi.org/10.2172/1366437

R.A. Esposito, J.C. Pashin, P.M. Walsh, Citronelle dome: A giant opportunity for multizone carbon storage and enhanced oil recovery in the Mississippi interior Salt Basin of Alabama. Environ. Geosci. **15**(2), 53–62 (2008)

E. Fathi, M.F. Adenan, N. Moryan, F. Belyadi, H. Belyadi, High-precision single-leak detection and localization in single-phase liquid pipelines using the negative pressure wave technique: an application in a real-field case study. SPE J. (2023a). https://doi.org/10.2118/217455-PA

S.E. Greenberg, Scaling up of deep saline storage in Illinois: From the Illinois Basin-Decatur project to the Illinois industrial CCS project, in *CCS Technical Workshop 2016*, (Tokyo, Japan, 2016)

J.A Hamling, C.D Gorecki, R.J Klapperich, D. Saini, E.N Steadman, Overview of the Bell Creek combined CO_2 storage and CO_2 enhanced oil recovery project. Energy Procedia **37**, 6402–6411, ISSN 1876-6102 (2013). https://doi.org/10.1016/j.egypro.2013.06.570. https://www.sciencedirect.com/science/article/pii/S1876610213008138

K. Hnottavange-Telleen, E. Chabora, R.J. Finley, S.E. Greenberg, S. Marsteller, Risk management in a large-scale CO_2 geosequestration pilot project, Illinois, USA. Energy Procedia **4**, 4044–4051 (2011). https://doi.org/10.1016/j.egypro.2011.02.346

S.D. Hovorka, T.A. Meckel, R.H. Trevino, J. Lu, J.-P. Nicot, J.-W. Choi, D. Freeman, P. Cook, T.M. Daley, J.B. Ajo-Franklin, B.M. Freifeld, C. Doughty, C.R. Carrigan, D. LaBrecque, Y.K. Kharaka, J.J. Thordsen, T.J. Phelps, C. Yang, K. Romanak, T. Zhang, R.M. Holt, J.S. Lindler, R.J. Butsch, Monitoring a large volume CO_2 injection: year two results from SECARB project at Denbury's Cranfield, Mississippi, USA. Energy Procedia **4**, 3478–3485 (2011). https://doi.org/10.1016/j.egypro.2011.02.274

S.D. Hovorka, T.A. Meckel, R.H. Treviño, Monitoring a large-volume injection at Cranfield, Mississippi—Project design and recommendations. Int. J. Greenh. Gas Con. **18**, 345–360 (2013). https://doi.org/10.1016/j.ijggc.2013.03.021

I. Iyegbekedo, E. Fathi, T.R. Carr, F. Belyadi, Machine learning-driven quantification of CO_2 plume dynamics at Illinois Basin Decatur project sites using microseismic data. Energies **17**(17), 4421 (2024)

E.J. Kutsienyo, M.S. Appold, M.E. Cather, Investigation of the effect of injected CO_2 on the morrow B sandstone through laboratory batch reaction experiments: Implications for CO_2 sequestration in the Farnsworth unit, northern Texas, USA. Energies **16**(12), 4611 (2023). https://doi.org/10.3390/en16124611

P. Larkin, W. Leiss, J. Arvai, M. Dusseault, M. Fall, R. Gracie, A. Heyes, D. Krewski, An integrated uncertainty and contingency planning and management framework for carbon capture and storage: a Canadian perspective. Int. J. Uncertain. Conting. Plan. Manag. **22**(1), 21–45 (2019). https://doi.org/10.1504/IJRAM.2019.103336

J.T. Litynski, S. Plasynski, H.G. McIlvried, C. Mahoney, R.D. Srivastava, The United States Department of Energy's Regional Carbon Sequestration Partnerships Program Validation Phase. Environ Int. **34**(1):127–38. Epub 2007 Oct 22. PMID: 17950875 (2008 Jan). https://doi.org/10.1016/j.envint.2007.07.005

J.T. Litynski, S.M. Klara, H.G. McIlvried, R.D. Srivastava, The United States Department of Energy's regional carbon sequestration partnerships program: a collaborative approach to carbon management. Environ. Int. **32**(1), 128–144 (2006). https://doi.org/10.1016/j.envint.2005.05.019

J.T. Litynski, T. Rodosta, D. Vikara, R. Srivastava, U.S. DOE'S R&D program to develop infrastructure for carbon storage: overview of the regional carbon sequestration partnerships and other R&D field projects. Energy Procedia **37**, 6527–6543 (2013). https://doi.org/10.1016/j.egypro.2013.06.584

E. Lewis, J. Moore, R. McDowell, K. Carter, P. Dinterman, W. Harrison, C. Medina, R. Ortt, B. Slater, M. Solis, T. Sparks, The Midwest regional carbon sequestration partnership petroleum fields database: compilation, utilization, and support for CCUS activities. Environ. Geosci. **28**(1), 25–32 (2021). https://doi.org/10.1306/eg.12172020008

A. Morgan, R. Grigg, W. Ampomah, A gate-to-gate life cycle assessment for the CO2-EOR operations at Farnsworth unit (FWU). Energies **14**(9), 2499 (2021). https://doi.org/10.3390/en14092499

B.J. McPherson, R.S. Balch, Lessons learned from the southwest regional partnership on carbon sequestration: a comprehensive, 22-year duration field demonstration program, in *Paper Presented at the SPE Europe Energy Conference and Exhibition*, (Turin, Italy, 2024). https://doi.org/10.2118/220068-MSr

T. Rodosta, W. Aljoe, G. Bromhal, D. Damiani, U.S. DOE regional carbon sequestration partnership initiative: new insights and lessons learned. Energy Procedia **114**, 5580–5592 (2017). https://doi.org/10.1016/j.egypro.2017.03.1698

L.H. Spangler, A. Cihan, B. Winkelman, B.C. De Vault, C. Chang, C.A. Shaw, C.M Oldenburg, D.W. Bowen, E.O. Adeniyi, H.P. Lisabeth, J.W. Carey, J. Blank, J.M. Antle, J.B. Ajo-Franklin, J.M. Matter, K. Barrett, L.M. Dobeck, L. Thomas, L. Pan, L. Huang, T.J.T. Kneafsey, A. Lindsey. L. Thrane, M. Stute, M. Leonti, O.M Omosebi, R. Perry, P.H. Stauffer, Q. Yang, Q. Zhou, R. Engel, R.O. Valdivia, R. Bricklemeyer, R. Feddema, S.M. Clegg, S.L. Codd, S.A. Mangini, S. Saltiel, S. Fairweather, S. Capalbo, T. Naberhaus, T.J. Kneafsey, Big Sky Regional Carbon Sequestration Partnership (Phase III Final Scientific/Technical Report) (2020)

M.R. Soltanian, M.A. Amooie, D.R. Cole, D.E. Graham, S.A. Hosseini, S. Hovorka, S.M. Pfiffner, T.J. Phelps, J. Moortgat, Simulating the Cranfield geological carbon sequestration project with high-resolution static models and an accurate equation of state. Int. J. Greenhouse Gas Control **54**(1), 282–296 (2016). https://doi.org/10.1016/j.ijggc.2016.10.002

R. Trautz, D. Thomas, D. Miller, M. Robertson, G. Koperna, D. Riestenberg, Geophysical monitoring using active seismic techniques at the Citronelle Alabama CO_2 storage demonstration site, Int. J. Greenh. Gas Con. **99**, 103084 (2020). ISSN 1750-5836. https://doi.org/10.1016/j.ijggc.2020.103084

Chapter 8
Site Selection and Characterization for Geological Carbon Storage

8.1 Overview and Objectives

As discussed in Chap. 7, Carbon capture and storage (CCS) project development follows a structured lifecycle, typically divided into four key phases: pre-development, pre-injection, injection, and post-injection site care (closure). This chapter outlined the main context for these phases, especially the pre-development (Site screening) and pre-injection (Site selection and characterization) phases, which are the basis for successful project execution. These early stages involve:

1. Site screening—Identifying prospective storage sites through regional assessments and basin-scale evaluations.
2. Site selection and characterization—Rigorously analyzing reservoir properties, seal integrity, and subsurface risks to qualify sites for commercial-scale storage.

These steps are critical for project feasibility and provide the foundation for Class VI well permit applications. The Class VI well permit mandates strict site evaluations to protect underground sources of drinking water (USDWs) and ensure long-term CO_2 containment.

8.1.1 Overview

This chapter outlines the key steps involved in identifying, assessing, and selecting suitable sites for geological CO_2 storage. By integrating guidelines from the *NETL Best Practices Manuals (BPMs)* and field-tested methodologies, this chapter demonstrates how systematic site assessment aligns with regulatory requirements and mitigates risks in CCS deployment.

"Site Screening, Site Selection, and Site Characterization for Geologic Storage Projects" is a comprehensive guide that outlines the necessary steps for identifying

E. Fathi and F. Belyadi, *Subsurface CO_2 Sequestration*,
https://doi.org/10.1007/978-3-032-28296-5_8

and developing a geologic storage site (Rodosta et al. 2017). The manual offers valuable recommendations starting from the early stages of regional exploration at the basin level to local site selection and characterization. It provides a systematic approach to assess the suitability of potential sites and ensures that only qualified locations are considered for commercial-scale storage operations. The manual emphasizes the importance of thorough site characterization, including data collection, geological analysis, and evaluation of key parameters such as reservoir capacity, permeability, and containment integrity. The discussion then shifts to downhole and laboratory evaluations essential for Class VI well permitting, addressing critical factors related to both the storage reservoir and caprock formations, as well as protecting underground sources of drinking water (USDW).

8.1.2 Learning Objectives

1. Applying the Best Practices Manuals (BPMs) provided by the National Energy Technology Laboratory (NETL) for Site Screening, Site Selection, and Site Characterization in geologic storage projects.
2. Evaluate the importance of adhering to best practices in ensuring safe and effective storage of carbon dioxide.
3. Implementing the site screening process for identifying suitable locations for geologic storage, considering factors such as geology, hydrogeology, and proximity to emission sources.
4. Evaluating the site selection and site characterization procedures, including data collection, monitoring, and uncertainty and contingency planning, for the successful implementation of carbon storage projects.

8.2 Site Screening for Geologic CO_2 Storage: Data Evaluation and Sub-Region Prioritization

The site screening phase systematically evaluates geologic, geographic, and social datasets to identify optimal areas for CO_2 storage within a target region that is in proximity to CO_2 emission sources. This process leverages existing data from public agencies (state geological surveys, oil and gas commissions), national resources (NETL reports, *Energy Data eXchange*™), and local stakeholders to efficiently prioritize areas for further investigation.

Key objectives of site screening include strategic elimination of the regions that are not viable for optimal storage sites, either because of technical or regulatory issues or social acceptance, and identifying the possible areas of interest. This can be done by assessing potential sub-regions across a basin to exclude non-viable zones while highlighting areas with optimal storage potential. This process implements multi-criteria ranking to identify and prioritize smaller selected areas using technical, practical, and social criteria established during project planning and development. Critical data required for this study include the geological data, site-specific data, and social data.

Geological data assess basin-scale suitability through formation properties, structural geology, seal integrity, and depositional history. The site-specific data evaluate the feasibility of the project through physical characteristics of the site, like topography, hydrology, and seismic hazards. Availability of the infrastructure like pipelines, roads, and electric supplies (access to grids) is a significant factor for the viability of the project. Next, land-use constraints such as protected areas and existing industrial activities are determined to ensure compliance with federal and state regulations. Finally, social data that analyzes the community characteristics such as demographic patterns and community structures, land ownership and traditional use patterns, public perception and acceptance of CCS technologies, and historical energy project interactions and equity considerations are studied to make sure long-term support and sustainability of the project.

Understanding community dynamics and stakeholder priorities helps to anticipate permitting challenges, optimize stakeholders engagement strategies, and foster long-term project support from the public through well designed community benefit plans. This social acceptance to operate in the area is increasingly recognized as equally critical as technical suitability for successful CCS deployment. Developing this multiple-dimensional data set will enable the team to identify the sweet spots where favorable geology, infrastructure, and community alignment intersect and develop risk-weighted rankings to guide Class VI permitting strategies. The screening procedure is presented as a workflow diagram for a systematic site screening framework for geologic CO_2 storage in Fig. 8.1.

The site screening process (Fig. 8.1) initiates with project definition, establishing technical thresholds and performance metrics for CO_2 storage suitability. Potential sub-regions are first analyzed through geologic studies, examining formation properties (porosity, thickness, depth), confining zone (Seal), and structural geology, including fault and fracture density. Depositional history analysis, including sediment source, rock composition, and fluid transport mechanism, further evaluates reservoir continuity and seal formation quality. This ensures the subsurface reservoir can safely store injected CO_2 over geologic timescales. Sub-regions failing to meet these geologic standards are immediately eliminated, while qualifying areas advance to subsequent feasibility assessments.

The second phase evaluates site-specific practical constraints, analyzing surface topography, hydrology, and infrastructure networks (pipelines and power access) alongside regulatory land-use limitations. This stage identifies operational restrictions such as protected ecosystems, national parks, or sensitive sites. Regions demonstrating both geologic integrity and development feasibility then undergo social context evaluation, where demographic patterns, native land rights, and community acceptance metrics are assessed through engagement records and survey data. This process ensures only locations with compatible technical, operational, and social conditions progress further.

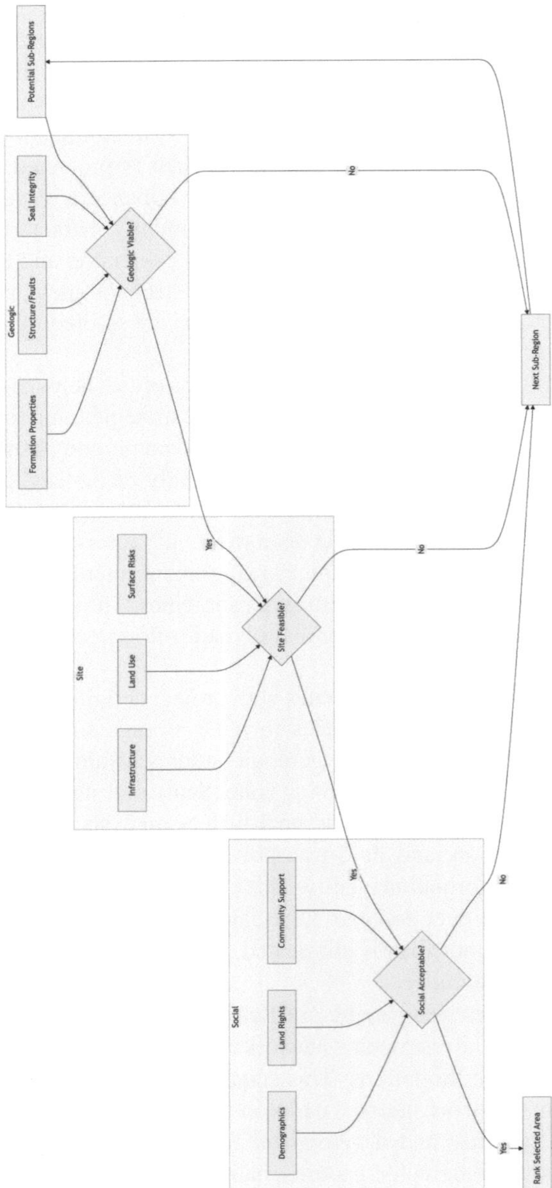

Fig. 8.1 Systematic site screening framework for geologic CO_2 storage: integrating geologic, infrastructure, and social criteria

Selected areas passing this multi-level screening are ranked using weighted criteria from the project definition phase, prioritizing sites with optimal storage security, development readiness, and community alignment. Top-ranked candidates advance to Site Selection for characterization through seismic surveys and reservoir modeling and simulation efforts. This workflow progressively evaluates remaining sub-regions. This iterative filtration methodology optimizes resource allocation by eliminating non-viable areas early while systematically identifying locations that satisfy the complete spectrum of geologic, engineering, and social requirements for permanent CO_2 storage.

Following, we discuss the details of each screening criterion, elaborating on the decision criteria for each subsection.

8.2.1 Geologic Assessment

The primary objective of geologic assessment is to evaluate the subsurface viability of potential sub-regions for CO_2 storage by evaluating three critical elements: formation properties, seal integrity, structure, and faults.

To enhance storage security and maximize capacity, it is preferable to inject CO_2 at depths where it becomes supercritical. Supercritical conditions for CO_2 occur at temperatures above 31.3 °C (88.3 °F) and pressures exceeding 7.4 MPa (1071 psi). These conditions are commonly found onshore at depths greater than 800 m (2,600 ft), where temperatures and fluid pressures (pore pressures) exceed these threshold values. At supercritical states, CO_2 exhibits a liquid-like density of approximately 500–800 kg/m^3 (31–50 lb/ft^3), resulting in significantly reduced volume compared to the gas phase at shallower depths.

Compliance with the Safe Drinking Water Act (SDWA) necessitates injection of CO_2 below underground sources of drinking water (USDWs) as well, although exemptions may be granted by the EPA in certain cases involving deep freshwater formations.

The sub-region analysis also needs to satisfy high porosity (>10%) and permeability (>10 mD) to ensure sufficient storage capacity and CO_2 injectivity. The low-salinity brines (typically <50,000 ppm total dissolved solids (TDS)) are also preferred since that enhances CO_2 dissolution trapping without compromising geochemical stability. Lower-salinity formations favor solubility trapping but require robust pressure management to avoid excessive brine acidification (pH <4.5), which could dissolve carbonate cements and pose integrity issues. It is also important to note that the TDS cannot be less than 10,000 ppm, as water sources with less than 10,000 ppm can be considered as USDW. Conversely, higher-salinity brines (>100,000 ppm) may limit dissolution but enhance mineral trapping through chlorite or feldspar reactions. Even though high salinity promotes the mineralization of CO_2, the dissolution trapping promoted at low salinity is preferred due to CO_2 storage at earlier stages of injection and at a higher rate than mineral trapping. It is also important to assess the in-situ pore pressure, which must be high enough (typically 1070 psi) to keep CO_2

in a dense supercritical state yet sufficiently below the fracture gradient to prevent caprock failure or induced seismicity. These combined criteria ensure the formation can securely store CO_2 while mitigating risks of buoyant migration or leakage.

The calculation of storage potentials involves determining the potential storage capacity of the selected formations. It considers the pore volumes within these formations and their ability to accommodate the planned injection volumes (Bachu et al. 2007). The calculation considers the results of the analyses conducted for the previous three elements.

8.2.1.1 Seal Integrity of Storage Sites

At supercritical conditions, CO_2 is less dense than both saline water and oil, making it buoyant and prone to migrating upward to the top of the reservoir. When injected at pressures above hydrostatic pressure, CO_2 can also drive saline water and other fluids laterally away from the injection point unless the reservoir has been sufficiently depressurized.

A seal (caprock), which is a protective barrier of low-permeability rock, prevents vertical migration of CO_2 and confines the CO_2 in the reservoir. Such seals typically consist of fine-grained materials like shale, mudstone, or well-cemented carbonate, as well as bedded salt or anhydrite, which effectively confine injected fluids.

During the site screening phase, regional geologic data are evaluated to identify potential confining zones within candidate sub-regions. This assessment considers the formation's type, depth, thickness, and lateral extent to ensure it can adequately contain CO_2. A suitable confining zone should extend well beyond the area of pressure influence from injection to protect underground sources of drinking water (USDWs).

To study the CO_2 confinement, focus is placed on the integrity of the confining zone, evaluating key parameters such as its thickness (generally >30 m), very low permeability (<0.01 mD), and lateral continuity to prevent vertical leakage. This analysis is complemented by structural geology studies, which examine fault and natural fracture density and orientation that could serve as potential leakage pathways. Finally, understanding the depositional history of the area helps to characterize reservoir heterogeneity and seal distribution. Together, these factors determine whether a sub-region has the mechanical and chemical capacity to securely contain CO_2 over timescales of 1000 years or more.

8.2.1.2 RCSP Regional Geologic Assessment for CO_2 Storage Potential

Regional Carbon Sequestration Partnerships (RCSP) has performed systematic geologic data analysis for CCS projects across the U.S. basins. These analyses quantify the alignment between stationary emission sources and subsurface storage resources. Figure 8.2 illustrates estimates of CO_2 stationary source emissions

RCSP or Geographic Region	CO_2 Stationary Sources		CO_2 Storage Resource Estimates (billion metric tons of CO_2)								
	CO_2 Emissions (million metric tons per year)	Number of Sources	Saline Formations			Oil and Gas Reservoirs			Unmineable Coal Areas		
			Low	Med***	High	Low	Med***	High	Low	Med***	High
BSCSP	115	301	211	805	2,152	<1	<1	1	<1	<1	<1
MGSC	267	380	41	163	421	<1	<1	<1	2	3	3
MRCSP	604	1,308	108	122	143	9	14	26	<1	<1	<1
PCOR*	522	946	305	583	1,012	2	4	9	7	7	7
SECARB	1,022	1,857	1,376	5,257	14,089	27	34	41	33	51	75
SWP	326	779	256	1,000	2,693	144	147	148	<1	1	2
WESTCARB*	162	555	82	398	1,124	4	5	7	11	17	25
Non-RCSP**	53	232	--	--	--	--	--	--	--	--	--
Total	3,071	6,358	2,379	8,328	21,633	186	205	232	54	80	113

Source: U.S. Carbon Storage Atlas – Fifth Edition (Atlas V); data current as of November 2014

* Totals include Canadian sources identified by the RCSP

** As of November 2014, "U.S. Non-RCSP" includes Connecticut, Delaware, Maine, Massachusetts, New Hampshire, Rhode Island, Vermont, and Puerto Rico

*** Medium = p50

Fig. 8.2 Estimates of CO_2 stationary source emissions and estimates of CO_2 storage resources for geological storage sites in RCSP geographic region DOE/NETL-2017/1844 https://www.netl.doe.gov/sites/default/files/2018-10/BPM-SiteScreening.pdf

and estimates of CO_2 storage resources for geological storage sites in the RCSP geographic region.

The Atlas V assessment reveals a connection between CO_2 emissions and storage potential across the United States. Annually, 3071 MMT (MMT) of CO_2 are emitted from 6358 stationary sources within the RCSP geographic regions, highlighting the critical need for allocating storage sites with the capacity to store this significant amount of emissions (DOE-NETL, Carbon Storage Atlas). The study reveals that saline formations dominate the geological storage sites, with a projected capacity of 2379–21,633 MMT, offering tremendous volumetric potential due to their widespread distribution and stable geologic confinement. These formations, particularly when coupled with optimal depth, salinity, and pressure conditions as outlined in prior screening criteria, represent the cornerstone of long-term CO_2 sequestration.

While oil and gas reservoirs present a more modest capacity (186–232 MMT), their strategic value lies in existing infrastructure and proven trapping mechanisms, enabling near-term deployment. Similarly, unminable coal seams (54–113 MMT) provide opportunities, leveraging adsorbed CO_2 storage in areas with mature coal basins.

The SECARB partnership leads with 1022 MMT of annual emissions and adequate storage potential, driven by extensive saline formations and concentrated industrial sources. The MRCSP (604 MMT) and PCOR (522 MMT) regions follow, indicating the interplay between emission density and subsurface suitability that defines scalable carbon management.

This geographic and technical stratification informs a tiered storage strategy. Saline formations serve as the backbone for bulk sequestration, while oil/gas reservoirs and coal seams address localized needs. Such prioritization is in line with the phased screening approach where geologic viability, infrastructure readiness, and social license mutually determine project viability and align with EPA Class VI permitting requirements. The Atlas V data not only quantifies resources but also emphasizes the need to match emission clusters with storage hubs, ensuring the technical and economic feasibility of the U.S. carbon management pipeline.

8.2.2 Site-Specific Data Analysis

The site-specific data analysis stage evaluates three critical factors to determine the suitability of potential sub-regions for CO_2 storage, integrating infrastructure, land use, and surface risk considerations.

Land-use studies include protected and sensitive areas that require careful consideration, such as legally designated wetlands, marine reserves, and culturally significant zones that impose strict limitations on subsurface activities. These areas are often governed by federal and state laws (e.g., Clean Water Act, Endangered Species Act) and require comprehensive environmental impact assessments to ensure compliance with regulations. For example, proximity to source water protection areas may necessitate additional monitoring plans to protect groundwater resources, while species habitats could restrict surface infrastructure placement (Fig. 8.3).

Population centers introduce multidimensional constraints, like standards for surface and subsurface operations. This includes surface pipeline or underground natural gas storage. Urban proximity complicates pore space acquisition, raises land access costs, and amplifies public engagement challenges. Permitting for seismic surveys or characterization wells in densely populated regions often faces lengthy approval timelines. Land-use conflicts may necessitate creative solutions such as deviated/directional or horizontal drilling.

Infrastructure studies, including existing hydrocarbon developments, present both benefits and concerns. While legacy oil and gas infrastructure offers valuable reservoir data (e.g., core samples, pressure histories), it also risks compromising storage integrity. Abandoned wells with inadequate cement bonding, which is not an uncommon issue in mature basins, require rigorous integrity testing to ensure CO_2 confinement. Some wells were also drilled a long time ago, and there is no documentation regarding their locations and conditions, that introduce immediate concern for the project. Conversely, active developments may compete for pore space or necessitate customized wellbore designs to avoid interference. In active oil and gas operation areas, it is important to select the storage site deeper than all the economically viable hydrocarbon resources to minimize the risk of drilling through the storage reservoir and possibly introducing integrity and leakage risks. The exact situation happened

Fig. 8.3 Assateague State Park, Maryland, USA. Fritz Geller-Grimm—2004

in 2026 in the West Virginia where the state mandates the CO_2 storage reservoirs located below the deepest oil and gas reservoirs in this case Utica-Point Pleasant Shale.

Pipeline Right-of-Ways (ROWs) are critical for project economics but require detailed evaluation. Existing CO_2 pipelines offer potential repurposing opportunities if their pressure ratings and materials align with supercritical CO_2 requirements. Non-CO_2 ROWs (e.g., decommissioned natural gas lines) may reduce surface disruption but often demand costly upgrades for corrosion resistance. Overall, these factors guide the ranking of sub-regions, where no single criterion serves as a disqualifier but rather flags the need for tailored mitigation strategies.

8.2.3 *Social Data Analysis*

The social data analysis phase is a critical component of the site screening workflow and is designed to systematically evaluate community perception and socio-political views on CO_2 storage activities within potential sub-regions. This analysis includes stakeholder engagement strategies, permitting activities, and long-term project support by examining three interconnected studies encompassing demographic, community support, and land rights.

Demographic studies form the first layer of this assessment, which can be obtained by mapping population distributions, employment sectors, and income discrepancies within the region using datasets from the U.S. Census Bureau, state agencies, and economic development portals. In regions with high industrial employment densities, or workforce in the energy sector, specially oil and gas industry, the infrastructure acceptance is expected to be high, while more tech-support communities might have more reservations.

Community support requires detailed investigation beyond surface-level surveys. This can include interviews with local leaders, analysis of public meeting records, and review of zoning board decisions. Public acceptance of CO_2 storage projects exists along a continuum ranging from active support to passive non-opposition (Wüstenhagen et al. 2007). Public support for CCS projects also depends on two key factors: how the technology is presented (as a temporary fix or long-term climate solution) (Whitmarsh et al., 2019), and how much the community already worries about climate change (Shackley et al. 2004).

Land-use histories and environmental legacies also need detailed assessment, particularly in regions where industrial, agricultural, or mixed-use are dominant. This will give an understanding of the level of familiarity and acceptance of activities like well-drilling, seismic acquisitions, and pipeline construction within the communities. The study should include identifying environmentally sensitive areas of concern within the region and investigating any history of environmental issues, particularly those associated with industrial projects. It is extremely important to identify the landowners, labor unions, or universities with prior CCS exposure. This can serve as credibility bridges for CCS projects during public hearings.

Combining the results of social studies from different layers falls under the social context analysis. Project developers can better understand the social dynamics, public perceptions, and potential challenges or opportunities associated with the sub-regions by having a good understanding of these studies. The social analysis provides valuable insights for effective community engagement, aligning project goals with local interests, and ensuring responsible and sustainable project development.

8.3 Site Selection for Geologic CO_2 Storage

During the site selection stage, the focus shifts to a detailed evaluation of the selected areas identified during the site screening process. The goal is to narrow down the options and identify potential sites suitable for further site characterization. This stage encompasses geologic storage evaluation, compliance assessment, reservoir modeling and simulation, storage viability assessment, and social baseline assessment.

Geologic storage evaluation involves a thorough examination of the geological data specific to each potential site, including formation porosity, permeability, mineralogy, mechanical and other characteristics critical to carbon storage. Key attributes of the storage reservoir, such as the presence of a suitable confining zone, trapping mechanisms to prevent CO_2 migration, and well injectivity potential, are studied. Using seismic data analysis, we estimate storage capacity and verify compliance with target injection volumes. If subsurface data confirm viable storage candidates, the process advances to compliance assessment analysis.

Compliance assessment analysis evaluates the selected areas against applicable local, state, and federal standards, including well categories, geologic trapping, formation pressure limit, and regulatory liability. Compliance with these regulations determines whether a site remains viable and progresses to reservoir modeling and simulation.

Reservoir modeling and simulation employ advanced computer models and simulations to predict the behavior of CO_2 storage at each potential site. This step integrates subsurface geological data and seismic information to define parameters, estimate data requirements and costs, and establish boundary conditions that account for uncertainties. If sufficient data support the construction of reliable models and initial results indicate effective injection zones, the analysis proceeds to storage viability assessment; otherwise, the site is eliminated, and the next selected area is evaluated.

In storage viability assessment, site-specific data such as surface access, infrastructure, area of review (AoR) requirements, and pore space ownership are examined to determine whether operational challenges can be managed. Following this, the social baseline assessment evaluates the compatibility of each potential site with local communities and stakeholders. This includes evaluating demographic trends, community engagement, public perceptions of climate and energy issues, and the region's land use and environmental history. The aim is to ensure community concerns can be addressed effectively and incorporated into an outreach plan.

Potential sites that pass the preceding evaluations are compiled into a prioritized list, and a site development plan is formulated. This plan details site-specific considerations, mitigation measures, and outlines the necessary steps for each site. Economic feasibility is then evaluated, considering cost-effectiveness, financial viability, and potential economic benefits. This assessment helps prioritize the sites for detailed site characterization. Based on subsurface characteristics, regulatory compliance, simulation results, site-specific attributes such as social considerations, and economic feasibility, the potential sites are ranked. The highest-ranked sites advance to the next phase, site characterization, which entails more detailed investigations and assessments.

The site selection workflow systematically guides decision-making by evaluating and narrowing selected areas. The goal is to identify and prioritize the most suitable sites for comprehensive site characterization, which will deliver the detailed information required for final site selection and successful implementation of carbon storage. Figure 8.4 presents the flowchart for integrated CO_2 storage site selection.

To better understand how each of these components contributes to identifying optimal storage sites, the following sections elaborate on each stage of the site selection workflow, describing how geological, regulatory, technical, social, and economic factors are systematically evaluated to narrow down and prioritize potential sites. This breakdown highlights the key considerations, methods, and decision points that guide the selection process.

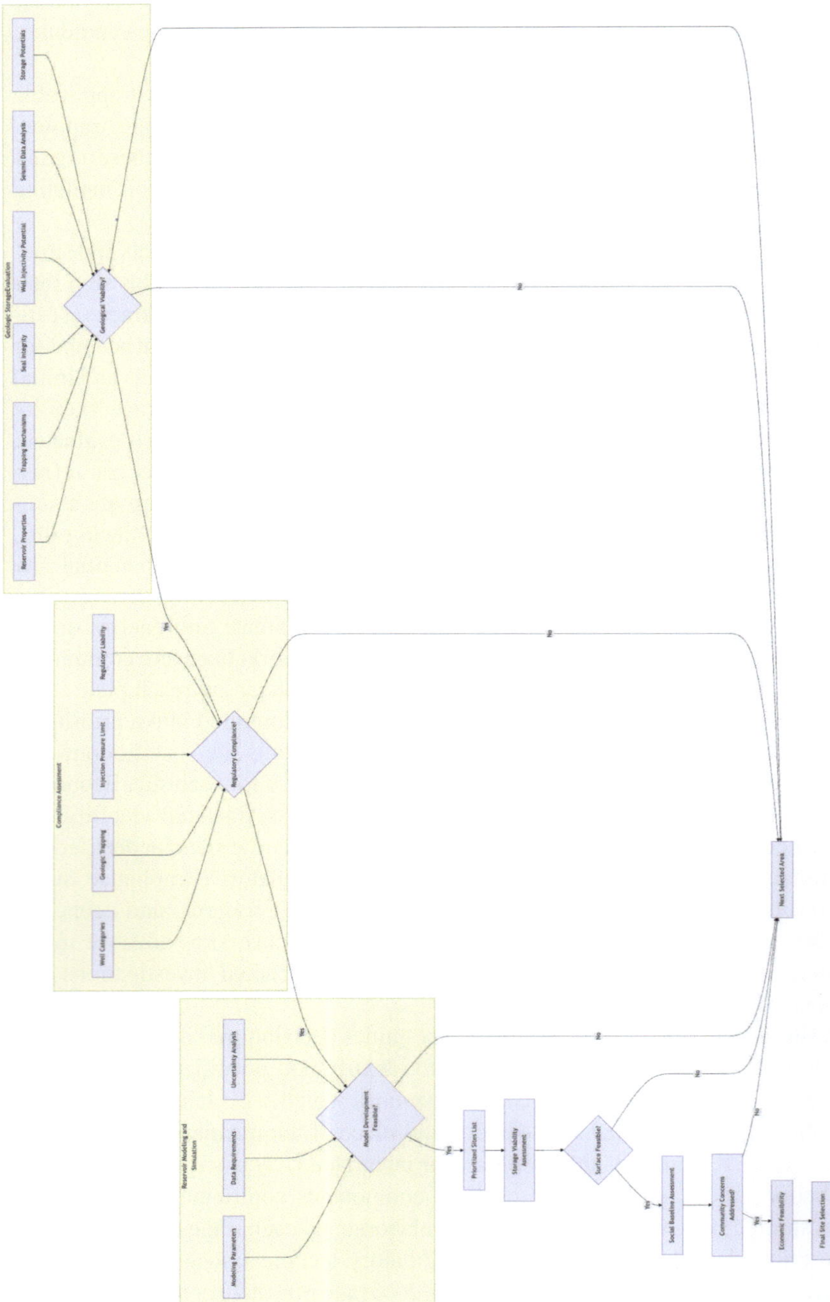

Fig. 8.4 Integrated CO_2 storage site selection workflow

8.3.1 Geologic Storage Evaluation

The geologic storage evaluation involves a detailed examination of the geological data specific to each potential site. This stage builds upon the geologic assessment conducted during site screening and aims to provide a more detailed understanding of the subsurface properties and structure within the selected areas. This evaluation helps reduce uncertainties associated with carbon storage operations. The geologic storage evaluation focuses on six primary elements, including reservoir properties, trapping mechanism, seal integrity, well injectivity potential, seismic data analysis, and storage potential.

8.3.1.1 Reservoir Properties

Reservoir property studies involve a systematic evaluation of potential CO_2 injection and storage sites within the selected areas. The process begins with the identification of specific reservoirs that are suitable for CO_2 storage. To achieve this, a preliminary stratigraphic column and structural map for the selected areas are developed to provide an initial overview of the subsurface. At this stage, it's required to utilize available well logs, core samples, and sidewall samples to map the upper and lower boundaries of storage reservoirs and to identify any heterogeneities within the layers. In cases where core and sidewall samples are limited, supplementing the data with adjacent wells or outcrop information is valuable. The next step involves determining the thickness and lateral extent of storage reservoirs that contain appropriate injection zones. Special attention is given to significant geological features, such as natural fractures, baffles, faults, and folds, which can influence fluid flow within the reservoir.

It's essential to thoroughly map and evaluate multiple potential injection zones that may exist at varying depths. Additionally, the feasibility of acquiring and analyzing existing seismic data to complement the well log data should be considered, weighing the associated costs and potential benefits of this additional information. The reservoir properties evaluation enhances the understanding of the subsurface conditions within the selected areas and provides valuable insights into the storage reservoirs, their properties, and the potential injection zones. The analysis incorporates available well log data, outcrop data and may consider the use of seismic data to refine the understanding of the subsurface structure. Figure 8.5 shows a combination of well logs, seismic, and outcrop studies performed at this stage. Mapping the storage reservoirs and assessing their characteristics (e.g., porosity, permeability, fluid saturations, thickness, clay content, mechanical properties, etc.) contributes to reducing uncertainties and improving the knowledge necessary for effective carbon storage site selection and subsequent operations. These studies, including creating the top and isopach maps, porosity and permeability maps, and natural fracture systems in the reservoir, are significantly important for later reservoir modeling and simulation.

Fig. 8.5 Storage reservoir characterization https://www.netl.doe.gov/sites/default/files/2018-10/BPM-SiteScreening.pdf

8.3.1.2 Seal Integrity

A reliable confining zone for CO_2 storage needs to be regional in scale and include at least one geologic seal that separates the CO_2 injection zones from the surface and underground sources of drinking water (USDWs). It should cover the entire area affected by the CO_2 plume pressure and saturations, and where pressure is elevated enough to potentially lift saline water to a USDW. The process involves developing stratigraphic columns and structural maps as is done for storage reservoirs and potential injection zones. To begin, the well logs used for storage reservoir analysis are utilized to map the top, base, thickness, and lateral extent of individual geologic seals or caprocks within the confining zone. The initial stratigraphic framework identifies significant stratigraphic units, such as sequences, unconformities, or folding surfaces, that may impact the confining zone's integrity.

The lithological analysis of the caprock, either using available well logs such as LithoScanner logging or XRD, XRF of core or sidewall samples, will reveal the mineralogy of the caprock that can help in quantifying the confining zone integrity and possible reactivity. Core samples can also be investigated for natural fracture counts and directions. Note that all available data from adjacent wells will be used for this purpose, as a limited budget is considered for this phase.

8.3.1.3 Well Injectivity Potential

Well injectivity potential refers to the ability of a rock formation to accept injected fluids without inducing fractures. It is a critical parameter in determining the number and design of injection wells and significantly influences overall project costs. During the site selection stage, injectivity can be estimated using the field tests such as injection step rate test, or well interference test, using production history data, hydrologic tests, and core plug analyses from adjacent wells. These methods provide valuable information regarding the formation's capacity to store CO_2 and help to confirm the integrity of the overlying caprock by managing pressure to stay below the breakdown pressure of the caprock.

The maximum allowable pressure for injection is regulated by the EPA's Underground Injection Control (UIC) Program, which establishes the Maximum Allowable Surface Injection Pressure (MASIP). MASIP accounts for factors such as reservoir, wellbore, and fluid conditions. Reservoir rock and mechanical properties define the formation breakdown pressure, natural fracture reactivation and propagation, and fault stability. Wellbore integrity, including casing collapse and burst pressures and cement integrity. It also considers the injected fluid density, viscosity, and frictional resistance of the injected fluid. MASIP is typically defined as a fraction of the formation's failure pressure, ensuring the integrity of both the injection and confining zones. Determining MASIP requires consideration of reservoir characteristics and wellbore flow dynamics to protect against mechanical failure.

Understanding the injectivity potential of a site is essential for effective planning and optimization of injection wells; it enables operators to efficiently design the CO_2 injection rate and safely store the CO_2. It also helps in better ultimate CO_2 injection volume, which helps in efficient techno-economic analysis of the project. A reliable assessment of the fracture gradient can be achieved through well logs, such as porosity, density, sonic, and geomechanical logs, and core analyses that measure properties like in-situ stresses, Young's modulus, and Poisson's ratio. Using well logs and core data, the frac gradient can be indirectly approximated by calculating the minimum horizontal stress. The density log provides the vertical stress, while porosity and resistivity logs can be used to estimate pore pressure (using empirical correlations such as Eaton's method, Zhang 2011); the mechanical properties are also obtained using geomechanical logs. Assuming the minimum horizontal stress as a good estimate of closure pressure, then closure pressure divided by depth gives estimates of the frac gradient. Note that for a more accurate frac gradient, one needs to use DFIT or diagnostic fracture injection test and obtain the instantaneous shut-in pressure that can be used to obtain the frac gradient. Additionally, estimating the surface treating pressure to avoid pipe failure involves evaluating expected well dimensions, casing specifications, anticipated injection rates, supercritical CO_2 properties, reservoir pressure, and fracture gradients (Belyadi et al. 2019).

8.3.1.4 Seismic Data Analysis

Seismic data analysis, such as 2D, 3D, and passive seismic, is necessary in validating the stratigraphic and structural framework in selected areas and bridging the information gap between existing wells (Fig. 8.6). It offers valuable insights into the confining zone's properties, reservoir storage capacity, and potential faulting that could lead to induced seismicity under pressure (Chen and Huang 2015).

To enhance the accuracy and reliability of seismic data, it is advisable to assess the cost and benefits of re-processing and re-interpreting the existing seismic data. This approach ensures that the most up-to-date and refined information is incorporated into the site evaluation. Furthermore, naturally occurring seismicity data can be

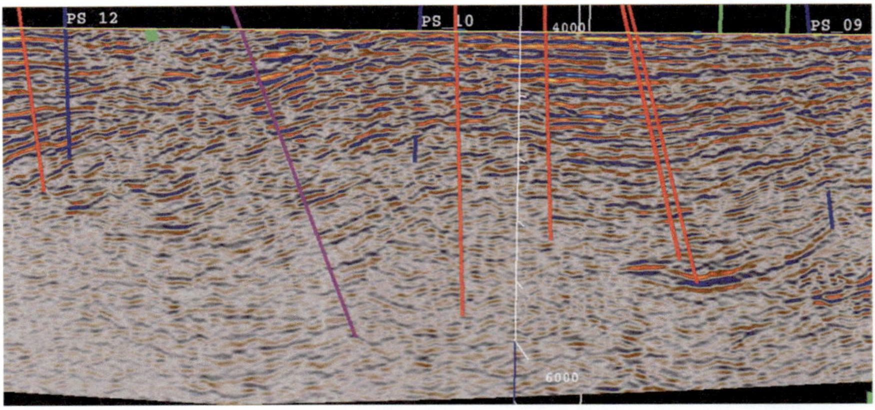

Fig. 8.6 Amplitude of reflected PS seismic wave (the converted shear wave) from survey line A (Traverse A) oriented in NE perspective from 4000 ms to 6000 ms PS time. Reference (Novel Use of P-wave and S-wave Seismic Attenuation for Deep Natural Gas Exploration and Development, Final Report DE-FC26-04NT42243)

utilized to analyze and identify the locations and characteristics of active faults. This data is mainly used for large-scale structures and regional analysis, where it helps to develop baseline geomechanics and aids in understanding the existing geological dynamics and potential risks associated with faulting in the selected areas. However, it is important to note that this passive seismic study needs to be integrated with active seismic, well logs, and core data for reservoir characterization.

8.3.1.5 Storage Potentials

The evaluation of storage potentials is an essential step in the site selection process. It involves refining the initial estimates (porosity, permeability, thickness, saturations) made during site screening by incorporating more detailed data on the structure and properties of the injection zones within the selected areas. By acquiring comprehensive information from adjacent wells about the storage reservoirs, including their geological characteristics and capacity for CO_2 storage (well logs, cores, seismic, and pressure tests), the estimates can be further validated and improved. This refinement process helps to ensure that the storage resource aligns with the carbon handling and storage approach where the CO_2 volume and rates are determined, and success criteria are set. Through the classification of the storage site, refined estimates are required in determining the suitability and potential of the site for CO_2 storage. This classification provides valuable insights into the status of the storage site that allows the project developers to make informed decisions based on the quality and compatibility of the storage potentials.

8.3.2 *Compliance Assessment*

In the United States, the Underground Injection Control (UIC) Program administered by the Environmental Protection Agency (EPA) regulates all six classes of underground injection wells under the Safe Drinking Water Act (SDWA). The primary goal of the UIC regulations is to safeguard Underground Sources of Drinking Water (USDWs) during CO_2 geologic storage. Specifically, the regulations aim to prevent:

1. Plume infiltration, which refers to the migration of CO_2 into the USDWs.
2. Brine intrusion caused by the increased pressures resulting from CO_2 injection.
3. Mobilization of potential subsurface contaminants such as trace metals and organics.

The UIC Program oversees the permitting, allocation, construction, monitoring and testing, closure, and post-closure care of injection wells that store or dispose of fluids (liquids, gases, semi-solids, or slurries) underground.

The compliance assessment is a critical component of site selection, as it assesses the potential regulatory requirements that the project must comply with. This analysis involves evaluating four key elements, including Well Categories, Geologic Trapping, Injection Pressure Limit, and Regulatory Liability. Conducting a thorough compliance assessment helps project developers to ensure compliance with relevant regulations.

8.3.2.1 Well Categories

The U.S. EPA's Underground Injection Control (UIC) program categorizes wells into six classes based on their purpose and operational risks. Class I wells handle hazardous and non-hazardous waste disposal into confined deep formations, while Class II wells manage fluids tied to oil and gas production, including wastewater and enhanced oil recovery (EOR) injectants. Class III wells support mineral extraction by injecting fluids to dissolve target resources, and Class IV wells were used for shallow hazardous/radioactive waste disposal near drinking water sources. Class V covers non-hazardous fluid injection, often near aquifers, and Class VI is dedicated to geologic CO_2 storage, with strict requirements for long-term containment.

For CO_2 storage projects, two relevant classifications are Class II, which governs CO_2-EOR operations, where injection boosts oil production while storing some CO_2, and Class VI, which applies to dedicated geologic sequestration, prioritizing permanent storage over hydrocarbon recovery.

A critical regulatory distinction arises when repurposing Class II wells for storage. If CO_2 storage becomes the primary function, operators must reclassify the well as Class VI, adhering to its enhanced safety protocols such as rigorous monitoring and extended post-injection oversight. This transition ensures alignment with the geologic sequestration framework and addresses risks distinct from traditional EOR activities.

8.3.2.2 Injection Pressure Limit

Evaluating the Class VI well injectivity is based on local geologic structures and relevant geomechanical data. Class VI wells have a maximum injection pressure limit of "90 % of the fracture pressure (fracture gradient×TVD) of the injection zone (s)." For Class VI wells, injection pressure should not cause new fractures or propagate existing fractures in the confining zone near the underground sources of drinking water (USDWs). Project developers must ensure that the subsurface geologic evaluation satisfies these regulatory requirements. For this purpose, diagnostic fracture injection test (DFIT) or minifrac if available in the adjacent wells in the region, can be used to have an estimate of the fracture gradient. The fracture gradient can be obtained from instantaneous shut-in pressure "ISIP" measured during the DFIT and used to calculate the bottomhole treating pressure. The bottomhole treating pressure minus the hydrostatic pressure of the sCO_2 plus all the pipe frictions will give a good estimate of maximum allowable surface injection pressure (MASIP).

8.3.2.3 Geologic Trapping

In the regulatory analysis, the assessment of geologic trapping is needed for both Class II and Class VI UIC wells. Project developers are required to provide evidence of the presence and adequacy of a reliable containment mechanism. This demonstration covers various aspects, including detailed information on local geologic structures, faults, and other geomechanical factors that are relevant to the site. Additionally, comprehensive maps and cross-sections depicting the lithology of the site and the locations of underground sources of drinking water (USDWs) must be included in the documentation. Additional reservoir pressure testing, such as injection pressure, the DFIT, or minifrac analysis, can be used to quantify the maximum pressure allowable for containment "Caprock" integrity. These measures ensure that the geologic trapping is robust and capable of effectively confining the injected fluids within the designated storage zones, mitigating the risk of potential leaks or migration into USDWs.

8.3.2.4 Regulatory Liability

Liability for CO_2 storage in the subsurface is a subject of ongoing discussion and debate. The uncertainty surrounding long-term regulatory liability and responsibility can impact the progress of a project. Currently, there is no universally accepted framework for assigning regulatory liability in CO_2 storage projects, although some states have passed or are considering legislative measures to address this issue. In Montana, for example, in May 2009, a law holds CO_2 injection operators fully responsible for storage safety, including long-term monitoring and verification. If the developer can demonstrate compliance with all UIC permit and regulatory requirements for CO_2 storage, the Montana Board of Oil and Gas has the authority to issue a "Certification of Completion" that provides some level of regulatory liability protection. In Kansas state is responsible for monitoring the CO_2 storage site. However, the operator must pay a fee per ton of CO_2 injected into the ground. The fee is collected in an account and will be used by the state for the purpose of monitoring and managing the storage site (Ingelson et al. 2010).

The issue of regulatory liability is complex and varies across jurisdictions. It is important for project developers to understand the regulatory landscape and potential regulatory liability implications associated with CO_2 storage. Clear guidelines and frameworks for regulatory liability assignment are necessary to provide confidence and clarity to stakeholders involved in CO_2 storage projects.

8.3.3 Reservoir Modeling and Simulation

During the reservoir modeling and simulation phase of site selection, the primary objective is to construct initial geological models that support subsequent numerical simulations. These models are essential for predicting the movement of the injected CO_2 plume and evaluating the extent and magnitude of the associated pressure front. The main goal of modeling is to assess the suitability of the storage reservoir to securely retain the targeted volume of CO_2 within the designated injection zone.

An important early step is to define the data requirements for reservoir modeling and simulation and evaluate the associated costs. This involves acquiring and analyzing relevant datasets, including well logs, core samples, seismic surveys, and other geological information, to ensure that the models are built upon reliable and comprehensive inputs. Next, boundary conditions must be established in the models by defining constraints on fluid flow, pressure, and temperature. This includes examples such as the presence of an aquifer connected to the reservoir and the boundary conditions it imposes in terms of pressure and rate. We also need to account for uncertainties inherent in the input data and evaluate their potential impact on simulation outcomes. Incorporating existing seismic data further enhances model accuracy by improving the characterization of subsurface structures, fault systems, and reservoir properties, resulting in an increase in the reliability of model predictions.

For reservoir model development, static models are constructed using storage reservoir and caprock properties identified during the earlier screening and subsurface analysis stages of site selection. Historical production data from nearby wells or results from injection pressure tests are then used to perform history matching. This step is the most important part of the model development as it requires detailed sensitivity analysis, identifying the major parameters impacting the reservoir model behavior (Key performance indicators) and history matching the actual reservoir response to create a more representative dynamic reservoir model. The history-matched model can subsequently be used to forecast CO_2 injection pressures and plume evolution under various operational scenarios, as well as to conduct sensitivity analyses and uncertainty quantification. Simulations are typically run for both

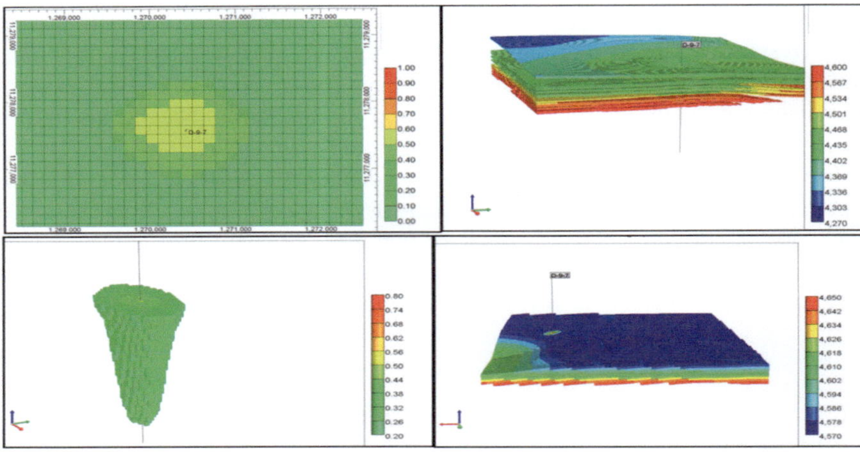

Fig. 8.7 CO_2 saturation distribution (left) and pressure distribution (right) at the end of 3 years of injection (top) and 30 years of injection (bottom) in Citronelle Dome, Alabama, Fathi et al. 2023a, b

the injection period to study area of review (AoR) and the post-injection period, post-injection site care (PISC).

Figure 8.7, for example, illustrates CO_2 saturation and pressure distributions within the Citronelle Dome, Alabama, for two scenarios of 3 years and 30 years of continuous CO_2 injection.

In addition to AoR and PISC studies, sensitivity analyses conducted using these models help quantify the influence of data uncertainties on CO_2 saturation and pressure predictions. This process involves identifying and defining key parameters, such as:

1. Rock properties: porosity, permeability, saturation, geomechanics, compressibility
2. Fluid properties: composition, salinity, PH
3. Rock–fluid interactions: relative permeability and capillary pressure curves
4. Reservoir: structural extension, presence of aquifers, initial pressure, and temperature
5. Operational conditions: CO_2 injection rates and total injection volumes

These parameters are critical to ensuring simulation accuracy and effectively predicting CO_2 behavior within the storage formation. Especially, CO_2 saturation plume expansion during and after injection has significant technical and nontechnical importance, as it identifies the pore spaces occupied by CO_2 and carries legal responsibilities regarding royalty payments to pore space owners. It also provides measures for safety considerations and for ensuring the containment of CO_2.

8.3.4 Storage Viability Assessment

The storage viability assessment is a major component of site selection and involves evaluating four key elements, including Infrastructure, Area of Review (AoR) Requirements, Surface Access, and Pore Space Ownership.

An infrastructure study involves assessing the availability and suitability of existing infrastructure necessary to support CO_2 geologic storage operations. This includes considerations such as transportation networks, pipelines, storage facilities, access roads, and electricity. The analysis aims to determine if the selected area has the required infrastructure in place or if additional investments are needed.

Area of Review (AoR) studies include quantifying three key metrics of CO_2 plume pressure differential extent, CO_2 plume saturation extent, and localized pressure differentials (at specified distances from the injection well). Given the risks associated with geologic carbon storage, particularly groundwater contamination and injection-induced seismicity, careful assessment of reservoir behavior under AoR is essential. Analyzing the area of review (AoR) requirements also involves assessing the distance and proximity of the storage site to various features such as drinking water sources, protected areas, sensitive ecosystems, and population centers. This evaluation ensures compliance with regulatory and environmental considerations and helps identify any potential conflicts or limitations.

The surface access study focus is on accessibility of the surface area above the storage site. This includes considerations such as land ownership, land-use restrictions, and potential conflicts with existing surface activities. Surface access is critical for construction, operation, and monitoring activities associated with CO_2 storage, and any limitations or constraints should be identified during this analysis.

Assessing the ownership rights and legal aspects related to the underground pore space where CO_2 will be stored is of great importance. This involves determining the ownership framework and legal regulations governing pore space rights in the specific jurisdiction. It is important to identify any potential conflicts or legal difficulties that may impact the use of the pore space for CO_2 storage. Pore space ownership, which refers to the ownership rights and legal framework surrounding the underground voids where CO_2 is stored, is primarily governed by state jurisdictions. However, the legal treatment of pore space ownership varies considerably from one state to another. Currently, most states have not passed specific legislation that directly addresses the issue of pore space ownership. This lack of specific laws leaves a gap in the regulatory framework that leads to uncertainty and potential challenges when it comes to determining ownership rights and responsibilities related to underground pore space. Due to the absence of clear guidelines, the legal treatment of pore space ownership may be subject to interpretation based on existing property laws, mineral rights regulations, and other relevant legal principles within each state. This variability in legal approaches across different states further adds to the complexity and lack of uniformity in the ownership of pore space. Some states like West Virginia, North Dakota, Montana, and Wyoming passed laws that the owner of the surface land automatically holds title to the pore space beneath it (in all subsurface layers, including deep saline formations suitable for CO_2 storage).

Although a selected area may possess favorable geologic characteristics, it may still face challenges that render it unsuitable for CO_2 storage. Infrastructure limitations, unresolved pore space ownership issues, surface access, or other factors can affect the overall suitability of a site. Therefore, it is significantly important to identify and address these issues during the site selection process to ensure the chosen site is suitable for successful CO_2 geologic storage operations.

8.3.5 Social Baseline Assessment

The social baseline assessment in the site selection process is necessary in understanding and addressing the socio-cultural factors that can influence the acceptance of the project by the society within the selected areas of interest. This component involves a direct investigation into the social dynamics surrounding the project. To conduct the social characterization, the project developer is required to gather and assess social data related to the project's potential impact on the local communities. This includes understanding the values, beliefs, and concerns of the community members through the different stakeholders involved. A community assessment helps in framing an effective outreach plan. Initially, the project developer can use readily accessible information such as local media coverage and websites addressing local opinions and engagements. However, in certain communities, a more direct approach may be necessary. This could involve conducting interviews with key stakeholders, organizing focus group discussions, and facilitating community discussions to gather firsthand perspectives. Furthermore, engaging with regulators and other relevant officials can provide valuable input and guidance. These interactions can help align the project with existing regulations and address potential concerns.

Once all the related information is collected, it can be used to assess the potential benefits the project may bring to the community. This assessment also allows for the identification and understanding of any concerns or issues that need to be addressed through the project's design and implementation. The gathered information is helping to develop an effective outreach plan that considers the specific needs, preferences, and priorities of the local community. Rothkirch and Ejderyan (2021) identified key place factors such as industrial familiarity, economic benefits, tourism reliance, environmental justice concerns, and community trust that influence public acceptance or opposition to carbon capture and storage (CCS) projects. Through a review of 38 CCS cases, the authors highlight that successful projects often align with local values (e.g., job creation in industrial areas) while failed ones conflict with community priorities (e.g., tourism or environmental equity). The study demonstrates how Geographic Information System (GIS) mapping can integrate these social factors into early-stage site selection, enabling developers to anticipate conflicts and prioritize locations with higher social compatibility.

8.4 Site Characterization

Site characterization is a critical phase in which selected CO_2 storage locations are thoroughly evaluated to determine their suitability. The goal is to assess the site's storage-related characteristics and determine if it is selected as a *Qualified Site*, indicating its readiness for development.

The site characterization process typically consists of two distinct stages: baseline reservoir assessment (BRA) and detailed characterization. During baseline reservoir assessment, existing data and information about the selected sites are utilized to gain an initial understanding of their storage potential. This includes analyzing available geological data, historical records, and any relevant studies conducted in the area. The purpose is to gather enough information to establish the BRA for the suitability of the sites for CO_2 storage. This will build on the site-specific characterization done in the site selection phase. Upon identifying candidates from selected sites during the BRA, a more comprehensive investigation is carried out for detailed characterization. This involves acquiring new and additional data through various methods such as site-specific surveys, drilling (such as ROP, WOB, drilling acceleration data, etc.), well logging, pressure testing, and laboratory analyses (Fathi et al. 2022). These detailed studies are used to refine the models built in site selection studies and update information regarding the subsurface conditions, reservoir properties, and potential risks associated with CO_2 storage.

The detailed characterization process finalizes the site characterization and identifies the qualified site. This detailed characterization indicates that the site has met the necessary criteria and is considered viable for CO_2 storage, further development, and implementation of carbon capture and storage projects.

Figure 8.8 shows the workflow for site characterization, focusing on key stakeholders and detailed characterization during site selection. Unlike previous screening and site selection phases, here the site characterization begins with assessing public outreach, including objectives and activities. At this stage, the public outreach team identifies the stakeholders and develops a public outreach program that supports the potential site's viability.

The next step involves analyzing regulatory requirements. This phase examines the regulatory landscape concerning CO_2 storage, assessing relevant regulations, including well plans and underground injection control (UIC) permit planning. If the potential site aligns with regulatory requirements, it's a green light to proceed. Otherwise, alternative potential areas are explored. The compliance assessment takes a close look at different activities, including operation, monitoring, and reporting.

Operating activities requirements include injection pressure requirements, well construction, and CO_2 steam purity analysis. Injection well operators must maintain injection pressure below 90% of the formation's fracture pressure to prevent creating new fractures or extending natural fractures in the injection zone. Injection between the outermost casing and wellbore is prohibited, and the tubing-casing annulus must be filled with an approved fluid, pressurized above injection pressure while ensuring continuous mechanical integrity (except during permitted maintenance). Operators

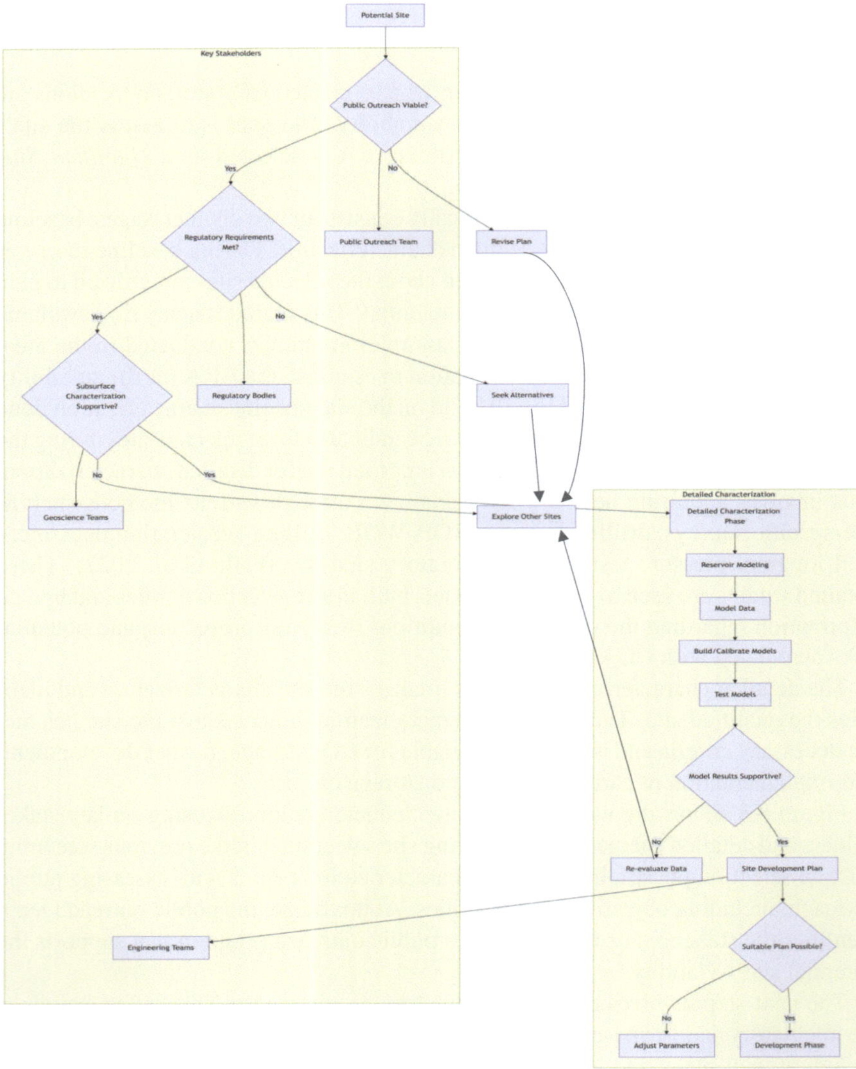

Fig. 8.8 Site characterization workflow

must install continuous monitoring systems for injection pressure, CO_2 flow rates, temperature, and annulus conditions, along with automatic shut-off systems and alarms for surface or downhole for onshore wells, and mandatory downhole for offshore wells (40 CFR § 146.88—Injection well operating requirements-ecfr.gov).

The fracture pressure can be obtained using the diagnostic fracture injection test (DFIT) for both reservoir and caprock. Well construction requirements mandate using coated casing, applying cement barriers, and demonstrating the mechanical integrity of the well for maximum allowable surface treating pressure plus a 20% safety factor.

CO_2 Stream Purity must also be ≥95% CO_2, and impurities must be below EPA thresholds. Monitoring activity requirements include obtaining the pre-injection baseline, active injection monitoring, and post-injection site care (PISC). Reporting requirements involve annual reports, incident reporting, and closure certification. Table 8.1 shows the details of regulatory activity requirements and key details.

After all the social and regulatory studies are completed, subsurface geology characterization comes next. In subsurface geology characterization, the focus is on understanding the reservoir framework and the properties of subsurface geology. If the baseline reservoir assessment (BRA) was supportive, then detailed characterizations will be performed. This phase includes in-depth geological, geophysical, geochemical, geomechanical, and hydrogeological assessments (Babarinde et al., 2023). These comprehensive studies lead to building a representative subsurface reservoir model that can be used for simulation and forecasting. Building and calibrating reservoir models is an essential step in the process. It involves collecting and analyzing data relevant to the potential site for modeling purposes. Models are developed and calibrated to simulate CO_2 behavior and its effects on the subsurface. These models are tested and verified by comparing their outputs with actual pressure tests performed in the reservoir, "History-matched model", If the modeling results confirm site viability, the process proceeds; otherwise, alternative potential areas are explored. Finally, the process concludes with the creation of an initial

Table 8.1 Regulatory activity requirements

Activities	Requirement	Key details
Operating	Injection pressure limits	≤90% of fracture pressure at the injection zone; real-time monitoring
	Well construction	CRA/epoxy-coated casing, cement barriers, demonstrate mechanical integrity
	CO_2 stream purity	≥95% CO_2; impurities below EPA thresholds,
Monitoring	Pre-injection baseline	3D seismic, USDW sampling, DFIT, groundwater chemistry, geochemical baseline
	Active injection monitoring	Annual 4D seismic, continuous DTS/DAS, microseismic arrays, annulus pressure monitoring
	Post-injection (PISC)	≥50-year monitoring until pressure < 10% of baseline and no USDW risk; site remains under EPA oversight until certified
Reporting	Annual reports	Injection volumes, pressure trends, integrity tests, plume evolution
	Incident reporting	Notify EPA within 24 h of leaks or endangerment; submit a 30-day corrective action plan
	Closure certification	Submit site closure plan; demonstrate plume stability and no USDW endangerment

site development plan. This phase involves formulating a comprehensive plan for site development based on the gathered information. The plan outlines the detailed characterization phase and subsequent development steps. Sites for which a suitable development plan can be established are designated as potential areas ready for development. This systematic approach streamlines the evaluation of potential sites and ensures a thorough assessment of key factors in the site characterization process.

8.4.1 *Baseline Reservoir Assessment (BRA) and Detailed Characterization*

During the baseline reservoir assessment and detailed characterization, the primary focus is on establishing the essential well-drilling requirements for the selected site. This requires careful well planning, covering appraisal, injection, and monitoring wells.

Appraisal wells are significant in the data gathering process to support the subsequent detailed characterization phase. These wells provide invaluable insights into the subsurface geology, reservoir properties, and the potential storage capacity. The data acquired from these wells is critical in further evaluating the site's suitability for CO_2 storage. Well planning also includes the identification of suitable locations for future injection wells, which serve as actual injection wells to store CO_2 in the reservoir.

Drilling of injection wells takes place as the project advances to the site development phase, indicating that the site meets the criteria for CO_2 storage. The drilling of monitoring wells is a critical component. These strategically positioned wells serve to monitor and assess the containment of CO_2 within the storage reservoir over time. Monitoring wells provide critical data regarding CO_2 behavior, potential migration, and the effectiveness of containment measures and can be equipped with downhole pressure gauges, downhole geophones, or retrievable fiber optic cables. Well planning during this phase ensures the identification and design of wells that are necessary for supporting the site's evaluation, detailed characterization, injection operations, and ongoing monitoring of CO_2 storage activities.

8.4.1.1 Well Planning

For each anticipated well at the potential site, a comprehensive well plan must be developed, including well design, formation evaluation, well testing, and injection tests. The well design involves determining the specifications and parameters for drilling, casing, and completion of the well. This includes considerations such as the wellbore trajectory, casing sizes and grade, and cementing techniques to ensure well integrity.

The formation evaluation focuses on gathering data about the subsurface formations encountered during drilling. Techniques such as logging, coring, and sidewall sampling are employed to assess the properties of the rock formations, such as porosity, permeability, mechanical properties, and fluid content. This information helps in understanding the potential for CO_2 storage within the reservoir.

Well testing involves conducting various tests to assess the performance and productivity of the well. These tests may include flow rate measurements (Drill stem test), pressure tests (Injection step rate test, DFIT), and interference tests to determine the behavior of the reservoir and assess the suitability of the well for CO_2 injection.

Injection tests are specifically designed to evaluate the injectivity and containment capabilities of the well. These tests include injecting fluids into the well at specified rates and pressures; the response of the formation and the wellbore can be observed and analyzed. This helps in understanding the potential of the reservoir to receive CO_2 and its ability for CO_2 storage and assessing any risks associated with injection operations. The detailed well plan helps the operator in proper design decisions, efficient evaluation and testing, and injection optimization of each anticipated well to make sure safe and efficient CO_2 storage operations at the site.

8.4.1.2 Formation Evaluation

The well plan should outline the data gathering activities or tests that will be conducted in a new wellbore to thoroughly characterize the injection and confining zones. These activities should be customized to the specific site, considering the level of certainty already achieved during the baseline reservoir assessment. At this stage, the following activities may be included:

1. Sampling whole cores of potential injection and confining formations: This involves extracting core samples of the rock formations to analyze their physical and chemical properties and investigate their suitability for injection and confinement, as shown in Fig. 8.9 (right).
2. Obtaining standard and advanced logging suites: Logging tools are deployed down the wellbore to measure various parameters, such as rock porosity, permeability, mineralogy, mechanical properties, fluid saturation, and natural fractures. Advanced logging techniques provide more detailed information about the subsurface formations, such as the presence of natural fractures using FMI logs. Figure 8.9 (left) shows a series of well logs obtained from the Plains CO_2 Reduction (PCOR) Partnership Program.
3. Obtaining sidewall cores to complement whole cores taken: Sidewall cores are small samples collected from specific intervals within the wellbore, providing additional data to supplement the analysis of whole cores and enhance the understanding of the subsurface geology. Sometimes, due to budget limitations or difficulties during the coring, sidewall cores are used instead.

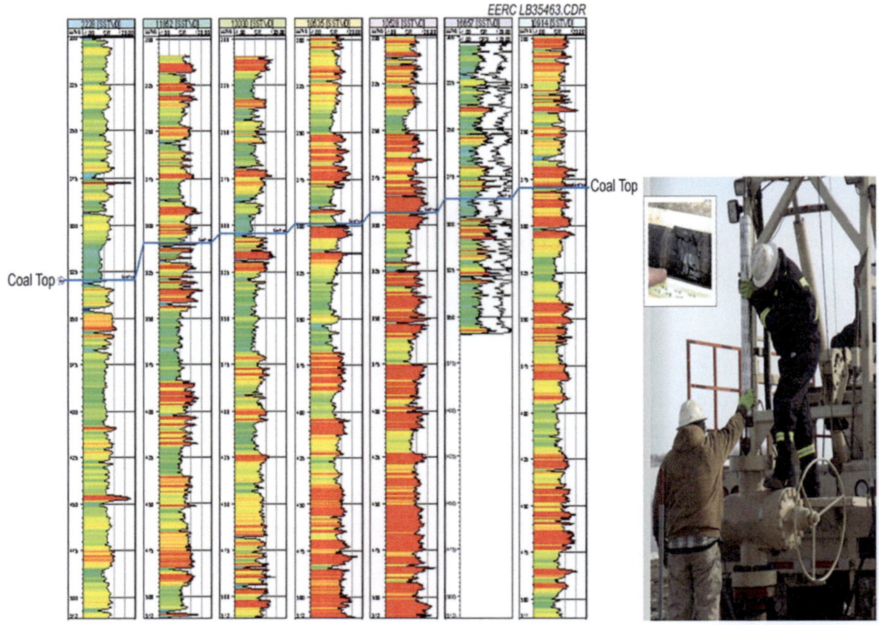

Fig. 8.9 Six geologic well logs from the Plains CO_2 Reduction (PCOR) Partnership Program. The correlated coal seam of interest is highlighted. https://www.netl.doe.gov/coal/carbon-storage/atlas/pcor/lignite

4. Performing geologic and hydrologic characterizations of above-zone monitoring intervals: This involves studying the geological and hydrological properties of intervals above the injection and confining zones to assess their potential impact on the storage operation and the effectiveness of monitoring activities.
5. Collecting fluid samples for geochemistry analysis: Fluid samples are collected from the wellbore to analyze their chemical composition. This analysis helps determine the potential interactions between the injected CO_2 and the formation fluids. This significantly helps in the assessment of storage suitability and potential risks.

The analysis of cores, formation fluids, pressure readings, and logs should be conducted with the aim of refining the understanding of both the injection and confining intervals at the site. This comprehensive formation evaluation provides key data for accurately characterizing the subsurface formations and making informed decisions about the site's suitability for carbon dioxide storage.

8.4.1.3 Well Testing-Drill Stem Test (DST)

It is important to perform well tests such as the Drill Stem Test (DST) to evaluate the formation of interest and determine its production potential. DSTs are widely used in the oil and gas industry to gather information about the fluids in the reservoir, its pressure, permeability, and flow characteristics (Marshall Black 1956; Ershaghi 2023).

A DST is typically conducted after the well has reached the predefined depth, targeting a specific formation. The goal is to assess whether the formation can produce fluids (oil, gas, or water) at economic rates and to better understand its properties. The DST is carried out using specialized equipment attached to the drill string. This equipment includes packers, which are rubber elements that expand to seal off (isolate) the formation from the rest of the wellbore, and a valve system, which controls the flow of fluids from the formation into the drill pipe.

The basic sequence of a DST includes several phases of initial hydrostatic pressure, setting the packers & initial flow, initial shut-in, final flow & final shut-in, and final hydrostatic (Motaei et al. 2024). While lowering the drill string into the well, the mud column exerts hydrostatic pressure on the formation. This baseline pressure is recorded before the test begins. Once the drill string reaches the target depth, the packers are set to isolate the zone of interest. The valve is then opened, allowing fluids from the formation to flow into the empty drill pipe. During this flow period, pressure and time are recorded. A drop in pressure indicates that the formation is producing fluids.

The next step is the initial shut-in, where a valve closes the chamber and stops the flow from the formation into the drill pipe. This leads to the pressure in the formation beginning to build back up as it equilibrates. This period is used to estimate the original reservoir pressure and assess how quickly the formation pressure recovers. The flow and shut-in sequence are repeated to confirm the earlier measurements and provide additional data on the formation's deliverability and pressure behavior. After the test is completed, tools are retrieved, and the mud column pressure is recorded again to confirm wellbore conditions. Throughout the test, pressure and time data are collected and plotted to create pressure-time behaviors (drawdown and buildup), which provide valuable insights into the reservoir.

The DST provides a variety of information about the formation, including fluid samples, reservoir pressure, formation properties, productivity estimates, and hydrodynamic information. Fluids from the reservoir (oil, gas, water) are collected for laboratory PVT analysis to determine their composition, density, viscosity, and other properties. Measurements of pressure during shut-in periods help determine the original (static) reservoir pressure and pressure changes during flow. The DST helps estimate properties such as permeability, skin factor (damage or stimulation near the wellbore), and radius of investigation (how far into the reservoir the test reaches). Engineers can estimate the reservoir's ability to produce fluids and evaluate its economic potential by studying the flow rate and pressure behavior during the test. The pressure response over time provides insights into reservoir boundaries, heterogeneity, and how fluids move through the rock. This is important for understanding the size and extent of the reservoir and planning future development. Figure 8.10 shows the schematic of the DST.

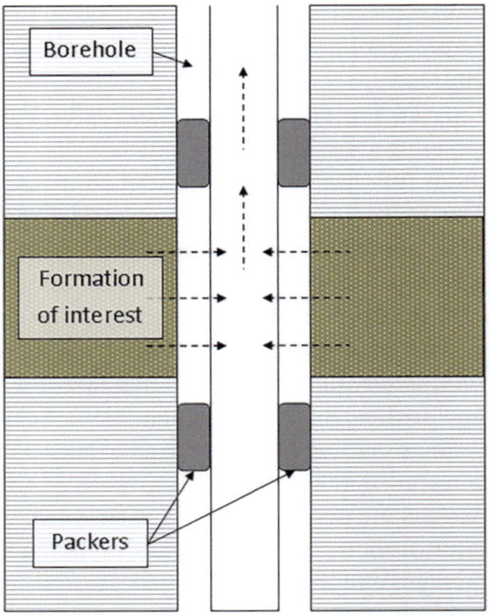

Fig. 8.10 Schematic of Drill Stem Test (DST), By Nwhit

8.4.1.4 Injection Tests and DFIT

Injection tests are an essential part of the well plan and involve the controlled injection of fluid into the reservoir. The purpose of a step-rate test is to evaluate formation injectivity and identify potential pressure limitations, such as fracture initiation or other flow barriers that may affect injection performance. During the test, fluid is injected at a series of increasing rates, with each rate maintained for a sufficient period to allow the pressure response to stabilize. Bottomhole pressure is continuously monitored throughout the test. The stabilized pressure corresponding to each injection rate is then plotted against the injection rate to evaluate the reservoir response. Changes in the slope of the pressure-rate relationship may indicate the onset of formation fracturing or other changes in flow behavior (Liu and Ehlig–Economides, 2019). Additionally, the well plan should outline the possibility of conducting injection tests using either brine or CO_2. These tests are to validate the presence of sufficient permeability within the reservoir and to identify any potential barriers that could impede the desired injection rates and pressures.

Another type of test that should be considered is a small-scale hydraulic fracture minifrac, known as a Diagnostic Fracture Injection Test (DFIT). DFITs are conducted to determine important reservoir rock properties such as fracture gradient, minimum/maximum horizontal stress, stress anisotropy, and reservoir properties like pressure and conductivity (Belyadi et al. 2017). DFIT measurements help engineers to gain a better understanding of the reservoir's fracture behavior and its response to injection pressures. DFIT is required to identify barriers and their integrity to ensure that the reservoir can support the required injection rates and pressures. Figure 8.11 shows an

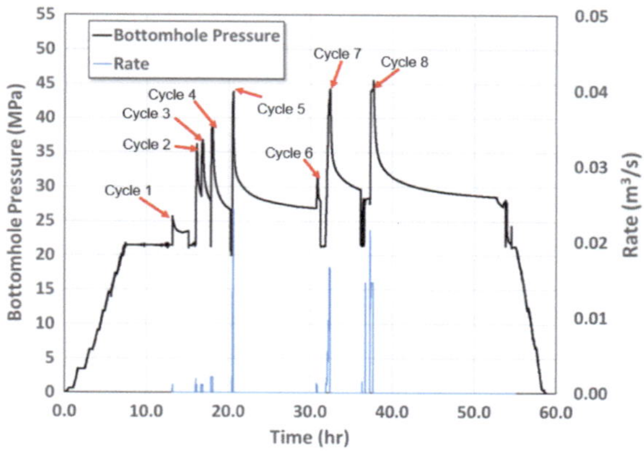

Cycles	Comment
Z1C1_17	•Injection fall-off •Packer Failure
Z1C2_17	•Micro HF
Z1C3_17	•Micro HF
Z1C4_17	•Micro HF
Z1C5_17	•DFIT •Extended shut-in
Z1C6_17	•Micro HF •Fracture not opened
Z1C7_17	•Step rate •Extended shut-in
Z1C8_17	•DFIT •Extended shut-in

Fig. 8.11 Well injection test in Zone 1 well 58–32 FORGE site, Xing et al. 2020

injection test performed in Well 58–32, which is a pilot well at the FORGE site. This analysis presents the pressure-rate relationships observed during the 2017 injection testing of Zone 1 across Cycles 1–8. Each cycle designation identifies the tested zone, sequence number, and measurement year. The dataset includes results from Hydraulic Fracturing (HF) and Diagnostic Fracture Injection Tests (DFIT) that provide insights into formation response and fracture behavior under controlled injection conditions (Xing et al. 2020).

8.4.1.5 DST Versus Injection Test

The main difference between a drill stem test (DST) and an injection test lies in their purpose and the direction of fluid flow. DST is performed to gather information about the reservoir properties, such as pressure, fluid composition, and permeability, before completing the well. In DST, fluid flows out of the reservoir into the wellbore, allowing data collection and evaluation of reservoir characteristics. An injection test is conducted to validate the existence of sufficient permeability in the reservoir and identify potential barriers to the required injection rates and pressures. In an injection test, fluid (typically brine or CO_2) is injected into the reservoir from the wellbore to assess the flow behavior, permeability, and the ability of the reservoir to accept injected fluids. DST focuses on fluid flow out of the reservoir to gather information, while injection tests involve fluid flow into the reservoir to evaluate its permeability and ability to accept injected fluids.

8.4.1.6 Reservoir Modeling and Simulation

The reservoir modeling and simulation component integrates analyses of the geological, geochemical, geomechanical, and hydrodynamic characteristics of the reservoir. Geological analysis examines the structure and composition of the reservoir rock, including rock types, stratigraphy, and the spatial distribution of formations. This helps define the physical framework of the reservoir.

Geochemical analysis focuses on the chemical properties of reservoir rocks and fluids. It assesses fluid–rock interactions, composition gradients, and potential reactions (including aqueous reactions between formation brine and sCO_2 and aqueous rock reactions that could result in rock dissolution or mineralization) that could influence reservoir performance (Roberts et al. 2017; Huq et al. 2015; Fathi et al. 2023a, b).

Geomechanical analysis evaluates the mechanical behavior of the reservoir, such as strength, stress distribution, and deformation potential. This is critical for ensuring reservoir stability and understanding its response to fluid injection (Fathi et al. 2024a).

Reservoir hydrodynamic analysis studies fluid movement and distribution within the reservoir, including fluid flow pathways, aquifer and boundary conditions, and pressure gradients.

These analyses draw on a wide range of datasets, including laboratory tests on core and sidewall samples, petrophysical data from well logs, and geophysical surveys. Data collected during site selection, such as reservoir fluid properties and boundary conditions, are also incorporated.

The results are used to construct static reservoir models and geomechanical earth models, which represent the reservoir's physical and mechanical properties. These static models are then integrated to develop a dynamic reservoir model, capable of simulating the pressure and saturation evolution of CO_2 within the *Area of Review (AoR)* during injection and post-injection site care (PISC), as illustrated in Fig. 8.12.

8.5 Downhole and Laboratory Activities for Science Well and Injection Well Permitting

Table 8.2 presents various data requirements and analysis techniques as per the US Environmental Protection Agency (EPA) requirements for different aspects related to injection confinement and Underground Sources of Drinking Water (USDW) zones. Downhole activity and lab activities are offered to meet the EPA requirement. Presented here is from Schlumberger lab and field services (www.slb.com); however, similar services are also provided by other vendors.

Analysis includes a litho scanner, XRD, XRF, and Fourier-transform infrared spectroscopy (FTIR) techniques for mineralogy assessment. Porosity and permeability analysis utilizes Conventional Magnetic Resonance (CMR-Plus) and Pulsed

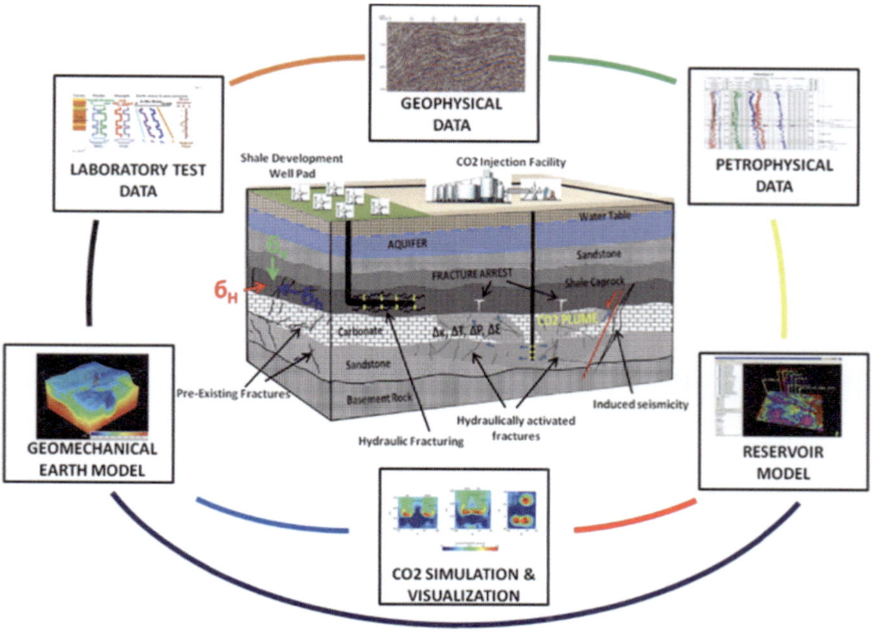

Fig. 8.12 The reservoir framework data NETL, US DOE https://www.netl.doe.gov/node/774

Neutron Capture (PEx) tools, along with Routine Core Analysis (RCAL) and Transient Radial Analysis (TRA) methods.

Capillary pressure analysis is conducted using CMR-Plus and Mercury Injection Capillary Pressure (MICP) techniques. Geological and facies characterization involves Formation Micro-Imager (FMI) and Quanta Geo tools, along with whole-core analysis for fracture, fault, and seal stability assessment.

Stress and mechanical properties are evaluated through FMI, core fracture analysis, and tests like triaxial stress testing, scratch tests, and Unconfined Compressive Strength (UCS) measurements.

Pore pressure analysis utilizes a Sonic scanner, Modular Formation Dynamics Tester (MDT), Minifrac, and data from pressure gauges.

Geochemical analysis is conducted through MDT water sampling and subsequent water analysis to understand CO_2 -brine interactions. Fracture pressure analysis involves MDT Minifrac tests and water analysis as well. Table 8.2 highlights the specific techniques and activities required for each parameter, whether it's for the injection zone, confining zone, or other aspects of reservoir characterization.

For the injection zone, specific activities include assessing salinity through water analysis and evaluating parameters such as resistivity, spontaneous potential (SP), and caliper measurements. Tools like Array Induction Tool (AIT), or lateraolog

Table 8.2 EPA data requirements & methods for CO_2 storage

Category	Parameters	Downhole methods	Laboratory methods	Relevant zones
Mineralogy and petrophysics	Mineralogy, porosity, permeability, capillary pressure	Litho scanner, CMR-plus, PEx	XRD, XRF, FTIR, RCAL, SCAL, MICP	Injection, confining
Geomechanics and structure	Fractures, faults, seal stability, stress, fracture pressure	FMI, sonic scanner, MDT Minifrac	Triaxial, scratch, UCS, Core fracture analysis	Confining
Fluid and geochemistry	Salinity, pore pressure, geochemical analysis, CO_2-brine interaction, DFA	MDT sampling, AIT, dielectric	Water analysis, Core flooding	Injection, confining, USDW
Well integrity	Cement integrity, mechanical integrity	Power eco, power flex, Optiq, pulsar WFL	–	Injection, confining, USDW
Reservoir dynamics	Pump tests, production forecasting, fluid pressure/temp/pH	PSP, Optiq, high-res well testing	–	Injection
Core and fluid sampling	Core samples, fluid samples	XL-rock, MDT water sampling	Whole Core analysis	Injection, confining

resistivity tools, depending on drilling fluid conditions, along with Modular Formation Dynamics Tester (MDT), and Dielectric+Litho Scanner are utilized for data acquisition.

Cement integrity is evaluated using tools like Power Eco, Power Flex, and optical tools to ensure the integrity of the cement barrier. Mechanical integrity assessment is performed through tools such as PowerEco, Optiq, and Pulsar WFL to examine the structural integrity of the well.

Core analysis is conducted to gather essential information for the injection permit. XL-Rock is Schlumbergers' sidewall coring tools used to aquire side wall cores, while fluid sampling using MDT is employed to measure fluid pressure, temperature, pH, conductivity, and contacts. Digital Fluid Analyzer (DFA) is used to analyze fluid samples.

Pump tests are performed to evaluate injection capabilities, and tools such as Packer Service Pump (PSP) and Optiq are employed. Production forecasting is conducted using high-resolution well testing methods to estimate future production performance. These additional requirements provide a comprehensive framework for evaluating various parameters and ensuring compliance with US EPA regulations for obtaining an injection permit.

These tests are selectively performed in three different zones, including the Injection zone, which is a formation targeted for CO_2 storage; the Confining zone, a layer above the injection zone that prevents upward migration; and the USDW zone, underground sources of drinking water, which is our most important zone and must remain protected under UIC rules and regulations.

Followings are the list of downhole and laboratory tests that are required selectively for each zone.

Downhole Methods:

- *Triple Combo logging: (Gamma Ray, Density amd Resistivity) AIT*: Array Induction Tool
- *CMR-Plus*: Magnetic Resonance Logging
- *FMI*: Formation Micro-Imager
- *MDT*: Modular Formation Dynamics Tester
- *PEx*: Porosity Exponent Logging
- *Power Eco/Flex*: Cement Evaluation Tools
- *Optiq, Pulsar WFL*: Integrity and Cement Bond Tools
- *XL-Rock*: Coring System
- *PSP*: Pump Test Tool
- *DFA*: Downhole Fluid Analysis
- *Quanta Geo and Litho Scanner*: Advanced Borehole Imaging
- *Sonic Scanner*: Stress and Mechanical Measurements

Laboratory Methods:

- *XRD, XRF, FTIR*: Mineralogy analyses
- *RCAL, SCAL, TRA (Tight Rock Analysis)*: Core analyses
- *MICP*: Capillary pressure
- *Core Flooding*: Fluid interaction tests

8.6 Conclusion

The successful deployment of carbon capture and storage (CCS) projects is defined as safe and permanent storage of the predesigned volume of CO_2 with a specified rate following the scheduled timeline. This heavily relies on detailed site screening, selection, and characterization. This chapter has examined the pre-development and pre-injection phases of CCS project lifecycles. We specially focused on emphasizing how geological, regulatory, and social assessments mitigate risks while ensuring regulatory compliance for Class VI well permits.

A key insight is that site screening and selection are iterative, science and data-driven processes. Initial basin-scale evaluations (regional assessments) identify promising storage formations, while site-specific analyses of reservoir properties, seal integrity, and injectivity determine technical feasibility. Advanced techniques

Fig. 8.13 Images designed by Freepik

from seismic surveys and well testing (DST, DFIT) to reservoir modeling enable us to predict CO_2 plume behavior and verify caprock integrity. The chapter also highlighted the growing importance of social characterization, as community acceptance and regulatory alignment are equally critical to project success.

Three major steps required for site qualification were

1. Geological suitability: Prioritizing formations with adequate porosity, permeability, and proven trapping mechanisms.
2. Regulatory readiness: Addressing Class VI requirements on injection pressures, Clean Water Act, Regulatory Liability frameworks, and long-term monitoring.
3. Stakeholder alignment: Integrating local socioeconomic data early in the life of the project to avoid delays.

Case studies from RCSP initiatives demonstrated how this phased approach reduces uncertainties.

Looking ahead, advancements in drilling, completions and monitoring of subsurface geological reservoirs integrated with AI-driven analytics could accelerate site characterization. However, challenges persist in balancing between the cost and benefit of the activities in supporting the economic viability of the project while ensuring safe and sustainable CO_2 sequestration (Fig. 8.13).

References

O. Babarinde, B. Schwartz, J. Meng, S. Kim, J.M. Segura, R.A. Schultz, H. Soroush, An overview of geological carbon sequestration and its geomechanical aspects. Enabling Secure Subsurface Storage in Future Energy Systems, J. M. Miocic, N. Heinemann, K. Edlmann, J. Alcalde, R. A. Schultz (2023)

S. Bachu, D. Bonijoly, J. Bradshaw, R. Burruss, S. Holloway, N.P. Christensen, CO_2 storage capacity estimation: methodology and gaps. Int. J. Greenh. Gas Con. **1**(4), 430–443 (2007). https://doi.org/10.1016/S1750-5836(07)00086-2

W.M. Black, A review of drill-stem testing techniques and analysis. J. Pet. Technol. **8**, 21–30 (1956). https://doi.org/10.2118/589-G

H. Belyadi, E. Fathi, F. Belyadi, Introduction to unconventional reservoirs, in *Hydraulic Fracturing in Unconventional Reservoirs*, (Elsevier, 2017), pp. 1–12

H. Belyadi, E. Fathi, F. Belyadi, *Hydraulic Fracturing in Unconventional Reservoirs: Theories, Operations, and Economic Analysis* (Gulf Professional Publishing, 2019)

T. Chen, L. Huang, Seismicity characterization around the Farnsworth field site for combined large-scale CO_2 storage and EOR. Int. J. Greenhouse Gas Control, **33**, 63–68 (2015). https://doi.org/10.1016/j.ijggc.2014.11.024

I. Ershaghi, *Solved Problems in Well Testing* (Springer International Publishing, 2023)

E. Fathi, G.B. Liwuitekong Dongho, B. Heidari, T.R. Carr, F. Belyadi, I. Bilgesu, Optimizing deep geothermal drilling for energy sustainability in the Appalachian Basin. Sustainability **16**(18), 8053 (2024a). https://doi.org/10.3390/su16188053

E. Fathi, D. Arcentales, F. Belyadi, Impacts of different operation conditions and geological formation characteristics on CO_2 sequestration in Citronelle dome, Alabama. Energies **16**(7) (2023b). https://doi.org/10.3390/en16073191

E. Fathi, M.F. Adenan, N. Moryan, F. Belyadi, H. Belyadi, High-precision single-leak detection and localization in single-phase liquid pipelines using the negative pressure wave technique: an application in a real-field case study. SPE J. (2023a). https://doi.org/10.2118/217455-PA

E. Fathi, T.R. Carr, M.F. Adenan, B. Panetta, A. Kumar, B.J. Carney, High-quality fracture network mapping using high frequency logging while drilling (LWD) data: MSEEL case study. Mach. Learn. Appl. **10**, 100421 (2022)

F. Huq, S.B. Haderlein, O.A. Cirpka, M. Nowak, P. Blum, P. Grathwohl, Flow-through experiments on water–rock interactions in a sandstone caused by CO_2 injection at pressures and temperatures mimicking reservoir conditions. Appl. Geochem. **58**, 136–146 (2015). https://doi.org/10.1016/j.apgeochem.2015.04.006

A. Ingelson, A. Kleffner, N. Nielson, Long-term liability for carbon capture and storage in depleted North American oil and gas reservoirs-a comparative analysis. Energy LJ **31**, 431 (2010)

G. Liu, C. Ehlig–Economides, Consistent Model for Injection and Falloff Pressure Match of Diagnostic Fracture Injection Tests (DFITs). SPE Drill & Compl **34** 345–355 (2019). https://doi.org/10.2118/187191-PA

E. Motaei, T. Ganat, K. Azhar, H. Al-Hadrami, Harnessing deep transient testing for reservoir characterization and CO_2 emission reduction in challenging geological settings. Pet. Res. **9**(3), 380–392 (2024). https://doi.org/10.1016/j.ptlrs.2024.01.014

J.J. Roberts, S.M.V. Gilfillan, L. Stalker, M. Naylor, Geochemical tracers for monitoring offshore CO_2 stores. Int. J. Greenh. Gas Con. **65**, 218–234 (2017). https://doi.org/10.1016/j.ijggc.2017.07.021

T. Rodosta, W. Aljoe, G. Bromhal, D. Damiani, U.S. DOE regional carbon sequestration partnership initiative: new insights and lessons learned. Energy Procedia **114**, 5580–5592 (2017). https://doi.org/10.1016/j.egypro.2017.03.1698

J.V. Rothkirch, O. Ejderyan, Anticipating the social fit of CCS projects by looking at place factors. Int. J. Greenhouse Gas Control, **110**, 103399 (2021). ISSN 1750-5836. https://doi.org/10.1016/j.ijggc.2021.103399

S. Shackley, C. McLachlan, C. Gough, The public perception of carbon dioxide captures and storage in the UK: Results from focus groups and a survey. Clim. Pol. **4**(4), 377–398 (2004). https://doi.org/10.1080/14693062.2004.9685538

L. Whitmarsh, D. Xenias, C.R. Jones, Framing effects on public support for carbon capture and storage. Palgrave Commun. **5**(1), 17 (2019). https://doi.org/10.1057/s41599-019-0217-x

R. Wüstenhagen, M. Wolsink, M.J. Bürer, Social acceptance of renewable energy innovation: An introduction to the concept. Energy Policy **35**(5), 2683–2691 (2007). https://doi.org/10.1016/j.enpol.2006.12.001

P. Xing, J. McLennan, J. Moore, In-situ stress measurements at the Utah frontier observatory for research in geothermal energy (FORGE) site. Energies **13**(21), 5842 (2020). https://doi.org/10.3390/en13215842

J. Zhang, Pore pressure prediction from well logs: Methods, modifications, and new approaches. Earth-Sci. Rev. **108**(1–2), 50–63 (2011)

Chapter 9
Carbon-Negative Solutions and Alternative Storage Options

9.1 Chapter Overview and Objectives

9.1.1 *Overview*

This chapter discusses the carbon-negative strategies and alternatives to underground geological CO_2 storage options with specific attention to analyzing markets, economics, and scalability of these technologies. Evolving negative emission technologies such as DAC, ocean alkalinity enhancement, and Bioenergy with Carbon Capture and Storage (BECCS) are discussed together with natural solutions like afforestation and soil carbon sequestration. This chapter also addresses plastic-related emissions and mitigation strategies and elaborates on the practical challenges and innovations in carbon capture and sequestration using alternative CO_2 storage sites.

9.1.2 *Learning Objectives*

1. Evaluate the economic and environmental feasibility of CO_2 utilization markets and storage permanence.
2. Analyze cost structures and policy incentives for large-scale CCS deployment.
3. Compare alternative CO_2 storage methods.
4. Explain the role of negative emission technologies in achieving carbon negativity.
5. Assess the impact of plastic production on GHG emissions and potential mitigation strategies.

E. Fathi and F. Belyadi, *Subsurface CO_2 Sequestration*,
https://doi.org/10.1007/978-3-032-28296-5_9

9.2 Permanence of CO_2 Storage

Before exploring alternatives to underground CO_2 sequestration, it is important to revisit the concept of permanence of CO_2 storage, as this concept can be used to quantify the quality and effectiveness of alternative CO_2 storage sites. The permanence of CO_2 storage refers to the long-term sequestration of carbon dioxide away from the atmosphere to effectively address climate change. The goal is to keep CO_2 stored for an extended period, ideally for at least 1000 years, without any human intervention. The significance of long-term storage is to prevent CO_2 from re-entering the atmospheric carbon cycle and contributing to climate change. If CO_2 is stored for only short periods, such as days, years, or decades, the problem of excess carbon dioxide is essentially only delayed and, in fact, transferred to future generations. This passing of the problem to future generations, including our children and grandchildren, is not a sustainable solution. To understand this concept, one can draw an analogy with the carbon dioxide adsorbed and stored in a tree. The CO_2 will be out of the carbon cycle if it is stored in the tree; however, when a tree dies and decomposes, the carbon dioxide will be released back into the atmosphere and contribute to CO_2 concentration in the atmosphere. Similarly, if CO_2 storage mechanisms are not designed for long-term permanence, the stored carbon may eventually be released, compromising the effectiveness of carbon dioxide mitigation efforts. To effectively mitigate climate change, it is critical to ensure that CO_2 storage solutions can maintain the sequestration of carbon dioxide for extended periods, minimizing the risk of re-entry into the atmosphere. The effectiveness of the alternative to underground geological storage of CO_2 can be investigated by first identifying the CO_2 storage mechanism and characterizing the storage properties. To ensure the effectiveness of storage, laboratory and field testing under controlled conditions, and long-term stability modeling of the stored CO_2 in the storage site are required. After storage, a routine monitoring and verification plan is required to ensure the effectiveness of the storage. For each stage, Uncertainty and Contingency Planning and proper mitigation planning are necessary for regulatory compliance.

9.3 Carbon Dioxide Utilization—Markets and Applications

The concept of converting all man-made carbon dioxide into valuable products or integrating it into other industrial processes, "Carbon Utilization" holds massive potential. It presents an opportunity to reduce the environmental impact of CO_2 emissions and create a more sustainable approach to managing carbon dioxide. It is also considered an alternative to underground geological CO_2 storage. CO_2 utilization occurs through two primary pathways, including direct use and indirect conversion (Valluri and Kawatra 2022). Direct applications involve implementing captured CO_2 without chemical transformation, such as in enhanced oil recovery (EOR) operations. Conversely, indirect utilization converts CO_2 into new products through chemical processes, yielding valuable outputs like methanol for fuel applications or urea for agricultural fertilizers. Value generated through CO_2 utilization can help offset some of the costs associated with carbon capture and transportation and improve the economic feasibility of CCUS projects. However, challenges remain when different CO_2 utilization projects are pursued, including the permanence of CO_2 storage that is long-term solutions like CO_2 storage in deep saline formations against short-term uses like methanol production.

While CO_2 can be transformed into various products, such as chemicals, fuels, or building materials, the sustainability of the CO_2 source and whether the product truly displaces higher-emission alternatives is a critical concern. The scalability and feasibility of these conversion processes also need careful consideration. The development of efficient and cost-effective conversion technologies is critical to ensure the practical implementation of such solutions. The environmental benefits of CO_2 utilization also depend on other factors, including the energy required for conversion and the carbon footprint of the energy source used.

Additionally, the size of the potential markets for CO_2 is another important factor. It is worth noting that the potential markets for CO_2 can vary depending on geographic location, industrial sectors, and policy frameworks (McLaughlin et al. 2023). In certain regions or industries, the availability of carbon dioxide as a feedstock for specific processes may be more significant. Exploring partnerships between different sectors and adopting innovative technologies can help identify new opportunities and expand the market for CO_2 utilization. While there are diverse applications for CO_2 utilization, including enhanced oil recovery, fertilizer production, food industry (carbonation of beverages), and others, the overall demand for CO_2 in these sectors may not be sufficient to accommodate all the carbon dioxide emissions generated globally. Currently, around 230 million tons (Mt) of CO_2 are utilized each year, primarily in fertilizer production (54%), enhanced oil recovery (35%), and food processing, cooling, and water treatment (11%) (IEA, Energy Technology Perspectives 2020). Scaling up these technologies to match global emissions will require significant advancements. Figure 9.1 shows an approximate comparison of CO_2 utilization in different sectors, where CO_2 EOR takes the majority of the CO_2 utilization market. In this study by NETL, fertilizer production is not included, as CO_2 is generated internally through the hydrogen production process from natural gas and consumed on-site to produce urea (also known as carbamide). Therefore, it's not counted as commercial CO_2 production.

Emerging applications such as CO_2-derived fuels, chemicals, and building materials also offer further potential but are still in the early stages of commercialization. For example, the George Olah Renewable Methanol Plant in Iceland became the world's first commercial facility to convert CO_2 into methanol, recycling 5500 tons of CO_2 annually (Carbon Recycling International 2019). In the chemical industry, companies like Covestro are using CO_2 as a feedstock for polymers, while others, such as Carbon8, are turning CO_2 and waste materials into construction products. These innovations demonstrate the flexibility of CO_2 utilization. However, the energy intensity of these conversions makes the widespread adoption of these technologies challenging.

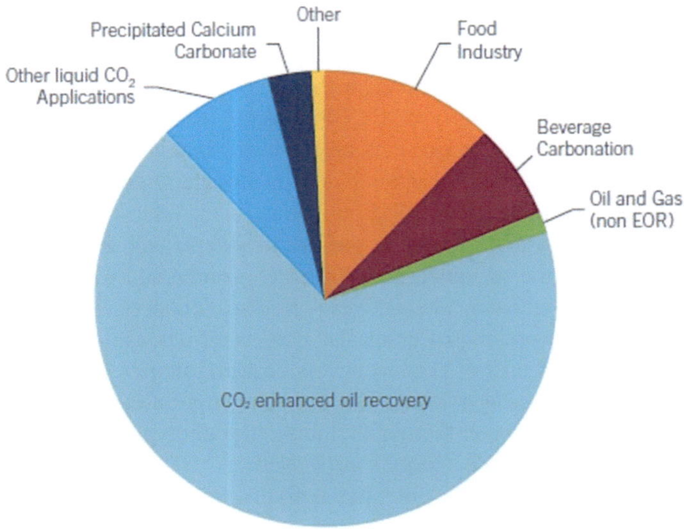

Fig. 9.1 Approximate application of CO_2 utilization in different sectors 9.2. Commercial carbon dioxide uses. www.netl.gov

Despite its potential, the long-term impact of CO_2 utilization on climate change mitigation is difficult to quantify. Many technologies are still in development, and their scalability, economic viability, and net emission reductions require further study. To create a carbon cycle where geological storage of carbon dioxide is no longer necessary, it is important to overcome these limitations and continue advancing research, development, and deployment of carbon capture, utilization, and storage (CCUS) technologies (Litynskia et al. 2013). This will require collaboration between governments, industries, and academia to drive innovation, scale up solutions, and create an enabling environment for the widespread adoption of CO_2 utilization practices. While the idea of converting all man-made carbon dioxide into valuable products and reusing CO_2 in industrial processes is fascinating, it faces significant challenges related to the fate of the captured CO_2 and the size of the potential markets. These limitations are a reminder that a combination of approaches, including carbon capture and storage, is necessary to address the full scale of CO_2 emissions. While utilization alone cannot solve the climate crisis, it can be integrated with permanent underground storage solutions as an effective carbon management strategy.

Enhanced hydrocarbon recovery bridges storage and utilization, creating value by additional oil production while sequestering some CO_2. Techniques like CO_2-EOR and CO_2-ECBM increase fossil fuel production while storing carbon dioxide. In oil reservoirs, CO_2-EOR is primarily applied to heavy oil production. The injected CO_2 reduces oil viscosity when dissolved in oil as a miscible injection, improving recovery rates. This process stores CO_2 while extending reservoir productivity; however, the main goal here is to increase oil recovery while achieving some degree of CO_2

storage capacity. To improve CO_2 storage efficiency and enhance oil recovery, water-alternating-gas (WAG) CO_2 injection can be used. In this approach, a slug of CO_2 is injected to reduce oil viscosity, followed by water injection to improve sweep efficiency and control CO_2 mobility. This process can be repeated using different injection strategies, such as a simple WAG approach, where the same volume of CO_2 and water is injected in each cycle; tapered WAG, where different volumes of CO_2 and water are injected over time; or hybrid WAG techniques. Other CO_2 EOR techniques, such as CO_2 huff-and-puff, can also be applied. In this method, CO_2 is injected into a single well, resulting in pressure maintenance and oil viscosity reduction during the soaking period. After the soaking period, the well is opened for production. This cycle of injection, shut-in/soaking, and production can be repeated multiple times. The selection of the most suitable CO_2 EOR technique highly depends on reservoir characteristics and fluid properties. Coalbed methane recovery operates differently, where coal shows higher affinity to adsorb CO_2 than methane. The CO_2 displaces methane through adsorption to the coal matrix and releases methane; this process simultaneously enhances gas recovery and stores CO_2. Even though these techniques are effective for hydrocarbon production, these formations offer limited storage capacity compared to dedicated geological sites. Between these two enhanced hydrocarbon recovery techniques, ECBM is not as common as CO_2-EOR due to the complexity of the process and the degree of uncertainty associated with highly heterogeneous formations like Coalbed methane.

The global market for CO_2 utilization for enhanced hydrocarbon recovery is relatively small compared to global emissions, estimated at around 0.5% of emissions. By 2040, the market for CO_2 utilization may reach 400 million tons per year, which is significantly smaller than the scale of carbon capture with underground geological storage projected by the International Energy Agency. Figure 9.2 illustrates the hybrid approaches that demonstrate how utilization can support CCS deployment through economic incentives for hydrocarbon production. The figure includes the schematic of the CO_2 storage process in deep saline formations, along with two key methods for CO_2 -enhanced hydrocarbon recovery, including CO_2 -enhanced coalbed methane and CO_2 -enhanced oil recovery.

Carbon utilization helps to reduce CO_2 emissions by repurposing CO_2; however, there is a need to also explore carbon-negative technologies that can go a step further by actively removing and permanently storing atmospheric CO_2 emissions.

9.4 Carbon-Negative Technologies

Carbon-Negative Technologies are innovative approaches and strategies designed to remove more carbon dioxide (CO_2) from the atmosphere than they release. This effectively reduces the overall concentration of CO_2 and contributes to mitigating climate change. These technologies are divided into ocean and land negative emission technologies, new technologies to remove CO_2 from the air, and hybrid approaches. Some of these technologies include:

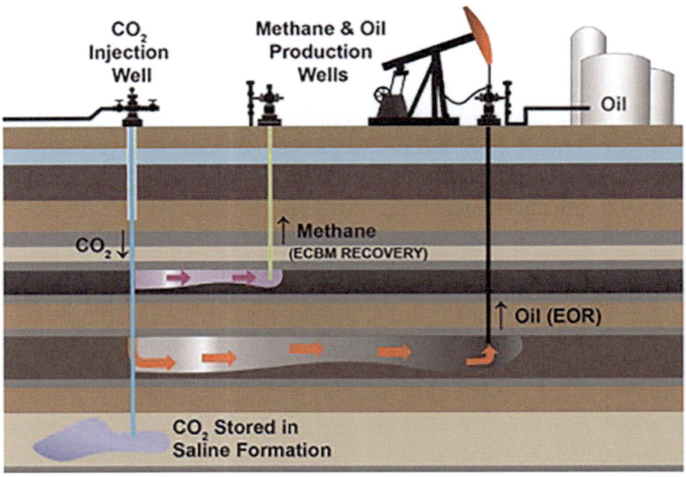

Fig. 9.2 CO_2 storage and enhanced hydrocarbon recovery process modified from https://netl.doe.gov/carbon-management/carbon-storage/faqs/carbon-storage-faqs

1- Ocean treatment to increase alkalinity: This involves adding alkaline substances like crushed limestone to the ocean to enhance its ability to absorb and store CO_2 as bicarbonate ions.
2- Ocean fertilization: Adding nutrients, such as iron, to stimulate phytoplankton growth in the ocean. Phytoplankton absorb CO_2 during photosynthesis and will sequester carbon for long periods after they die and are buried under the ocean floor.
3- Blue carbon habitat restoration focuses on restoring coastal ecosystems like mangroves, seagrasses, and salt marshes. These habitats have high carbon sequestration potential and can store significant amounts of carbon in their biomass and sediments.
4- Enhanced weathering on land: This technique accelerates the natural process of rocks weathering to absorb CO_2. It typically involves grinding and spreading certain rocks on land surfaces to increase their surface area for faster CO_2 absorption.
5- Afforestation and reforestation: Planting new forests (afforestation) or restoring existing forests (reforestation) effectively absorb CO_2 through photosynthesis and store it in trees and forest ecosystems.
6- Biochar with carbon storage in soil: Biochar, produced from biomass, when added to soil, enhances soil fertility and long-term carbon storage.
7- Soil carbon sequestration: Various agricultural practices like conservation tillage, cover cropping, and rotational grazing can increase carbon sequestration in soil by promoting the accumulation of organic matter.

8- Direct air capture: Using large-scale machines to directly capture CO_2 from ambient air. The captured CO_2 can then be stored underground or utilized in various industrial processes.
9- Bioenergy with CCS: Combining bioenergy production (burning biomass) with Carbon Capture and Storage (CCS) to capture and store CO_2 emissions generated during the bioenergy process.
10- Converting CO_2 to long-lived products: Certain technologies aim to capture CO_2 and convert it into stable, long-lived products like building materials or mineral carbonates, which can be stored or utilized in various industries.

These carbon-negative technologies hold significant promise in addressing climate change by actively removing CO_2 from the atmosphere and enabling net-negative emissions. However, their large-scale viability depends on critical factors such as technology readiness, permanence of carbon storage, atmospheric CO_2 removal capacity, cost-effectiveness, and full lifecycle emissions. A thorough assessment of these aspects is required to determine their true potential in global climate mitigation strategies.

9.4.1 Ocean Treatment to Increase Alkalinity

Oceans are the Earth's largest natural carbon dioxide sink. Approximately 30% of atmospheric CO_2 emissions are absorbed by oceans and dissolve in ocean water (Gruber et al. 2019). This uptake occurs through the diffusion of CO_2 from the atmosphere into the ocean, where it reacts with seawater to form carbonic acid. Carbonic acid then dissociates into bicarbonate (HCO_3^-) and hydrogen ions (H^+). The increase in hydrogen ions lowers the ocean's pH, a process known as ocean acidification. As ocean pH decreases, the water becomes more acidic, which in turn reduces its capacity to absorb additional CO_2 from the atmosphere. This weakens the ocean's role as a carbon sink over time. Enhancing oceanic CO_2 uptake by implementing natural processes or engineered solutions provides a pathway to mitigate atmospheric CO_2 levels and climate change.

To enhance the uptake of CO_2 by the oceans, one approach is to introduce an alkaline substance or base to the water. In this ocean treatment method, a large quantity of a base, such as lime (calcium oxide or CaO), is spread over extensive areas of the ocean. Figure 9.3 depicts the schematic of the ocean alkalinity enhancement, where a massive amount of alkalinity is added to the ocean using ship deployment. The purpose is to increase the ocean's capacity to absorb CO_2 and neutralize the effects of ocean acidification. When a base reacts with an acid, they neutralize each other, creating a more balanced pH level. While this approach shows potential for generating negative emissions, it is important to consider the potential disadvantages of this technique. The large-scale addition of alkaline substances to the ocean raises important concerns from both environmental and legal perspectives. This approach may lead to unintended and potentially irreversible ecological consequences. The full Project

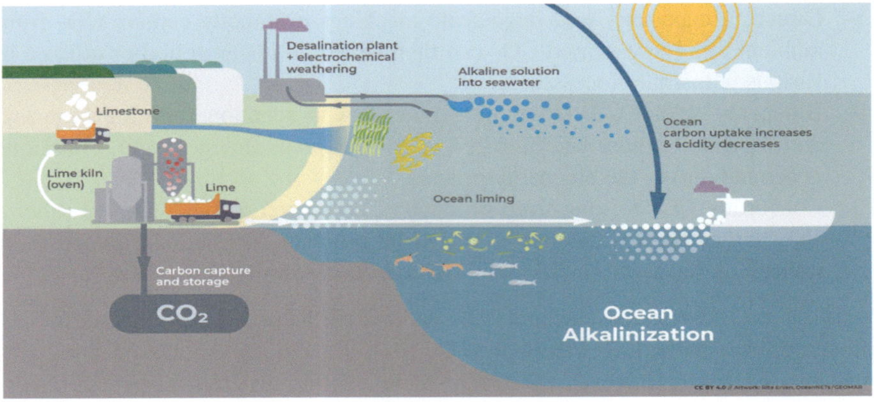

Fig. 9.3 Schematic of ocean alkalinity enhancement (Image: CC By 4.0. Acknowledgement: Rita Erven, GEOMAR/OceanNETs)

Boundaries and Objectives of its environmental impacts are poorly understood and are not truly quantified. Given the scale and non-reversible nature of such interventions to the ecosystem of the ocean, it is essential to thoroughly evaluate potential ecological risks and ensure strict compliance with international environmental regulations. Additionally, the effectiveness of this method is uncertain. It remains unclear how much alkaline material would be required to meaningfully restore ocean CO_2 absorption capacity. A detailed cost-benefit analysis is also needed, as lime extraction, grinding, transportation, and deployment are extremely energy intensive, and associated CO_2 emissions from this process also need to be quantified, which might exceed the improvement in CO_2 adsorption via ocean treatment.

9.4.2 *Ocean Fertilization*

Like land-based plants, marine plants also absorb CO_2 during the process of photosynthesis. The concept of ocean fertilization aims to artificially enhance the photosynthetic activity of microscopic marine plants like phytoplankton to enhance CO_2 uptake and potentially accelerate the removal of atmospheric CO_2. This can be done by adding nitrogen/phosphorous fertilizers in nutrient-poor parts of oceans, adding iron sulfates to enhance the phytoplankton growth, or Artificial upwelling in the ocean, where pumping the nutrient-rich deep ocean water upward to the shallower depths results in stimulation of phytoplankton growth and enhances CO_2 uptake. This process is also called biological pumping, where deep ocean water (100–1000 m depth) rich in nitrates, phosphates, and micronutrients (e.g., iron) is pumped to the surface to act as an internal nutrient supply for phytoplankton blooms rather than being prepared for shipping and dumping into the ocean. Figure 9.4

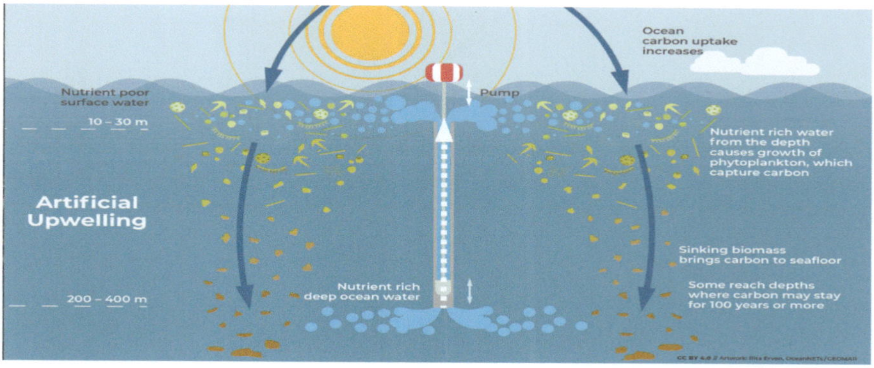

Fig. 9.4 Artificial upwelling of highly nutrient-rich deep ocean water to the surface for a phytoplankton bloom. Image: CC By 4.0. Acknowledgement: Rita Erven, GEOMAR/OceanNETs https://www.oceannets.eu/infographics/

shows the schematic of providing nutrient-rich water from deep ocean water to feed phytoplankton in the surface water, enhancing their CO_2 absorption capacity.

This process has several advantages, including supporting commercial fishing by adding potential ecosystem benefits, supporting ocean productivity, and increasing the fish supply.

However, despite several experiments conducted (Hernández et al. 2018), there are significant concerns regarding the large-scale implementation of ocean fertilization, as less than 1% of carbon stored by phytoplankton reaches the ocean floor for long-term storage, and the rest will be recycled at the surface (Ducklow et al. 2001). There is also a risk of CO_2 release back to the atmosphere even before mineralization (Robinson et al. 2014).

There are also concerns regarding the ecological disturbance by the production of excess harmful algal blooms and deoxidation of the ocean, expanding the dead zones in the ocean (Parry 2012). Key considerations also include the immense quantity of fertilizers required, the regulatory framework under international law to govern such interventions, and the potential unintended consequences of disrupting the natural marine carbon cycle. These challenges emphasize the need for further research and careful evaluation before considering widespread adoption of ocean fertilization as a carbon-negative technology. Figure 9.5 shows the schematic of adding nutrients like iron sulfate to grow phytoplankton that results in enhanced CO_2 capture, generating biomass that sinks to the sea floor and CO_2 permanently stored.

9.4.3 *Blue Carbon Habitat Restoration*

"Blue carbon" refers to the carbon that is stored in coastal or marine ecosystems. These ecosystems, such as mangroves, salt marshes, and seagrasses, have the unique

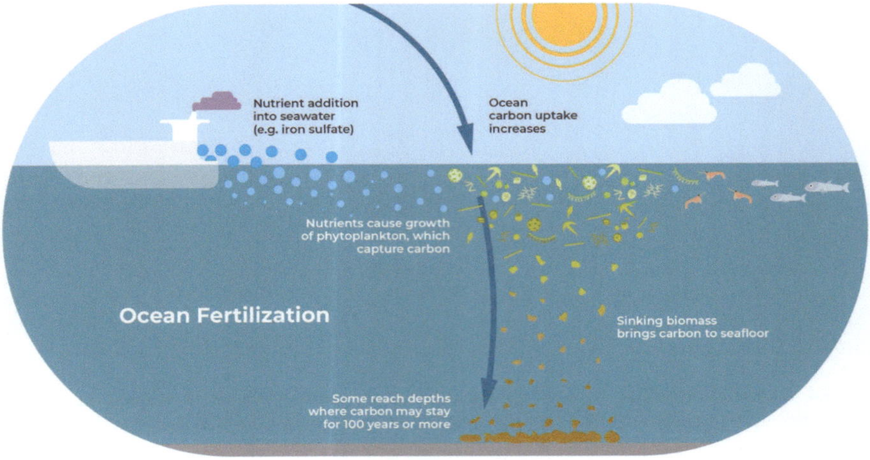

Fig. 9.5 Schematic of ocean fertilization process leading to permanent CO_2 sequestration at sea floor. Image: CC By 4.0. Acknowledgement: Rita Erven, GEOMAR/OceanNETs. https://www.oceannets.eu/infographics/

ability to draw in and sequester CO_2 from the atmosphere. This property makes them valuable resources in mitigating climate change (Bromberg Gedan et al. 2009). Mangroves are tropical coastal forests dominated by mangrove trees. They are found in bays, where rivers meet the sea. They provide ecosystem services such as protecting coastlines from erosion and providing habitats for diverse marine life. Mangroves are estimated to sequester 6–8 metric tons of carbon dioxide equivalent per hectare each year. However, 30–50% of the mangroves have been lost in the past 50 years due to natural causes, repurposing for agricultural use, such as land or shrimp farming. Salt marshes, on the other hand, are coastal wetlands that experience periodic flooding and draining by saltwater brought in by tides (Mcowen et al. 2017). They serve as important environments for various plant and animal species while effectively capturing and storing carbon dioxide. They have been found to have a similar capacity for CO_2 storage as mangroves. Studies by Crooks et al. 2011 and Duarte et al. 2008, indicated that the global coverage of saltmarsh areas has been lost by 25–50%. Some were lost due to repurposing for agricultural activities, industrial usage, or other natural events (Gedan et al. 2009).

Seagrasses are underwater flowering plants with extensive root systems that can store carbon dioxide in deep seabed soils (Montero-Hidalgo et al. 2023). Even though the total coverage of Seagrasses is limited to less than 0.2% of the ocean floor, they are significant in CO_2 sequestration by roughly 27.4 million metric tons of carbon each year (Duarte et al. 2013). Studies have shown that Seagrasses have been lost by around 29% globally due to pollution and coastal development (Waycott et al. 2009).

Restoring these habitats can significantly contribute to climate change mitigation and coastal protection. Conserving and restoring these coastal ecosystems is a

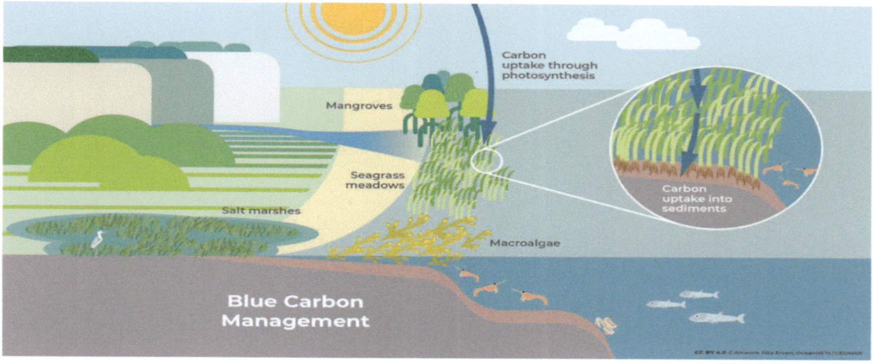

Fig. 9.6 Schematic of blue carbon habitats including Mangroves, Seagrass meadows, and salt marshes and their carbon uptake processes Image: CC By 4.0. Acknowledgement: Rita Erven, GEOMAR/OceanNETs. https://www.oceannets.eu/infographics/

suggested strategy to maintain their capacity to remove CO_2 from the air and help mitigate climate change. Protecting these blue carbon habitats can preserve their carbon storage potential and the ecological benefits they provide. Successful restoration examples include Mikoko Pamoja (Kenya), which is a community-driven project by Lingocal, including seedling planting and forest monitoring (https://mikokopamoja.org/).Other successful examples of restoration of seagrass are the one performed over the last 20 years in Virginia's coastal bays, where scientists and volunteers have seeded more than 70 million eelgrass seeds across four once-lifeless lagoons (Orth et al. 2020). Figure 9.6 shows the schematic of blue carbon habitats, including Mangroves, Seagrass meadows, and salt marshes, and their carbon uptake processes.

9.4.4 Enhanced Weathering

Weathering is a natural process that breaks down rocks and contributes to carbon absorption. Currently, weathering absorbs approximately 0.3% of global fossil fuel emissions (Beerling et al. 2018). The process starts with rainwater absorbing CO_2 from the atmosphere during its descent. When this acidic rainwater meets rocks, it gradually breaks them down, generating bicarbonate ions. Over time, these bicarbonate ions are carried by water into the oceans, where carbon is sequestered on the seafloor or dissolved in ocean water. Enhanced weathering aims to accelerate this process. Crushing rocks into a fine powder and spreading that over land can accelerate natural weathering action. This resulting powder can be applied to large agricultural areas and utilized in conjunction with soil microbes to expedite the chemical reactions involved (Bach et al. 2019). This approach enhances the rate of carbon absorption from rain and offers potential for carbon mitigation in agricultural settings. Figure 9.7 shows the schematic of the enhanced weathering process where

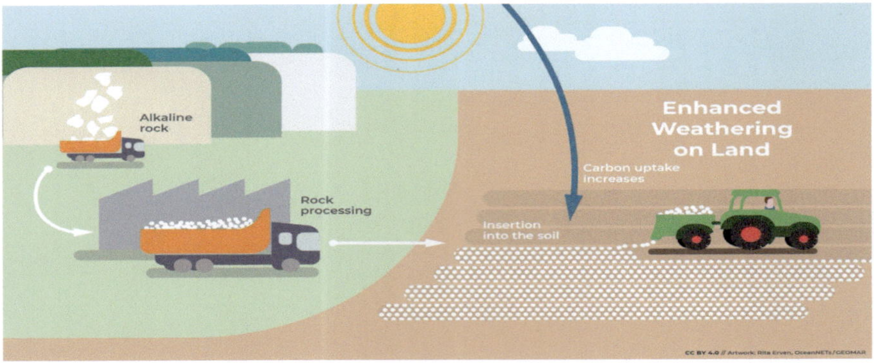

Fig. 9.7 Schematic of enhanced weathering Image: CC By 4.0. Acknowledgement: Rita Erven, GEOMAR/OceanNETs https://www.oceannets.eu/infographics/

rock powders are trucked and distributed in farmland to increase the CO_2 adsorption capacity of the soil. The mineral dissolution products were then transferred to the ocean for permanent storage. This method offers several advantages, particularly its focus on CO_2 sequestration through mineralization, which is a process that enables permanent carbon storage. One significant co-benefit of enhanced weathering is its compatibility with agricultural practices. The application of crushed silicate rocks, such as basalt, does not interfere with normal farming operations (Kristen Coyne 2023). In fact, it may improve soil quality by releasing beneficial nutrients like potassium (K^+), magnesium (Mg^{2+}), and silicon (Si), potentially enhancing soil fertility and crop yields.

However, despite these advantages, the technology remains immature for large-scale commercial deployment. A major limitation is the slow kinetics of the reaction between rock powder and atmospheric CO_2, which significantly constrains the amount of carbon that can be sequestered annually. In addition, the process is highly energy-intensive, requiring significant resources for mining, grinding, and transporting rock material. The net carbon benefit is uncertain, as the CO_2 emissions associated with these operations have not been comprehensively quantified. Lastly, while the addition of rock powder can enrich soils, it also raises environmental concerns. The potential release of heavy metals and the risk of soil and water contamination must be carefully evaluated before widespread implementation of this technology. Depending on the location where the technology is planned to be deployed, there also might be federal, state, and local regulations that require further analysis and quantification of rock powders that could potentially limit the widespread addition of crushed rock in soil.

9.4.5 *Afforestation and Reforestation*

Afforestation and reforestation are two approaches to increasing forest cover and restoring degraded areas. Afforestation involves planting trees in areas where there were no trees before, while reforestation focuses on restoring areas where trees have been cut down or degraded (Cho et al. 2019). Young forests typically store more CO_2 per hectare per year due to their rapid growth and high rates of biomass accumulation. In contrast, mature forests absorb less CO_2, reflecting slower growth but sustained long-term carbon storage in biomass and soils (Pan et al. 2011).

Both afforestation and reforestation are considered feasible options for addressing climate change, but they also come with certain drawbacks and uncertainties. Reforestation is widely regarded as beneficial due to its positive ecological impact, especially when native tree species are used. However, one potential challenge for afforestation is the availability of suitable land and the need to consider biodiversity when selecting tree species. Careful planning and consideration of ecological factors are necessary to ensure successful afforestation and reforestation efforts. While the common goal of both afforestation and reforestation is to plant enough trees in an area to call it a forest, afforestation specifically relates to areas where there have never been trees before, while reforestation is the planting of new trees where most trees have been felled. Figure 9.8 shows sample plot landscape of artificial afforestation at different times, where the evolution of artificial afforestation for 5 different cases in different time intervals including view (a) and (b) after 2 years, (c) and (d) after 5 years, (d) and (e) view after 11 years, (f) and (g) view after 15 years and finally (i) and (j) view after 31 years (Yin et al. 2022).

Afforestation and reforestation present a proven low-tech solution for carbon sequestration. It is natural carbon capture systems that do not require expensive infrastructure. In addition to storing CO_2, this process also offers significant benefits, including soil stabilization and water regulation. Economically, tree-based carbon removal is a relatively cost-effective and more sustainable solution compared to most engineered alternatives like direct air capture (DAC). However, with all the advantages it comes with some challenges, including the long time required to complete the afforestation or reforestation, typically 20–50 years. The permanence of CO_2 is also a concern, as it will not store CO_2 for 1000 years. Events such as wildfires, pests, and deforestation can reverse these gains, as seen in the 2023 Canadian wildfires, which released an estimated 1.7 gigatons of CO_2 based on (GFAS) v1.2 data, 2023. Finally, the recent studies reveal that up to 90% of rainforest carbon credits issued by Verra may have overstated their actual carbon benefits due to inadequate monitoring and verification, which highlights challenges in quantifying the capacity and effectiveness of this process (The Guardian/Die Zeit/Source Material (2023)).

Fig. 9.8 Evolution of artificial afforestation view **a** and **b** After 2 years **c** and **d** View after 5 years **d** and **e** View after 11 years, and **f** and **g** view after 15 years, and finally (i) and (j) view after 31 years. (Yin et al. 2022)

9.4.6 *Direct Air Capture*

Direct Air Capture (DAC) is a technology that involves the development of devices designed to extract carbon dioxide directly from the atmosphere and store it in a permanent manner. Two distinct methods currently employed for capturing carbon dioxide (CO_2) from the atmosphere include solid and Liquid direct air capture. Solid DAC uses solid adsorbents at ambient to low pressure and medium temperatures (80–120 °C); these are flexible designs that make easy deployment, transportation, and maintenance of the system (Zentou et al. 2025). The solid DAC is a scalable design that broadens its application. The challenges with solid DAS are the slow CO_2 capture and release and more frequent maintenance due to filter clogging and a decrease in efficiency. A great example of a commercial solid DAC project is the ORCA located in Iceland.

Liquid DAC, on the other hand, relies on an aqueous basic solution such as potassium hydroxide and operates at high temperatures (300–900 °C) to release captured CO_2. The liquid DAC offers much higher CO_2 capture efficiency as it operates in a continuous loop and takes advantage of mature chemical handling that has been practiced for decades in the chemical industry. However, as it operates at high temperature, it's much more energy intensive and requires large volumes of water. The facility is also much bigger and requires large reactors. One great example of liquid DAC is the STRATOS project by Carbon Engineering (Chopra et al. 2026).

In general, the process of capturing CO_2 from ambient air poses certain challenges due to the low concentration of CO_2 molecules present in the air (Mikulčić et al. 2019). Roughly one molecule of CO_2 needs to be separated from approximately 2500 molecules of air, making it an energy-intensive process compared to capturing CO_2 from concentrated sources. As a result, there are financial costs associated with implementing DAC technology. However, one notable advantage of Direct Air Capture is that it requires minimal land usage. Unlike other carbon mitigation strategies that rely on extensive land areas or forests, DAC technology can be implemented in compact spaces. This aspect makes it a particularly attractive option in scenarios where land availability is limited or when alternative land use is not feasible. Despite the energy intensity and financial costs, Direct Air Capture introduces a practical option for actively removing CO_2 from the atmosphere. It provides a complementary approach to other carbon reduction measures and has the potential to significantly contribute to mitigating climate change by directly addressing atmospheric CO_2 levels. Twenty-seven Direct Air Capture (DAC) plants are currently operational worldwide, collectively capturing nearly 0.01 million tons of CO_2 annually (https://www.iea.org, Budinis et al. 2023).

Figure 9.9 illustrates a schematic of the solid sorbent direct air capture (DAC) process powered by renewable solar energy. In this system, ambient air is drawn in using large fans and passed through filters that selectively capture CO_2 and release the carbon dioxide-free gas back to the air. Once saturated, the filters are replaced with clean filters to continue the CO_2 capture processes. The removed saturated filters are then heated to release the captured CO_2, after which they are cleaned and prepared for reinstallation. The released CO_2 is then compressed to a supercritical state for either permanent geological storage or for use in various industrial applications, such as in the food and beverage industry.

9.4.7 Bioenergy with CCS

Bioenergy with Carbon Capture and Storage (BECCS) combines bioenergy production from biomass, such as agricultural residues, forestry byproducts, or dedicated energy crops, with carbon capture and storage (CCS) technology. As a negative emissions technology, BECCS benefits from the natural ability of biomass (e.g., trees, crops, or algae) to absorb CO_2 from the atmosphere through photosynthesis. When this biomass is combusted or converted to produce energy (e.g., electricity or biofuels), the resulting CO_2 emissions are captured and permanently stored in deep underground geological formations (Davoodi et al. 2023). Because the CO_2 absorbed during growth is removed from the atmosphere and not re-released, the overall process results in net-negative CO_2 emissions.

Figure 9.10 presents a conceptual illustration of two BECCS pathways: biomass absorbing atmospheric CO_2 and subsequently being used for either electricity generation or biofuel production. In both cases, the CO_2 released during conversion

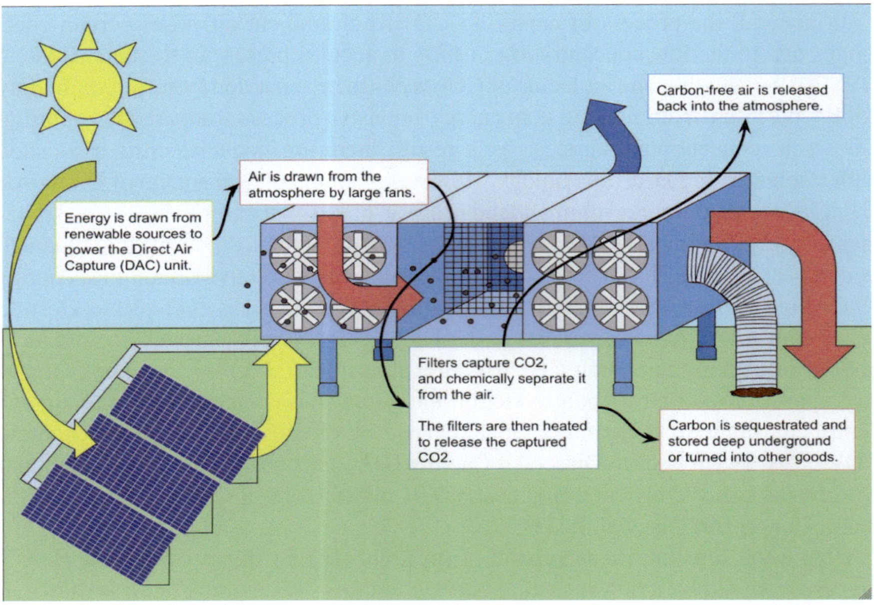

Fig. 9.9 Direct Air Capture (DAC) Cartoon by Steven Gandhi

is captured and stored underground. Typically, 85–95% of CO_2 emissions from bioenergy plants can be captured through these processes.

While BECCS offers the dual benefit of generating renewable energy and achieving carbon removal, its deployment must be carefully managed to avoid adverse environmental and social impacts. Key considerations include sustainable biomass sourcing, responsible land and water use, and the protection of biodiversity (Zheng and Xu 2014). Like afforestation, BECCS poses challenges related to land competition and ecosystem impacts. However, like Direct Air Capture, BECCS offers

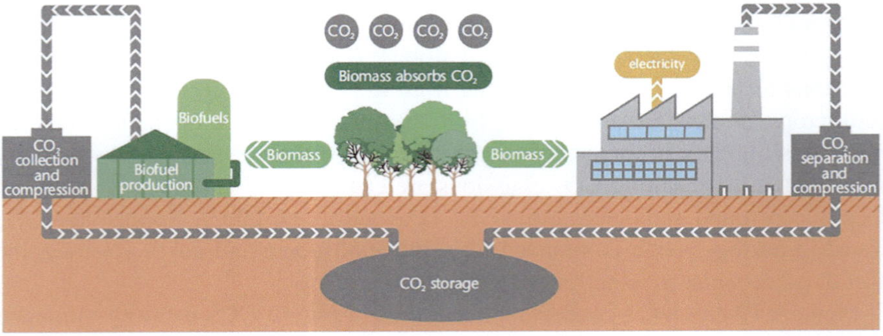

Fig. 9.10 Schematic view of Bio energy with carbon capture and sequestration. Source: IEA technology roadmap, delivering sustainable bioenergy (2017)

the critical advantage of permanent geological CO_2 storage, effectively removing carbon from the atmospheric cycle.

As of now, approximately two million metric tons of biogenic CO_2 are captured annually, mainly from bioethanol production. However, with around 20 new BECCS facilities announced since 2022, global capture capacity is projected to grow to 15 million metric tons of biogenic CO_2 per year (Mathilde Fajardy and Carl Greenfield 2023 www.iea.org).

9.4.8 Biochar with Carbon Storage in Soil

Biochar is a form of charcoal generated by subjecting organic materials like wood, crop residues, or animal manure to thermal decomposition in a process called pyrolysis (Zakaria et al. 2023). This method takes place in the absence of oxygen, involving the heating of biomass at relatively low temperatures. This results in the carbon from the biomass being locked into the biochar, which has a slow decomposition rate. Biochar is then applied to soil, providing several benefits for soil health and regeneration. Figure 9.11 shows the biochar produced from woody biomass.

One advantage of biochar is its ability to improve soil fertility and structure, enhancing nutrient retention and water-holding capacity. It can also promote microbial activity and reduce soil erosion. Additionally, biochar has the potential to sequester carbon in the soil for an extended period, contributing to carbon storage. The advantage of biochar is its ability to remain stable in soils and permanently store absorbed CO_2. Figure 9.12 shows the role of biochar in carbon sequestration

Fig. 9.11 Biochar produced from woody biomass. By Oregon Department of Forestry

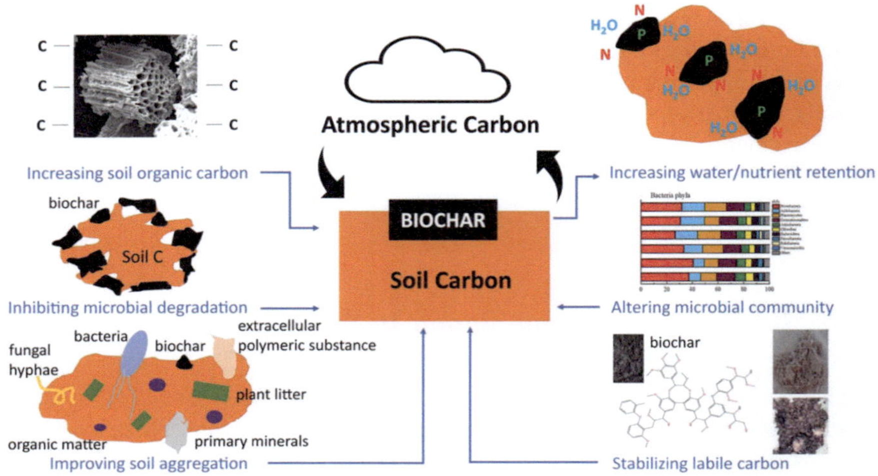

Fig. 9.12 Different biochars' mechanisms on soil health and carbon dioxide sequestration (Li and Tasnady 2023)

in soil by multiple processes, including increasing the soil organic content that helps increase the absorption capacity of soil, inhibiting microbial degradation to help stabilize the stored CO_2, stabilizing labile carbon, improving soil aggregation, and water and nutrient retention. These mechanisms work synergistically to increase long-term carbon storage in soils, with their combined influence leading to greater overall carbon retention. It is estimated that 50 to 90% of biomass carbon is stabilized in solid form in biochar (Li and Tasnady 2023).

However, there are still uncertainties surrounding the long-term stability of biochar in soils. Some concerns revolve around the potential for biochar to dissolve over time and be transported to water bodies like rivers and wetlands, eventually reaching the oceans. This could limit its effectiveness as a permanent carbon storage solution. There are also concerns about land and biomass competition where large-scale deployment needs to compete with agriculture and the food industry. It is also possible that poor-quality biochar shows a reverse impact on soil health due to the generation of toxic components. To fully employ the benefits of biochar and ensure its sustainable use, further research is needed to better understand its interactions with soil and the environment. Addressing these uncertainties in the application of biochar can potentially play a valuable role in mitigating climate change and supporting soil health.

Fig. 9.13 Tidal wetlands of Pinckney Island National Wildlife Refuge By Gentry George, U.S. Fish and Wildlife Service—http://www.public-domain-image.com/public-domain-images-pictures-free-stock-photos/wallpapers-public-domain-images-pictures/a-scenic-view-of-the-landscape-at-the-pinckney-national-wildlife-refuge.jpg, Public Domain. https://commons.wikimedia.org/w/index.php?curid=24919954

9.4.9 Soil Carbon Sequestration

The concept suggests implementing straightforward modifications to farming practices to transform agricultural soils, which have suffered degradation due to modern farming methods, into carbon sinks. Actions like restoring grasslands and establishing wetlands and ponds have the potential to absorb significant quantities of carbon dioxide from the atmosphere. Nevertheless, there is uncertainty regarding how soils would respond to a warming climate and their capacity to store carbon, especially in regions where water availability could be compromised. Further investigation is required to better understand these dynamics and potential risks. Figure 9.13 shows the Pinckney Island National Wildlife Refuge (NWR) identified as a significant site for soil carbon sequestration, especially within its salt marshes and tidal wetlands.

9.4.10 Converting CO_2 to Long-Lived Products

Converting CO_2 into long-lived products is another way of storing carbon and reducing atmospheric CO_2 levels (Varshney et al. 2022). One approach involves utilizing plant-based materials in construction, effectively sequestering carbon in durable structures. Timber and bamboo can be employed as structural elements, while wool can provide insulation. These alternatives serve as environmentally friendly substitutes for traditional construction materials like steel and concrete. Furthermore, advancements in technology have led to the development of CO_2 -absorbing cements and construction materials. These innovative materials, such as aggregates, can actively sequester carbon dioxide for extended periods, potentially spanning decades or even centuries. By incorporating these carbon-negative materials into construction practices, we can contribute to long-term carbon storage and mitigate the impacts of climate change. Other examples of converting carbon dioxide to useful products are to produce Diamond from CO_2 captured from direct air capture, such as those commercially produced by the AETHER company (https://aetherdiamonds.com/), or using small post-combustion CO_2 capture technology to capture CO_2 and convert that to soap, such as those provided by the Clean O2 company. There are significantly more applications of converting captured CO_2 to useful products; however, one major issue remains the permeability of CO_2, where different products show significantly different sequestration time frames.

9.5 The Potential Capacity of Negative Emission Technologies

The potential capacity of negative emission technologies varies across different options and their maturity levels. Currently, some options are not included in the Intergovernmental Panel on Climate Change (IPCC) pathways due to their limited development. Agriculture, forestry, and land-use measures have the potential to contribute between 8 and 14 billion tons of CO_2 per year by 2050 (IPCC, Sixth Assessment Report, 2021). This encompasses various practices related to land management and carbon sequestration. Bioenergy with CCS (BECCS) has the potential to contribute from 0.5 to 11.3 billion tons of CO_2 per year by 2050. This approach combines biomass energy production with carbon capture and storage.

Direct Air Capture is considered a "wild card" technology, as its capacity is not constrained by identified limitations. In theory, it has the potential for almost infinitely scalable carbon removal. However, the cost of Direct Air Capture technologies could influence the overall global cost of climate change mitigation efforts.

It's important to note that these figures should be considered in the context of the 33 billion tons of global energy-related CO_2 emissions in 2021, indicating the scale of the challenge in achieving significant negative emissions.

9.6 The Impact of Plastics on Greenhouse Gas Emissions

Over the past 70 years, global plastic production has significantly increased from two million tons of plastic to more than 450 million tons annually (OECD 2022). Only in the past 20 years, plastic production has doubled, indicating the increasing demand

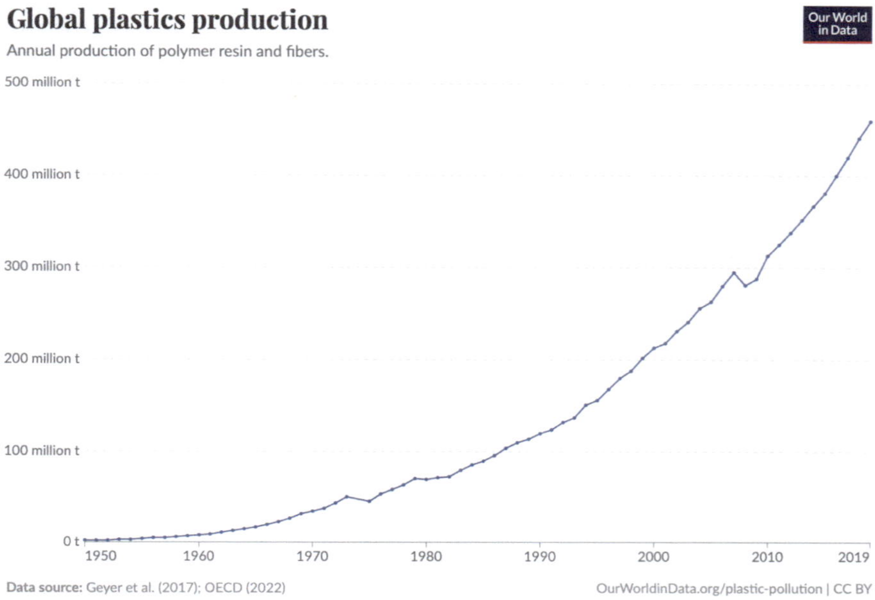

Fig. 9.14 Global plastics production between 1950 and 2019. Geyer et al. (2017), OECD (2022)—with major processing by Our World in Data

for plastic use on a daily basis. Plastic has many applications in different industries, from construction and healthcare to food packaging and electronics. This wide range of applications necessitates efficient plastic waste management. Figure 9.14 shows an exponential increasing trend of annual global plastic production from the 1950s to 2019 (Geyer et al. 2017).

The entire life cycle of plastics from production to disposal generates significant greenhouse gas (GHG) emissions. According to the Organization for Economic Co-operation and Development (OECD 2022), global plastics were responsible for 1.8 billion tons of CO_2-equivalent emissions in a recent assessment. Most plastics' carbon dioxide emissions are produced during the manufacturing phase. Figure 9.15 shows the greenhouse gas emissions from the plastic life cycle from 2015 to 2019 (Ritchie et al. 2023). It is estimated that 3.3% of global emissions are produced during the life cycle of plastic production (Jones et al. 2023).

9.6.1 Strategies for Mitigation

The impact of plastics on greenhouse gas emissions is significant due to their inefficient management and disposal practices. Here are the key points: A large portion of plastic waste, roughly more than 70%, ends up in landfills or escapes into the

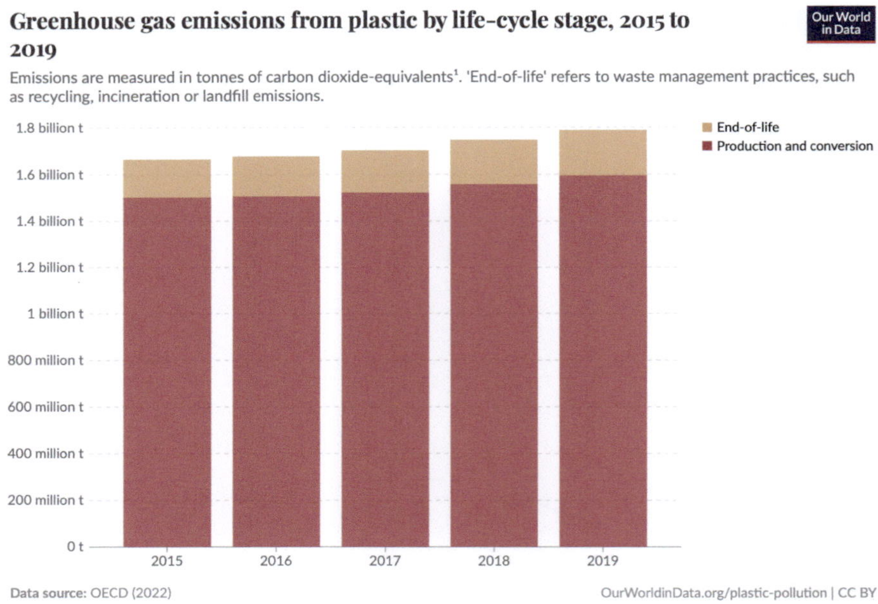

Fig. 9.15 Greenhouse gas emissions from plastic by life cycle stage (OECD (2022)—processed by Our World in Data)

natural environment. This amount significantly contributes to environmental pollution. Figure 9.16 shows the distribution of plastic waste around the world between 2000 and 2019, where landfill and mismanagement take the bulk part, as recycling is a minute part of the waste management portfolio (OECD 2023). Figure 9.16 also shows that in 2019 only 9% of the world's plastic waste was recycled, while the vast majority, 50%, ends up in landfills, where it sits indefinitely and slowly breaks down; 20% was mismanaged, which is neither recycled, safely incinerated, nor properly contained. Most of the plastic waste often pollutes rivers, lakes, and oceans. Waste management practices vary dramatically by region. For example, Europe relies heavily on incineration for plastic disposal, while the United States sends 75% of its plastic waste to landfills. These differences highlight the urgent need for better recycling systems, waste policies, and global cooperation to tackle plastic pollution effectively (OECD 2023).

When plastics are recycled or transformed into other products, the carbon they contain is temporarily embedded in the economy. The fate of this carbon depends on the subsequent use or disposal of the recycled plastic. If plastics are incinerated after recycling, the carbon in them is released back into the atmosphere as CO_2, which directly contributes to greenhouse gas emissions. Plastics that are not recycled and are buried in landfills can temporarily store carbon in the upper layers of the Earth. However, there is no guarantee that this carbon will remain stored for the minimum term of 1000 years. When plastics are not recycled or properly disposed of, they

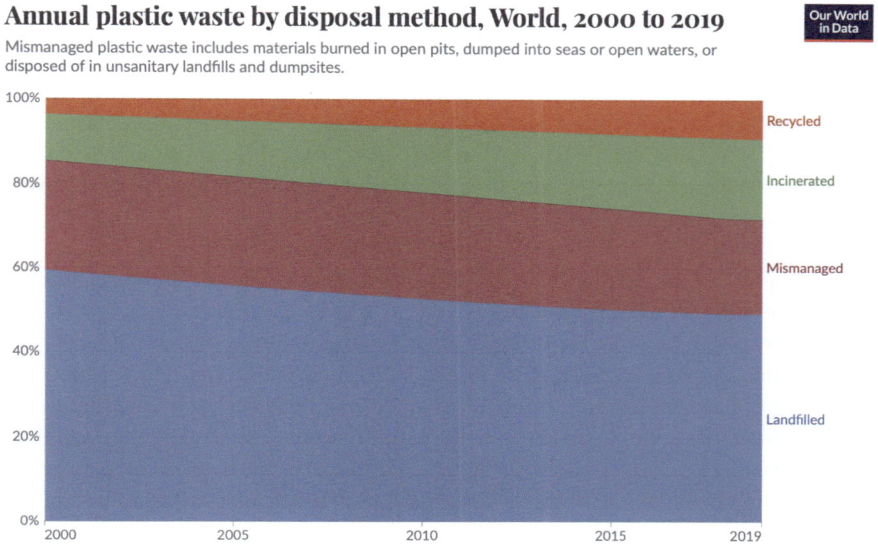

Fig. 9.16 Distribution of plastic waste around the world between 2000 and 2019 (OECD, 2023, processed by Our World in Data)

accumulate in the natural environment, where they degrade slowly over time and release CO_2 back into the atmosphere.

The improper management of plastic waste meaningfully contributes to carbon emissions and environmental pollution. Therefore, there is a critical need for effective recycling and waste management strategies. This is the most important mitigation strategy to reduce the negative impacts of plastics. The waste management strategies can focus on expanding the recycling infrastructure, as less than 9% of plastics are properly recycled. It is also important to scale up the collection systems to properly and safely collect and separate the plastics for recycling. Advancing the disposal technologies to increase the efficiency and reduce the carbon footprint is another topic that requires special attention, as incineration will simply release greenhouse gases into the atmosphere. This can be done by pyrolysis, where plastic is converted to fuel (Hasan et al. 2025), or by gasification with lower emissions for synthesis gas production (Saebea et al. 2020). Shift in using renewable energies during plastic production and disposal also significantly reduces the carbon footprint in the life cycle of plastic. This can also improve if coupled with carbon capture technologies in production and disposal facilities. Plastic production from alternative materials such as biodegradable plant-based polymers can also significantly help to mitigate plastic pollution (Samir et al. 2022). Governments can also contribute to reducing plastic pollution by imposing taxes or banning the use of items like plastic bags, containers, etc. They can also help with initiatives on public awareness campaigns or implement extended producer responsibilities where the manufacturers are held accountable for plastic disposal management.

9.7 Conclusion

This chapter has explored a diverse portfolio of carbon-negative strategies that complement traditional geological CO_2 storage. These solutions, ranging from direct air capture (DAC) and bioenergy with carbon capture and storage (BECCS) to nature-based approaches such as afforestation, soil carbon sequestration, and blue carbon restoration, demonstrate that effective climate mitigation requires a multidimensional approach.

The main consideration is the trade-off between scalability, permanence, and cost across different methods. Engineered solutions like DAC offer precise, measurable CO_2 removal but face high energy demands and economic barriers. In contrast, natural climate solutions (forests, soils, wetlands) provide co-benefits for biodiversity and local communities but face challenges like high cost and uncertainties in the quantification of their impact in reducing emissions and unseen environmental consequences. The chapter also highlighted emerging marine-based strategies (ocean alkalinity enhancement, fertilization), which could unlock vast storage potential but require careful assessment of ecological impacts.

The key insights from this chapter include:

1. BECCS and DAC are among the most scalable technological options but depend on policy support (e.g., carbon pricing) to become viable. Projects like Orca DAC in Iceland prove technical feasibility, while BECCS deployments in bioenergy plants show integration potential.
2. Nature-based solutions must be studied in more detail and supported by regulations to ensure safe and sustainable CO_2 storage; for instance, soil carbon gains can reverse if land management practices change.
3. CO_2 utilization, such as converting CO_2 into building materials or plastics, presents dual opportunities by reducing emissions while creating markets for captured carbon. However, most CO_2 utilization technologies remain energy-intensive, and the permanency of CO_2 storage is also a concern.

The chapter also addressed plastic-related emissions, emphasizing that mitigation requires both advanced recycling technologies and finding replacements for plastic production, which is a complex but necessary transition for a low-carbon future.

Looking ahead, achieving gigaton-scale carbon removal will demand detailed policy frameworks that incentivize high-permanence solutions (e.g., DAC). Cross-sector collaboration is necessary to drive down costs (e.g., renewable-powered DAC plants). Monitoring technologies also required to accurately verify and quantify the impact of carbon sequestration projects.

In conclusion, based on the current state of technology and numerous successful examples, carbon capture and storage (CCS) technologies hold significant potential for effectively mitigating climate change through the capture and sequestration of carbon dioxide (López et al. 2023). These proven technologies have demonstrated their ability to capture CO_2 emissions from various sources, such as power plants and industrial facilities, and store them underground in geological formations. With

Fig. 9.17 Images designed by Freepik

proper implementation and continued advancements, CCS has the capacity to achieve large-scale carbon reduction and contribute to global emission reduction targets. While carbon-negative technologies offer innovative approaches, they are still in the early stages of development and face challenges in terms of scalability, environmental impact, and cost-effectiveness. Therefore, the focus should be on advancing and expanding the deployment of carbon capture and storage technologies, as they offer a more viable and immediate solution for reducing greenhouse gas emissions and addressing climate change (Fig. 9.17).

References

L.T. Bach, S.J. Gill, R.E.M. Rickaby, S. Gore, P. Renforth, CO_2 removal with enhanced weathering and ocean alkalinity enhancement: potential risks and CO-benefits for marine pelagic ecosystems. Front. Clim. **1**, 7 (2019). https://doi.org/10.3389/fclim.2019.00007

D.J. Beerling, J.R. Leake, S.P. Long, et al. Farming with crops and rocks to address global climate, food and soil security. Nat. Plants 4, 138–147 (2018). https://doi.org/10.1038/s41477-018-0108-y

K. Bromberg Gedan, B.R. Silliman, M.D. Bertness, Centuries of human-driven change in salt marsh ecosystems. Annu. Rev. Mar. Sci. **1**(1), 117–141 (2009). https://doi.org/10.1146/annurev.marine.010908.163930

S. Budinis et al., *Direct Air Capture: A Key Technology for Net Zero* (International Energy Agency, 2023)

S.-H. Cho, M. Soh, B.C. English, T.E. Yu, C.N. Boyer, Targeting payments for forest carbon sequestration given ecological and economic objectives. Forest Policy Econ. **100**, 214–226 (2019). https://doi.org/10.1016/j.forpol.2018.12.003

H. Chopra, C.H. Stanfield, Y. Teng, P.K. Thallapally, R. Kahsar, K. Khandelwal, et al., Scrubbing the sky: reimagining the air as a resource. ACS Sustain. Resour. Manag. **40** (2026)

K. Coyne, Crushing it: can ground-up rock help save the earth. **CSA News, 68**(4), 4–11 (2023). https://doi.org/10.1002/csan.20986

S. Crooks, D. Herr, J. Tamelander, D. Laffoley, J. Vandever, *Mitigating Climate Change through Restoration and Management of Coastal Wetlands and near-Shore Marine Ecosystems: Challenges and opportunities (Environment Department Papers No. 121)* (World Bank, 2011)

S. Davoodi, M. Al-Shargabi, D.A. Wood, V.S. Rukavishnikov, K.M. Minaev, Review of technological progress in carbon dioxide capture, storage, and utilization. Gas Sci. Eng. **117**, 205070 (2023). https://doi.org/10.1016/j.jgsce.2023.205070

C.M. Duarte, W.C. Dennison, R.J.W. Orth, T.J.B. Carruthers, The charisma of coastal ecosystems: addressing the imbalance. Estuar. Coasts **31**(2), 233–238 (2008). https://doi.org/10.1007/s12237-008-9038-7

C.M. Duarte, I.J. Losada, I.E. Hendriks, I. Mazarrasa, N. Marbà, The role of coastal plant communities for climate change mitigation and adaptation. Nat. Clim. Chang. **3**(11), 961–968 (2013). https://doi.org/10.1038/nclimate1970

H. Ducklow, D. Steinberg, K. Buesseler, Upper ocean carbon export and the biological pump. Oceanography **14**(4), 50–58 (2001). https://doi.org/10.5670/oceanog.2001.06

K.B. Gedan, B.R. Silliman, M.D. Bertness, Centuries of human-driven change in salt marsh ecosystems. Annu. Rev. Mar. Sci. **1**, 117–141 (2009). https://doi.org/10.1146/annurev.marine.010908.163930

R. Geyer, J.R. Jambeck, K.L. Law, Production, use, and fate of all plastics ever made. Sci. Adv. **3**(7), e1700782 (2017). https://doi.org/10.1126/sciadv.1700782

N. Gruber, D. Clement, B.R. Carter, R.A. Feely, S. van Heuven, M. Hoppema, M. Ishii, R.M. Key, A. Kozyr, S.K. Lauvset, C.L. Monaco, J.T. Mathis, A. Murata, A. Olsen, F.F. Perez, C.L. Sabine, T. Tanhua, R. Wanninkhof, The oceanic sink for anthropogenic CO_2 from 1994 to 2007. Science **363**(6432), 1193–1199 (2019). https://doi.org/10.1126/science.aau5153

M.M. Hasan, R. Haque, M.I. Jahirul, M.G. Rasul, Pyrolysis of plastic waste for sustainable energy recovery: technological advancements and environmental impacts. Energy Convers. Manag. **326**, 119511 (2025). https://doi.org/10.1016/j.enconman.2025.119511

N. Hernández-Hernández, L.T. Bach, M.F. Montero, J. Taucher, I. Baños, W. Guan, M. Espósito, A. Ludwig, E.P. Achterberg, U. Riebesell, J. Arístegui, High CO_2 Under Nutrient Fertilization Increases Primary Production and Biomass in Subtropical Phytoplankton Communities: A Mesocosm Approach. Front. Mar. Sci. **5**, 213 (2018). https://doi.org/10.3389/fmars.2018.00213

IEAGHG, The status and challenges of CO_2 shipping infrastructures (2020). https://ieaghg-publications.s3.eu-north-1.amazonaws.com/Technical+Reports/2020-10+The+Status+and+Challenges+of+CO2+Shipping+Infrastructures.pdf

M.W. Jones, G.P. Peters, T. Gasser, R.M. Andrew, C. Schwingshackl, J. Gütschow, R.A. Houghton, P. Friedlingstein, J. Pongratz, C. Le Quéré, National contributions to climate change due to historical emissions of carbon dioxide, methane and nitrous oxide. Earth System Science Data **15**(12), 5301–5369 (2023). https://doi.org/10.5194/essd-15-5301-2023

S. Li, D. Tasnady, Biochar for soil carbon sequestration: Current knowledge, mechanisms, and future perspectives. **9**, 67 (2023). https://doi.org/10.3390/c9030067

J.T. Litynski, T. Rodosta, D. Vikara, R. Srivastava, U.S. DOE'S R&D program to develop infrastructure for carbon storage: overview of the regional carbon sequestration partnerships and other R&D field projects. Energy Procedia **37**, 6527–6543 (2013). https://doi.org/10.1016/j.egypro.2013.06.584

L.R. López, P. Dessì, A. Cabrera-Codony, L. Rocha-Melogno, B. Kraakman, V. Naddeo, M.D. Balaguer, S. Puig, CO_2 in indoor environments: From environmental and health risk to potential renewable carbon source. Sci. Total Environ. **856**(2), 159088 (2023). https://doi.org/10.1016/j.scitotenv.2022.159088

C.J. Mcowen, L.V. Weatherdon, J.V. Bochove, E. Sullivan, S. Blyth, C. Zockler, D. Stanwell-Smith, N. Kingston, C.S. Martin, M. Spalding, S. Fletcher, A global map of saltmarshes. Biodivers. Data J. **5**, e11764 (2017). https://doi.org/10.3897/BDJ.5.e11764

H. McLaughlin, A.A. Littlefield, M. Menefee, A. Kinzer, T. Hull, B.K. Sovacool, M.D. Bazilian, J. Kim, S. Griffiths, Carbon capture utilization and storage in review: Sociotechnical implications for a carbon reliant world. Renew. Sust. Energ. Rev. **177**, 113215 (2023). https://doi.org/10.1016/j.rser.2023.113215

H. Mikulčić, I.R. Skov, D.F. Dominković, S.R.W. Alwi, Z.A. Manan, R. Tan, N. Duić, S.N.H. Mohamad, X. Wang, Flexible carbon capture and utilization technologies in future energy systems and the utilization pathways of captured CO_2. Renew. Sust. Energ. Rev. **114**, 109338 (2019). https://doi.org/10.1016/j.rser.2019.109338

M. Montero-Hidalgo, F. Tuya, F. Otero-Ferrer, R. Haroun, F. Santos-Martín, Mapping and assessing seagrass meadows changes and blue carbon under past, current, and future scenarios. Sci. Total Environ. **872**, 162244 (2023). https://doi.org/10.1016/j.scitotenv.2023.162244

H. Ritchie, P. Rosado, M. Roser, CO_2 and greenhouse gas emissions. Our world in data (2023).

J. Robinson, E.E. Popova, A. Yool, M. Srokosz, R.S. Lampitt, J.R. Blundell, How deep is deep enough? Ocean iron fertilization and carbon sequestration in the Southern Ocean. Geophys. Res. Lett. **41**(7), 2489–2495 (2014). https://doi.org/10.1002/2013GL058799

D. Saebea, P. Ruengrit, A. Arpornwichanop, Y. Patcharavorachot, Gasification of plastic waste for synthesis gas production. Energy Rep. **6**(Suppl. 1), 202–207 (2020). https://doi.org/10.1016/j.egyr.2019.08.043

A. Samir, F.H. Ashour, A.A.A. Hakim, et al., Recent advances in biodegradable polymers for sustainable applications. NPJ Mater. Degrad. **6**(1), 68 (2022). https://doi.org/10.1038/s41529-022-00277-7

K. Varshney, M. Pedersen Zari, N. Bakshi, Carbon sequestration and habitat provisioning through building-integrated vegetation: A global survey of experts. Buildings **12**, 1458 (2022). https://doi.org/10.3390/buildings12091458

Organisation for Economic Co-operation and Development, Global plastics outlook (2022). https://stats.oecd.org/viewhtml.aspx?datasetcode=PLASTIC_WASTE_5&lang=en

R.J. Orth, J.S. Lefcheck, K.S. McGlathery, L. Aoki, M.L. Luckenbach, K.A. Moore, M.P.J. Oreska, R. Snyder, D.J. Wilcox, B. Lusk, Restoration of seagrass habitat leads to rapid recovery of coastal ecosystem services. Sci. Adv. **6**(41), eabc6434 (2020). https://doi.org/10.1126/sciadv.abc6434

Parry, W. (2012, July 18). Could fertilizing the oceans reduce global warming? Live Science. https://www.livescience.com/21684-geoengineering-iron-fertilization-climate.html

Y. Pan, R.A. Birdsey, J. Fang, R. Houghton, P.E. Kauppi, W.A. Kurz, O.L. Phillips, A. Shvidenko, S.L. Lewis, J.G. Canadell, P. Ciais, R.B. Jackson, S.W. Pacala, A.D. McGuire, S. Piao, A. Rautiainen, S. Sitch, D. Hayes, A large and persistent carbon sink in the world's forests. Science **333**(6045), 988–993 (2011). https://doi.org/10.1126/science.1201609

S. Valluri, V. Claremboux, S. Kawatra, Opportunities and challenges in CO_2 utilization. J. Environ. Sci. **113**, 322–344 (2022). https://doi.org/10.1016/j.jes.2021.05.043

M. Waycott, C.M. Duarte, T.J.B. Carruthers, R.J. Orth, W.C. Dennison, S. Olyarnik, A. Calladine, J.W. Fourqurean, K.L. Heck, A.R. Hughes, G.A. Kendrick, W.J. Kenworthy, F.T. Short, S.L. Williams, Accelerating loss of seagrasses across the globe threatens coastal ecosystems. Proc. Natl. Acad. Sci. **106**(30), 12377–12381 (2009). https://doi.org/10.1073/pnas.0905620106

H. Yin, B. Yu, Z. Zhang, L. Jia, J. Xue, X. Chen, S. Wang, J. Han, Z. Xing, Evaluating the impact of different afforestation timescales on the soil quality in a typical watershed valley using the criteria importance through intercriteria correlation method and gray target model. Front. For. Glob. Chang. **5**, 997949 (2022). https://doi.org/10.3389/ffgc.2022.997949

M.R. Zakaria, M.A.A. Farid, Y. Andou, I. Ramli, M.A. Hassan, Production of biochar and activated carbon from oil palm biomass: Current status, prospects, and challenges. Ind. Crop. Prod. **199**, 116767 (2023). https://doi.org/10.1016/j.indcrop.2023.116767

H. Zentou, B. Hoque, M.A. Abdalla, A.F. Saber, O.Y. Abdelaziz, M. Aliyu, A.M. Alkhedhair, A.J. Alabduly, M.M. Abdelnaby, Recent advances and challenges in solid sorbents for CO_2 capture. Carbon Capture Sci. Technol. **15**, 100386 (2025). https://doi.org/10.1016/j.ccst.2025.100386

B. Zheng, J. Xu, Carbon capture and storage development trends from a techno-paradigm perspective. Energies **7**(8), 5221–5250 (2014). https://doi.org/10.3390/en7085221

Glossary

Acronym	Full name	Category	Notes
ACR	American carbon registry	Carbon markets	
ADM	Archer Daniels Midland company	Projects	IBDP participant
AIT	Array induction tool	Well logging	
API	American petroleum institute	Standards	
AoR	Area of review	Regulations	EPA requirement
ASU	Air separation unit	Technology	CO_2 capture
BECCS	Bioenergy with carbon capture and storage	Technology	Negative emissions
BIL	Bipartisan infrastructure Law	Policy	US (2021)
BPA	Bonneville power administration	Organizations	US Pacific NW
BRA	Baseline reservoir assessment	Reservoir engineering	Pre-injection
BPMs	Best practices manuals	Standards	NETL publications
BSCSP	Big sky carbon sequestration partnership	Projects	Montana/ Wyoming
BTU	British thermal units	Measurements	Energy
CAPEX	Capital expenditure	Economics	
CAR	Climate action reserve	Carbon markets	
CARB	California air resources board	Regulations	LCFS administrator
CAT	Climate action tracker	Monitoring	NGO initiative
CCER	China certified emission reduction	Carbon markets	
CCS	Carbon capture and storage	Technology	
CCUS	Carbon capture, utilization and storage	Technology	

(continued)

E. Fathi and F. Belyadi, *Subsurface CO_2 Sequestration*,
https://doi.org/10.1007/978-3-032-28296-5

(continued)

Acronym	Full name	Category	Notes
CDM	Clean development mechanism	Carbon markets	UN program
CERs	Certified emission reductions	Carbon markets	CDM credits
CFCs	Chlorofluorocarbons	Chemicals	Ozone-depleting
CHP	Combined heat and power	Technology	Efficiency
CMG GEM	Computer modelling group—Generalized equation-of-state model	Software	Reservoir Sim
CMIP5	Coupled model Intercomparison project phase 5	Climate science	IPCC models
CMR-plus	Conventional magnetic resonance	Well logging	Schlumberger tool
CO_2-ESGR	CO_2 enhanced shale gas recovery	Technology	Unconventional
CO_2e	Carbon dioxide equivalents	Measurements	GHG metric
COC	Cost of CO_2 capture	Economics	$/ton
COP28	28th conference of the parties (UNFCCC)	Policy	2023 UAE
COP29	29th conference of the parties (UNFCCC)	Policy	2024 Azerbaijan
CORSIA	Carbon offsetting and reduction scheme for international aviation	Policy	ICAO
CPU	CO_2 processing unit	Technology	Purification
CSR	Corporate social responsibility	Business	
DAS	Distributed acoustic sensing	Monitoring	Fiber optic
DCS	Dedicated carbon storage	Projects	Non-EOR
DEA	Diethanolamine	Chemicals	Solvent
DFA	Digital fluid analyzer / downhole fluid analysis	Well testing	Schlumberger
DFIT	Diagnostic fracture injection test	Well testing	Fracture pressure
DIC	Dissolved inorganic carbon	Chemistry	
DMG	Downhole memory gauges	Equipment	Pressure/temp
DOE	Department of energy	Organizations	US
DOT	Department of transportation	Organizations	US (PHMSA)
DST	Drill stem test	Well testing	
DSS	Distributed strain sensing	Monitoring	Fiber optic
DTS	Distributed temperature sensing	Monitoring	Fiber optic
ECBM	Enhanced coalbed methane recovery	Technology	
EEA	European Environment Agency	Organizations	EU
EERC	Energy & environmental research center	Organizations	North Dakota
EEX	European energy exchange	Carbon markets	Germany
EIA	Energy Information Administration	Organizations	US
EOR	Enhanced oil recovery	Technology	

(continued)

(continued)

Acronym	Full name	Category	Notes
EPA	Environmental Protection Agency	Regulations	US
EPS	Emission performance standards	Regulations	
ETS	Emissions trading system	Carbon markets	EU, China, etc.
EU	European Union	Organizations	
FEED	Front-end engineering design	Engineering	Pre-construction
FMI	Formation micro-imager	Well logging	Schlumberger
FTIR	Fourier-transform infrared spectroscopy	Analysis	
FWU	Farnsworth unit	Projects	EOR site
GAO	Government accountability office	Organizations	US
GCF	Green climate fund	Finance	UN
GFAS	Global fire assimilation system	Monitoring	NASA
GHCN	Global historical climatology network	Data	NOAA
GHG	Greenhouse gas	Climate science	
GIS	Geographic information system	Technology	Spatial analysis
GtC	Gigatons of carbon	Measurements	
GWP100	Global warming potential (100-yr)	Climate science	IPCC metric
HCFCs	Hydrochlorofluorocarbons	Chemicals	
HSE	Health, safety, environment	Standards	
IEA	International energy agency	Organizations	OECD
IBDP	Illinois Basin–Decatur project	Projects	ADM CCS
IGCC	Integrated gasification combined cycle	Technology	Power plants
IPCC	Intergovernmental panel on climate change	Organizations	UN
ISIP	Instantaneous shut-in pressure	Well testing	DFIT analysis
ITC	Investment tax credit	Policy	US (26 USC §48)
KRX	Korea exchange	Carbon markets	
kWh	Kilowatt-hour	Measurements	Energy
LCFS	Low carbon fuel standard	Regulations	CA, OR
LCOE	Levelized cost of energy	Economics	\$/MWh
LCOS	Levelized cost of storage	Economics	
LNG	Liquefied natural gas	Energy	
LDS	Leak detection system	Monitoring	Pipelines
MASIP	Maximum allowable surface injection pressure	Engineering	UIC limit
MDEA	Methyldiethanolamine	Chemicals	Solvent
MEA	Monoethanolamine	Chemicals	Solvent
MGSC	Midwest geological sequestration consortium	Projects	
MICP	Mercury injection capillary pressure	Analysis	Porosity

(continued)

(continued)

Acronym	Full name	Category	Notes
MMV	Measurement, monitoring & verification	Regulations	
MMT	Million metric tons	Measurements	
MOU	Memorandum of understanding	Agreements	
MRCSP	Midwest regional carbon sequestration partnership	Projects	
MVA	Monitoring, Verification & Accounting	Regulations	
NA	Not applicable		
NASA	National aeronautics & space administration	Organizations	US
NB	Not build		Alternative scenario
NDCs	Nationally determined contributions	Policy	Paris agreement
NETL	National Energy Technology Laboratory	Organizations	US DOE
NGO	Non-governmental organization	Organizations	
NOAA	National oceanic & atmospheric administration	Organizations	US
NPW	Negative pressure wave	Monitoring	Leak detection
NWR	National Wildlife Refuge	Organizations	US
OECD	Organisation for economic co-operation and development	Organizations	
O&M	Operations & maintenance	Economics	
OPEX	Operational expenditure	Economics	
PCC	Post-combustion capture	Technology	
PCOR	Plains CO_2 reduction partnership	Projects	
PHMSA	Pipeline & hazardous materials safety administration	Regulations	US DOT
PISC	Post injection site care	Regulations	UIC requirement
PNL	Pulsed neutron logging	Well logging	
PTC	Production tax credit	Policy	US (26 USC §45)
PVT	Pressure-volume-temperature	Reservoir engineering	Fluid properties
QBTU	Quadrillion British thermal units	Measurements	Energy
R&D	Research & development		
RCAL	Routine Core analysis	Analysis	
RCPs	Representative concentration pathways	Climate science	IPCC scenarios
RCSP	Regional carbon sequestration partnership	Projects	US DOE
REDD	Reducing emissions from deforestation/ degradation	Policy	UN
RGGI	Regional greenhouse gas initiative	Policy	US northeast

(continued)

(continued)

Acronym	Full name	Category	Notes
ROP	Rate of penetration	Drilling	Ft/hr
ROWs	Rights-of-way	Land use	Pipelines
RPS	Renewable portfolio standard	Policy	State-level
SACROC	Scurry area canyon reef operators committee	Projects	Texas EOR
SCADA	Supervisory control & data acquisition	Technology Analysis	
SCAL	Special Core analysis		
SECARB	Southeast regional carbon sequestration partnership	Projects	
SIGTTO	Society of Int'l gas Tanker & Terminal Operators	Organizations	
SMR	Steam methane reforming	Technology	Hydrogen prod.
SP	Spontaneous potential	Well logging	
SWP	Southwest regional partnership	Projects	
TCM	Technology Centre Mongstad	Facilities	Norway
TDS	Total dissolved solids	Measurements	Water quality
TEA	Triethanolamine	Chemicals	Solvent
TOB	Torque on bit	Drilling	
TRA	Transient radial analysis	Analysis	Well testing
UCS	Unconfined compressive strength	Geomechanics	
UIC	Underground injection control	Regulations	EPA program
UK	United Kingdom		
UNFCCC	UN framework convention on climate change	Policy	
USDA	US Department of Agriculture	Organizations	
USDW	Underground source of drinking water	Regulations	SDWA
UT	Ultrasonic testing	Inspection	Pipeline integrity
VOCs	Volatile organic compounds	Chemicals	Air pollutants
WOB	Weight on bit	Drilling	
XRD	X-ray diffraction	Analysis	Mineralogy
XRF	X-ray fluorescence	Analysis	Composition

Appendix A

Knowledge Check Chap. 1

1. How much have global temperatures risen since the Industrial Revolution?

   (a) 1.04 °C
   (b) 2.2 °C
   (c) 4.1 °C
   (d) 10 °C.

   Answer: (a) 1.04 °C
2. How much has sea ice cover in the Arctic shrunk between 1980 and 2020?

   (a) 10%
   (b) 25%
   (c) 38%
   (d) 51%.

   Answer: (d) 51%
3. By how much has the mean sea level risen since 1992?

   (a) 100 cm
   (b) 50 cm
   (c) 35 cm
   (d) 10.10 cm

   Answer: (d) 10.10 cm
4. What is the primary source of energy consumed globally?

   (a) Coal.
   (b) Natural gas.
   (c) Petroleum.
   (d) Renewable energy.

E. Fathi and F. Belyadi, *Subsurface CO_2 Sequestration*,
https://doi.org/10.1007/978-3-032-28296-5

Answer: (c) Petroleum

5. Which of the following countries is the largest emitter of greenhouse gases?

   (a) China.
   (b) United States.
   (c) India.
   (d) European Union.

   Answer: (a) China

6. Which sector consumes the largest share of primary energy worldwide?

   (a) Industrial sector.
   (b) Residential sector.
   (c) Transportation sector.
   (d) Commercial sector.

   Answer: (a) Industrial sector

7. When did the Paris Agreement enter into force?

   (a) 2015
   (b) 2016
   (c) 2017
   (d) 2018

   Answer: (b) 2016

8. Which country formally announced its withdrawal from the Paris Agreement but later rejoined?

   (a) China.
   (b) Germany.
   (c) United States.
   (d) India.

   Answer: (c) United States

9. What percentage of global emissions did the United States account for when it initially withdrew from the agreement?

   (a) 5%
   (b) 10%
   (c) 15%
   (d) 20%

   Answer: (c) 15%

10. What does "neutral carbon balance" mean?

   (a) Balancing carbon emissions with carbon offsets.
   (b) Achieving zero carbon emissions.
   (c) Reducing carbon emissions by a certain percentage.
   (d) Maintaining the current level of carbon emissions.

   Answer: (b) Balancing carbon emissions with carbon offsets

11. Why is neutral carbon balance important?
    (a) It helps to mitigate the impacts of climate change.
    (b) It ensures sustainable and cleaner energy sources.
    (c) It promotes environmental conservation and biodiversity.
    (d) All the above.

    Answer: (a) It helps to mitigate the impacts of climate change.
12. How can CCUS contribute to improving the quality of life?
    (a) By reducing greenhouse gas emissions.
    (b) By providing cleaner air and water.
    (c) By supporting sustainable economic growth.
    (d) All the above.

    Answer: (d) All of the above
13. How can CCUS contribute to the fight against poverty?
    (a) By creating job opportunities in the clean energy sector.
    (b) By providing access to affordable and reliable energy.
    (c) By supporting economic development in communities.
    (d) All the above.

    Answer: (d) All of the above
14. How can CCUS help alleviate environmental concerns?
    (a) By capturing and storing carbon dioxide emissions.
    (b) By reducing air pollution from fossil fuel combustion.
    (c) By preventing carbon dioxide from entering the atmosphere.
    (d) All the above.

    Answer: (d) All of the above.

Knowledge Check Chap. 2

1. Which of the following is a carbon negative CCS facility?
   (a) Boundary Dam CCS facility.
   (b) Illinois Industrial CCS Project.
   (c) Al Reyadah facility.
   (d) Norcem Brevik facility.

   Answer: (b) Illinois Industrial CCS Project.
2. What is the main objective of carbon negative CCS facilities?
   (a) To increase the efficiency of energy production.
   (b) To produce renewable energy.
   (c) To reduce overall greenhouse gas emissions from carbon cycle.

(d) To extract carbon from the atmosphere.

Answer: (c) To reduce overall greenhouse gas emissions from carbon cycle

3. Which process is involved in achieving carbon negativity in CCS facilities?

   (a) Conversion of biomass into ethanol while CCS.
   (b) Transformation of CO_2 into biomass.
   (c) Injection and storage of CO_2 in geological formations.
   (d) Combustion of fossil fuels.

   Answer: (a) Conversion of biomass into ethanol while CCS

4. What is the primary characteristic of a Near Carbon Neutral CCS facility?

   (a) It captures and stores 100% of CO_2 emissions.
   (b) It utilizes renewable energy sources for CO_2 capture.
   (c) It significantly reduces carbon emissions but still has a small fraction entering the atmosphere.
   (d) It focuses on capturing CO_2 emissions from specific industries only.

   Answer: (c) It significantly reduces carbon emissions but still has a small fraction entering the atmosphere

5. Which facility is mentioned as an example of a Near Carbon Neutral CCS facility?

   (a) Boundary Dam CCS facility.
   (b) Quest CCS facility.
   (c) CarbFix project.
   (d) Orca by Climeworks (Iceland).

   Answer: (d) Orca by Climeworks (Iceland)

6. What is the main objective of Near Carbon Neutral CCS facilities?

   (a) To eliminate carbon emissions.
   (b) To offset carbon emissions through carbon capture and storage.
   (c) To convert captured carbon into usable products.
   (d) To achieve carbon neutrality through renewable energy integration.

   Answer: (b) To offset carbon emissions through carbon capture and storage

7. What is the primary characteristic of a Carbon Positive CCS facility?

   (a) It captures and stores 100% of CO_2 emissions.
   (b) It produces additional carbon emissions in the process.
   (c) It utilizes renewable energy sources for CO_2 capture.
   (d) It focuses on capturing CO_2 emissions from specific industries only.

   Answer: (b) It produces additional carbon emissions in the process

8. Which facility is mentioned as an example of a Carbon Positive CCS facility?

    (a) Quest CCS facility.
    (b) Norcem Brevik facility.
    (c) Al Reyadah facility.
    (d) All of the above.

    Answer: (d) All of the above.

9. What is the main challenge associated with Carbon Positive CCS facilities?

    (a) Achieving complete carbon neutrality.
    (b) Finding suitable geological storage sites.
    (c) Balancing carbon emission reduction with energy production needs.
    (d) Eliminating the need for fossil fuel use.

    Answer: (c) Balancing carbon emission reduction with energy production needs

10. Blue hydrogen is produced by.

    (a) Electrolysis of water.
    (b) Reforming natural gas with carbon capture.
    (c) Biomass gasification.
    (d) Direct capture from the atmosphere.

    Answer: (b) Reforming natural gas with carbon capture

11. Which industry is the largest global emitter of carbon dioxide (CO_2)?

    (a) Steel industry.
    (b) Cement industry.
    (c) Chemical industry.
    (d) Automotive industry.

    Answer: (b) Cement industry

12. The production of cement results in the release of CO_2 primarily from.

    (a) Combustion of fossil fuels.
    (b) Decomposition of limestone.
    (c) Chemical reactions during the curing process.
    (d) Waste disposal and treatment processes.

    Answer: (b) Decomposition of limestone

13. Which of the following is a significant challenge in achieving carbon neutrality in the chemical industry?

    (a) Limited availability of renewable energy sources.
    (b) Lack of efficient carbon capture technologies.
    (c) High dependence on fossil fuels as feedstock.
    (d) Inadequate government regulations and policies.

    Answer: (c) High dependence on fossil fuels as feedstock

14. Blue hydrogen can be used as a low-carbon alternative in.

(a) Steel production.
(b) Cement production.
(c) Chemical production.
(d) All of the above.

Answer: (d) All of the above

Knowledge Check Chap. 3

1. What is the purpose of a carbon tax?

   (a) To distribute carbon credits among participating companies.
   (b) To incentivize companies to reduce their greenhouse gas emissions.
   (c) To generate revenue for the government through emissions trading.
   (d) To establish international standards and maximum caps for emissions.

   Answer: (b) To incentivize companies to reduce their greenhouse gas emissions
2. Which countries have implemented cap-and-trade programs?

   (a) India, China, and Brazil.
   (b) United States, European Union, and Mexico.
   (c) United Kingdom, Canada, and Australia.
   (d) Japan, Russia, and South Korea.

   Answer: (b) United States, European Union, and Mexico
3. What is a potential challenge of cap-and-trade systems?

   (a) Insufficient participation from companies and industries.
   (b) Difficulty in accurately monitoring and determining emissions levels.
   (c) Lack of incentives for companies to invest in cleaner technologies.
   (d) Inability to achieve an international consensus on emissions and caps.

   Answer: (b) Difficulty in accurately monitoring and determining emissions levels
4. What is a carbon credit?

   (a) A tax imposed on carbon-intensive industries.
   (b) A financial incentive for reducing greenhouse gas emissions.
   (c) A unit representing one metric ton of CO_2 emissions reduced or avoided.
   (d) A penalty imposed on companies exceeding their emission limits.

   Answer: (c) A unit representing one metric ton of CO_2 emissions reduced or avoided
5. Which sectors are major buyers of carbon offset credits in the voluntary market?

   (a) Agricultural and forestry industries.
   (b) Cryptocurrency platforms and airlines.
   (c) Oil companies and car manufacturers.
   (d) Renewable energy and technology companies.

Answer: (b) Cryptocurrency platforms and airlines

6. How do companies participate in the voluntary market to offset their emissions?

   (a) By purchasing carbon credits from regulated exchanges.
   (b) By investing in renewable energy projects.
   (c) By implementing internal emission reduction strategies.
   (d) By participating in government-mandated carbon markets.

   Answer: (a) By purchasing carbon credits from regulated exchanges.

Knowledge Check Chap. 4

1. Which of the following technologies involves capturing CO_2 from the combustion gases before they are released into the atmosphere?

   (a) Oxyfuel combustion.
   (b) Chemical looping combustion.
   (c) Post-combustion capture.
   (d) Pre-combustion capture.

   Answer: (c) Post-combustion capture

2. Which capture technology relies on chemical solvents to absorb CO_2 from flue gas?

   (a) Oxyfuel combustion.
   (b) Chemical looping combustion.
   (c) Post-combustion capture.
   (d) Pre-combustion capture.

   Answer: (c) Post-combustion capture

3. Which capture technology can be retrofitted to existing power plants without major modifications?

   (a) Pre-combustion capture.
   (b) Chemical looping combustion.
   (c) Oxyfuel combustion.
   (d) Post-combustion capture.

   Answer: (d) Post-combustion capture

4. Which of the following is a common solvent used for post-combustion carbon capture?

   (a) Water.

(b) Nitrogen.
(c) Oxygen.
(d) Amine-based solvents.

Answer: (d) Amine-based solvents

5. What is the purpose of compressing the separated CO_2 in post-combustion capture?

(a) To increase its purity.
(b) To reduce its volume for transportation.
(c) To remove impurities.
(d) To facilitate storage in underground reservoirs.

Answer: (b) To reduce its volume for transportation

6. Which membrane separation mechanism relies on the size difference of molecules to separate them?

(a) Size sieving.
(b) Surface diffusion.
(c) Solution diffusion.
(d) Facilitated transport.

Answer: (a) Size sieving

7. Which membrane separation mechanism involves the movement of molecules across the membrane surface?

(a) Size sieving.
(b) Surface diffusion.
(c) Solution diffusion.
(d) Facilitated transport.

Answer: (b) Surface diffusion

8. Complete combustion of hydrocarbon fuels produces which of the following gases?

(a) Carbon monoxide (CO)
(b) Carbon dioxide (CO_2)
(c) Nitrogen dioxide (NO2)
(d) Sulfur dioxide (SO2)

Answer: (b) Carbon dioxide (CO_2)

9. Incomplete combustion of hydrocarbon fuels can result in the formation of which harmful gas?

(a) Oxygen (O2)
(b) Carbon dioxide (CO_2)
(c) Nitrogen dioxide (NO2)
(d) Carbon monoxide (CO)

Answer: (d) Carbon monoxide (CO)

10. Which of the following technologies involves burning fuel with nearly pure oxygen instead of air?

    (a) Post-combustion capture.
    (b) Oxyfuel combustion.
    (c) Pre-combustion capture.
    (d) Chemical looping combustion.

    Answer: (b) Oxyfuel combustion

11. What is the primary purpose of flue gas processing in oxyfuel combustion?

    (a) To control flame temperature.
    (b) To reduce the volume of flue gas.
    (c) To remove pollutants from the flue gas.
    (d) To separate CO_2 from the flue gas.

    Answer: (c) To remove pollutants from the flue gas

12. What is the primary purpose of an air separation unit (ASU) in oxyfuel combustion?

    (a) To produce oxygen for combustion.
    (b) To control flame temperature.
    (c) To remove pollutants from the flue gas.
    (d) To separate CO_2 from the flue gas.

    Answer: (a) To produce oxygen for combustion

13. What is the primary purpose of using oxyfuel combustion in power plants?

    (a) To increase the flame temperature.
    (b) To reduce the concentration of CO_2 in the flue gas.
    (c) To generate a flue gas with high CO_2 concentration.
    (d) To improve combustion efficiency.

    Answer: (c) To generate a flue gas with high CO_2 concentration

14. What is the main advantage of oxyfuel combustion compared to air-fired combustion?

    (a) Lower capital costs.
    (b) Higher energy efficiency.
    (c) Reduced emissions of pollutants.
    (d) Faster combustion reactions.

    Answer: (c) Reduced emissions of pollutants.

15. How is the flame temperature controlled in oxyfuel combustion?

    (a) By adjusting the oxygen flow rate.
    (b) By recycling part of the flue gas.
    (c) By increasing the fuel supply.
    (d) By adjusting the air-fuel ratio.

    Answer: (b) By recycling part of the flue gas

16. What is the effect of using oxyfuel combustion on the volume of flue gas produced?

    (a) It increases the volume of flue gas.
    (b) It decreases the volume of flue gas.
    (c) It has no significant effect on the volume of flue gas.
    (d) It depends on the type of fuel used.

    Answer: (b) It decreases the volume of flue gas
17. What is one of the disadvantages of oxyfuel combustion technology?

    (a) Increased combustion efficiency.
    (b) Higher capital costs.
    (c) Lower CO_2 capture efficiency.
    (d) Reduced emissions of greenhouse gases.

    Answer: (b) Higher capital costs
18. What is the main characteristic of Direct-Fired Oxy-Fuel Combustion?

    (a) It utilizes supercritical carbon dioxide as the working fluid.
    (b) It achieves nearly 100% CO_2 capture efficiency.
    (c) It involves the combustion of fuel with pure oxygen.
    (d) It relies on the use of air-fuel ratio control for temperature regulation.

    Answer: (a) It utilizes supercritical carbon dioxide as the working fluid.
19. Which capture technology involves gasification of fuel to produce a syngas containing mainly hydrogen and carbon monoxide?

    (a) Oxyfuel combustion.
    (b) Pre-combustion capture.
    (c) Chemical looping combustion.
    (d) Post-combustion capture.

    Answer: (b) Pre-combustion capture
20. Which of the following is an advantage of pre-combustion capture?

    (a) Lower capital cost compared to conventional power plants.
    (b) Higher CO_2 concentration in flue gas streams.
    (c) Simplicity of the capture process.
    (d) Availability of multiple technology options.

    Answer: (b) Higher CO_2 concentration in flue gas streams
21. What is a potential disadvantage of pre-combustion capture?

    (a) Limited availability of Integrated Gasification Combined Cycle (IGCC) plants.
    (b) Low efficiency in capturing CO_2.
    (c) Lower capital cost compared to conventional power plants.
    (d) Ease of integration with existing power plants.

Answer: (a) Limited availability of Integrated Gasification Combined Cycle (IGCC) plants.

22. Which of the following is an advantage of pre-combustion capture over post-combustion capture?

(a) Lower operating costs.
(b) Higher purity of captured CO_2.
(c) Higher CO_2 removal efficiency.
(d) Compatibility with existing power plants.

Answer: (b) Higher purity of captured CO_2

23. What is a potential disadvantage of pre-combustion capture technology?

(a) Higher capital cost compared to conventional power plants.
(b) Ease of retrofitting to existing power plants.
(c) Lower efficiency in capturing CO_2 compared to post-combustion capture.
(d) Availability of multiple technology options.

Answer: (a) Higher capital cost compared to conventional power plants

24. Which of the following is a challenge associated with pre-combustion capture?

(a) Limited storage options for captured CO_2.
(b) Higher emissions of NOx and SOx.
(c) Difficulty in achieving high CO_2 concentrations in flue gas streams.
(d) Limited availability of coal feedstock.

Answer: (c) Difficulty in achieving high CO_2 concentrations in flue gas streams

25. What is an advantage of pre-combustion capture technology in terms of environmental impact?

(a) Reduced emissions of greenhouse gases.
(b) Decreased water usage in the capture process.
(c) Improved air quality due to reduced particulate matter emissions.
(d) Lower land requirements for power plant installations.

Answer: (a) Reduced emissions of greenhouse gases

26. Which carbon capture technology can achieve the highest CO_2 capture efficiency?

(a) Pre-combustion capture.
(b) Post-combustion capture.
(c) Oxyfuel combustion.

Answer: (c) Oxyfuel combustion

27. Which carbon capture technology requires the least retrofitting of existing power plants?

(a) Pre-combustion capture.
(b) Post-combustion capture.

(c) Oxyfuel combustion.

Answer: (b) post-combustion capture

28. Which carbon capture technology is most suitable for new power plant constructions?

    (a) Pre-combustion capture.
    (b) Post-combustion capture.
    (c) Oxyfuel combustion.

    Answer: (c) Oxyfuel combustion

Knowledge Check Chap. 5

1. Which of the following is NOT an internal leak detection technique for CO_2 transport pipelines?

    (a) Mass balance.
    (b) Tracer gas injection.
    (c) Fiber optic cables.
    (d) Acoustic sensors.

    Answer: (c) Fiber optic cables

2. What is the primary advantage of using external leak detection techniques for CO_2 pipelines?

    (a) Early detection of leaks.
    (b) Accurate localization of leaks.
    (c) Non-intrusive monitoring.
    (d) Cost-effectiveness.

    Answer: (a) non-intrusive monitoring

3. Which technique utilizes laser light to detect leaks by measuring the amount of energy absorbed?

    (a) Infrared cameras.
    (b) Fiber optic cables.
    (c) Leak sniffers.
    (d) Laser-based absorption.

    Answer: (d) Laser-based absorption

4. What type of signals can be used for leak detection using the NPW technique?

    (a) Visual signals.
    (b) Pressure signals.
    (c) Acoustic signals.

(d) Temperature signals.

Answer: (b) Pressure signals

5. How does the NPW technique detect leaks?

(a) By analyzing pressure drop patterns over time.
(b) By measuring temperature changes in the pipeline.
(c) By detecting acoustic signals emitted by leaks.
(d) By monitoring flow rates and pump statuses.

Answer: (a) By analyzing pressure drop patterns over time

6. Which of the following is a challenge associated with the NPW technique?

(a) High equipment cost.
(b) Noisy transducer data.
(c) Limited sample size of leak data.
(d) Difficulty in analyzing pressure derivatives.

Answer: (b) Noisy transducer data

7. Which technique relies on the analysis of noise levels to detect and localize leaks?

(a) Fiber optic cables.
(b) Infrared cameras.
(c) Acoustic sensors.
(d) Mass balance calculations.

Answer: (c) Acoustic sensors

8. Which API standard addresses computational pipeline monitoring for liquids?

(a) API 1130.
(b) API 1149.
(c) API 1155.
(d) API 1166.

Answer: (a) API 1130

9. What is the primary objective of an effective Leak Detection System (LDS)?

(a) Minimizing fluid loss due to leaks.
(b) Generating accurate flow and location measurements.
(c) Providing reliable and timely alarms.
(d) Operating under non-ideal circumstances.

Answer: (a) Minimizing fluid loss due to leaks

10. Which surveillance technique involves the use of drones or aircraft to monitor CO_2 pipeline emissions?

(a) Visual inspections.
(b) Aerial monitoring.

(c) Remote sensing.
(d) Mass balance calculations.

Answer: (b) Aerial monitoring.

Knowledge Check Chap. 6

1. What is the primary aim of CO_2 storage?

(a) To generate electricity.
(b) To reduce greenhouse gas emissions.
(c) To increase oil production.
(d) To enhance coal bed methane recovery.

Answer: (b) To reduce greenhouse gas emission

2. At surface conditions, what is the approximate density of CO_2 in its supercritical state?

(a) 0.1 kg/m^3.
(b) 1 kg/m^3.
(c) 50 kg/m^3.
(d) 500 kg/m^3.

Answer: (d) 500 kg/m^3

3. What is the main trapping mechanism for CO_2 storage in geological formations?

(a) Structural trapping.
(b) Solubility trapping.
(c) Capillary trapping.
(d) Capillary trapping.

Answer: (a) Structural trapping

4. Which of the following is a geological storage site for CO_2?

(a) Depleted oil and gas reservoirs.
(b) Deep ocean trenches.
(c) Urban areas.
(d) Agricultural fields.

Answer: (a) Depleted oil and gas reservoirs

5. How does CO_2 move within a geological storage site?

(a) By convection.
(b) By diffusion.
(c) By advection.

(d) All of the above.

Answer: (b) All of the above

6. Which technique involves injecting CO_2 into a reservoir to increase oil production?

   (a) CO_2 storage.
   (b) CO_2 trapping.
   (c) CO_2-EOR.
   (d) CO_2 sequestration.

   Answer: (c) CO_2-EOR

7. In CO_2-enhanced coal bed methane recovery, what happens when CO_2 is injected into a deep coal seam?

   (a) CO_2 adsorbs onto the coal, releasing methane.
   (b) CO_2 dissolves in the coal, enhancing methane production.
   (c) CO_2 displaces methane from the coal, increasing methane recovery.
   (d) CO_2 reacts with coal, forming methane.

   Answer: (a) CO_2 adsorbs onto the coal, releasing methane

8. Which trapping mechanism is related to the higher affinity of coal for adsorbing CO_2?

   (a) Solubility trapping.
   (b) Capillary trapping.
   (c) Capillary trapping.
   (d) Adsorption trapping.

   Answer: (d) Adsorption trapping

9. What is the primary objective of CO_2 injection in the Mt. Simon Sandstone at the IBDP site?

   (a) CO_2 storage.
   (b) CO_2-EOR.
   (c) CO_2 trapping in coal beds.
   (d) CO_2 sequestration in saline aquifers.

   Answer: (a) CO_2 storage

10. Which type of rock is often considered suitable for carbon dioxide storage due to its high porosity and permeability?

    (a) Sandstone.
    (b) Granite.
    (c) Shale.

(d) Basalt.

Answer: (a) Sandstone

11. What is the importance of natural traps in carbon dioxide storage?

    (a) They enhance the permeability of the reservoir rock.
    (b) They provide a source of carbon dioxide for injection.
    (c) They prevent the upward escape and lateral movement of injected gas.
    (d) They facilitate the migration of carbon dioxide to adjacent formations.

    Answer: (c) They prevent the upward escape and lateral movement of injected gas

12. Which of the following mechanisms creates barriers that limit lateral fluid flow and is considered a primary trapping mechanism?

    (a) Capillary trapping.
    (b) Dissolution trapping.
    (c) Structural traps.
    (d) Mineral trapping.

    Answer: (c) Structural traps

13. How does capillary trapping contribute to CO_2 storage?

    (a) It immobilizes CO_2 in smaller pores, preventing its free flow.
    (b) It dissolves CO_2 into formation fluids, reducing its mobility.
    (c) It reacts with minerals, forming stable carbonate minerals.
    (d) It creates barriers that limit lateral fluid flow.

    Answer: (a) It immobilizes CO_2 in smaller pores, preventing its free flow

14. What is the role of mineral trapping in CO_2 storage?

    (a) It creates barriers that limit lateral fluid flow.
    (b) It immobilizes CO_2 in smaller pores, preventing its free flow.
    (c) It dissolves CO_2 into formation fluids, reducing its mobility.
    (d) It reacts with minerals, forming stable carbonate minerals.

    Answer: (d) It reacts with minerals, forming stable carbonate minerals that act as long-term storage.

Knowledge Check Chap. 7

1. What is the primary purpose of conducting an Uncertainty and Contingency Planning in a carbon management project?

    (a) To identify potential project risks and uncertainties.
    (b) To secure surface rights for the project.
    (c) To estimate the cost of CO_2 transportation.
    (d) To determine the optimal site for CCS implementation.

Answer: (a) To identify potential project risks and uncertainties

2. Why is it important to consider the cost of CO_2 transportation in CCS site selection?

   (a) It determines the availability of CO_2 sources.
   (b) It affects the project's financial viability.
   (c) It determines the required staffing and competence levels.
   (d) It ensures compliance with regulatory requirements.

   Answer: (b) It affects the project's financial viability

3. What are some potential risks that need to be considered in a project's Uncertainty and Contingency Planning?

   (a) Geotechnical issues and community acceptance.
   (b) Availability of skilled personnel and funding.
   (c) Compliance with federal, state, and local regulations.
   (d) All the above.

   Answer: (d) All the above

4. What factors influence the economic viability of a CCS project site?

   (a) CO_2 transportation costs and terrain characteristics.
   (b) Skilled personnel and funding availability.
   (c) Uncertainty and Contingency Planning and mitigation plans.
   (d) Regulatory compliance and public acceptance.
   (e) All of the above.

   Answer: (e) All of the above

5. How can a variable CO_2 supply impact the cost of a CCS project?

   (a) It can result in delays in project completion.
   (b) It may require additional data collection and analysis.
   (c) It affects the efficiency and reliability of pipelines and compressors.
   (d) It increases the risk of mechanical failures in equipment.
   (e) All of the above.

   Answer: (e) All of the above

6. What are some considerations in determining the scope of a carbon management project?

   (a) Geology, community, and modeling issues.
   (b) Costs, schedule, and resources.
   (c) CO_2 supply, delivery system options and chemical properties.
   (d) All of the above.

   Answer: (d) All of the above

7. How does the choice of a CCS site relate to the cost of CO_2 transportation?

(a) A closer CO_2 supply reduces transportation costs.
(b) Sensitive areas increase transportation costs.
(c) Rugged topography results in higher transportation costs.
(d) All of the above.

Answer: (d) All of the above

8. What are the key factors to consider when planning for resource needs in a carbon management project?

(a) Availability of federal subsidies and tax offsets.
(b) Sizing of pipelines and compressors for variable supply.
(c) The staffing and skillsets required for each project stage.
(d) The financial viability and profitability of the project.
(e) All of the above.

Answer: (e) All of the above.

Knowledge Check Chap. 8

1. What is the main objective of Geologic Storage Evaluation during site screening?

(a) Evaluate social data.
(b) Identify potential sub-regions.
(c) Analyze demographic trends.
(d) Assess land use patterns.

Answer: (b) Identify potential sub-regions

2. What is one of the key elements evaluated in the Geologic Storage Evaluation during site screening?

(a) Public outreach implications.
(b) Storage Formation.
(c) Pipeline right-of-ways.
(d) Social context.

Answer: (b) Storage Formation

3. What type of areas are considered in the protected and sensitive areas analysis?

(a) Industrial areas.
(b) Urban areas.
(c) Wetlands and sea areas.
(d) Agricultural areas.

Answer: (c) Wetlands and sea areas

4. Why is it important to consider existing resource development during site screening?

(a) It helps in determininbg demographic trends.
(b) It provides information about land ownership patterns.
(c) It allows for the evaluation of potential storage reservoirs.
(d) It can have benefits and concerns for CO_2 storage projects.

Answer: (d) It can have benefits and concerns for CO_2 storage projects

5. What is one factor to evaluate when considering pipeline right-of-ways?

(a) Age of the pipelines.
(b) Climate change attitudes.
(c) Socio-economic issues.
(d) Demographic trends.

Answer: (a) Age of the pipelines

6. What is the purpose of public outreach implications assessment during site screening?

(a) To analyze the regional geological data.
(b) To determine suitable confining zones.
(c) To understand the socio-economic issues of communities.
(d) To identify potential sub-regions.

Answer: (c) To understand the socio-economic issues of communities

7. What are the temperature and pressure conditions required for carbon dioxide (CO_2) to be in its supercritical state?

(a) 25.3 °C (77.5 °F) and 5.4 MPa (785 psi)
(b) 31.3 °C (88.3 °F) and 7.4 MPa (1071 psi)
(c) 20.0 °C (68.0 °F) and 10.0 MPa (1450 psi)
(d) 35.5 °C (95.9 °F) and 3.2 MPa (465 psi)

Answer: (b) 31.3 °C (88.3 °F) and 7.4 MPa (1071 psi)

8. What type of rock units are typically part of a suitable confining zone in geological storage sites?

(a) Igneous rocks and sandstones.
(b) Sandstones and conglomerates.
(c) Shales and mudstones.
(d) Limestones and dolomites.

Answer: (c) Shales and mudstones

9. Which of the following types of data can be used in Geologic Storage Evaluation?

(a) Seismic data.
(b) Core and side wall samples.
(c) Well log data.
(d) Outcrop.

All of the above.

Answer: (e) All of the above

10. What is the purpose of Geologic Storage Evaluation in site selection?

   (a) To analyze community demographics.
   (b) To assess economic feasibility.
   (c) To understand the subsurface structure and properties.
   (d) To evaluate regulatory requirements.

   Answer: (c) To understand the subsurface structure and properties

11. How is seismic data used in Geologic Storage Evaluation?

   (a) To identify top and bottom of formation.
   (b) To detect change in fluid saturations.
   (c) To map the faults and major discontinuities.
   (d) All of the above.

   Answer: (d) All of the above

12. What are the benefits of Geologic Storage Evaluation in site selection?

   (a) Reducing social risks.
   (b) Identifying market trends.
   (c) Enhancing environmental sustainability.
   (d) Reducing uncertainty and improving decision-making.

   Answer: (d) Reducing uncertainty and improving decision-making

13. Which regulatory program is responsible for overseeing underground injection wells in the United States?

   (a) EPA's Clean Air Act program.
   (b) USDA's Natural Resources Conservation Service.
   (c) EPA's Underground Injection Control (UIC) Program.
   (d) DOE's Office of Environmental Management.

   Answer: (c) EPA's Underground Injection Control (UIC) Program

14. What are the primary objectives of the UIC Program in relation to CO_2 geologic storage?

   (a) Protecting underground sources of drinking water (USDWs)
   (b) Promoting community engagement.
   (c) Ensuring financial sustainability.
   (d) Monitoring air quality.

   Answer: (a) Protecting underground sources of drinking water (USDWs)

15. Which class of UIC wells is used for injection of carbon dioxide (CO_2) into underground subsurface rock formations for long-term storage?

   (a) Class I.

(b) Class II.
(c) Class III.
(d) Class VI.

Answer: (d) Class VI

16. What is a key concern in terms of Regulatory Liability for CO_2 geologic storage?

(a) Ensuring community participation.
(b) Evaluating economic feasibility.
(c) Managing technical challenges.
(d) Assigning long-term Regulatory Liability and responsibility.

Answer: (d) Assigning long-term Regulatory Liability and responsibility

17. What is the purpose of developing geologic models in site selection?

(a) Assessing community demographics.
(b) Estimating project costs.
(c) Predicting natural fracture patterns.
(d) Performing numerical simulations and predicting CO_2 movement.

Answer: (d) Performing numerical simulations and predicting CO_2 movement

18. Which parameter is fundamental in modeling the movement of injected CO_2?

(a) Injection rate.
(b) Presence of baffles and natural fractures.
(c) Reservoir rock and fluid properties.
(d) All of the above.

Answer: (d) All of the above

19. What role do models play in assessing the suitability of a storage reservoir for CO_2?

(a) Estimating pressure sidtributions.
(b) Estimating saturation distributions.
(c) Testing assumptions about CO_2 retention.
(d) Predicting CO_2 dynamics after injection.
(e) All of the above.

Answer: (e) All of the above

20. How can models be useful in site selection?

(a) Identifying infrastructure requirements.
(b) Analyzing emission supply trends.
(c) Evaluating economic impact.
(d) Assessing the importance of data uncertainty.

Answer: (d) Assessing the importance of data uncertainty

21. What does pore space ownership refer to in site suitability analysis?

(a) Ownership of land surface.
(b) Ownership of underground minerals.

(c) Ownership of underground voids for CO_2 storage.
(d) Ownership of airspace above the site.

Answer: (c) Ownership of underground voids for CO_2 storage

22. What is the focus of social characterization in site selection?

(a) Assessing geological suitability.
(b) Evaluating financial viability.
(c) Analyzing community demographics.
(d) Examining regulatory requirements.

Answer: (c) Analyzing community demographics

23. Why is infrastructure an important element in site suitability analysis?

(a) It determines social acceptability.
(b) It affects project costs and feasibility.
(c) It influences regulatory requirements.
(d) It impacts geological characteristics.

Answer: (b) It affects project costs and feasibility

24. What is the purpose of a community assessment in social characterization?

(a) Identifying potential social risks.
(b) Assessing financial viability.
(c) Evaluating technical feasibility.
(d) Analyzing market trends.

Answer: (a) Identifying potential social risks

25. What type of data is typically gathered during formation evaluation in site characterization?

(a) Reservoir pressure and fluid samples.
(b) Porosity and permeability measurements.
(c) Injection rates and pressures.
(d) Geological and geophysical data.
(e) All of the above.

Answer: (e) All of the above

26. What is the purpose of conducting a drill stem test (DST) during well testing?

(a) To collect fluid samples for analysis.
(b) To estimate the flow rate of the reservoir.
(c) To measure the formation properties such as permeability and skin.
(d) All of the above.

Answer: (d) All of the above

27. Which of the following information can be obtained from a DST?

(a) Fluid composition analysis.
(b) Reservoir pressure.
(c) Formation permeability.

(d) All of the above.

Answer: (d) All of the above

28. What is the purpose of conducting injection tests during well characterization?

    (a) To measure the pressure of the injected fluid.
    (b) To assess the reservoir's permeability.
    (c) To determine the reservoir boundaries.
    (d) To analyze the fluid composition of the reservoir.

    Answer: (b) To assess the reservoir's permeability

29. What is the purpose of conducting a small-scale hydraulic fracture stress measurement (DFIT)?

    (a) To measure the fracture gradient.
    (b) To formation breakdown pressure.
    (c) To validate the presence of sufficient permeability within the reservoir.
    (d) To determine the magnitudes of in situ stress within the reservoir.
    (e) All of the above.

    Answer: (e) All of the above

Knowledge Check Chap. 9

1. Which of the following is a carbon negative technology that involves adding a base material to the ocean to increase CO_2 absorption and counteract ocean acidification?

    (a) Ocean fertilization.
    (b) Biochar production.
    (c) Direct Air Capture.
    (d) Enhanced weathering.

    Answer: (a) Ocean fertilization

2. What is the primary advantage of Direct Air Capture as a carbon negative technology?

    (a) It requires minimal land use.
    (b) It can be implemented at a large scale.
    (c) It has no environmental impact.
    (d) It is cost-effective.

    Answer: (b) It can be implemented at a large scale.

3. Which carbon negative technology involves the conversion of biomass into biochar through the process of burning with limited oxygen supply?

    (a) Ocean fertilization.
    (b) Enhanced weathering.

(c) Bioenergy with CCS.
(d) Biochar production.

Answer: (d) Biochar production

4. Blue carbon refers to the carbon stored in which type of ecosystem?

(a) Forests and woodlands.
(b) Coastal or marine ecosystems.
(c) Grasslands and prairies.
(d) Wetlands and peatlands.

Answer: (b) Coastal or marine ecosystems

5. Reforestation and afforestation are examples of carbon negative technologies that involve.

(a) Storing carbon in the oceans.
(b) Capturing CO_2 from the air directly.
(c) Converting CO_2 into long-lived products.
(d) Increasing the carbon uptake by planting trees.

Answer: (d) Increasing the carbon uptake by planting trees.

References

1. K.M. Abbott, P.A. Zaidel, A.H. Roy, K.M. Houle, K.H. Nislow, Investigating impacts of small dams and dam removal on dissolved oxygen in streams. PLoS One **17**(11), e0277647 (2022). https://doi.org/10.1371/journal.pone.0277647
2. M.F. Adenan, E. Fathi, T.R. Carr, B. Panetta, Advancing subsurface analysis: Integrating computer vision and deep learning for the near real-time interpretation of borehole image logs in the Illinois Basin-Decatur project. Interpretation **12**(2), SB19–SB27 (2024)
3. R. Ahmed, G. Liu, B. Yousaf, Q. Abbas, H. Ullah, M.U. Ali, Recent advances in carbon-based renewable adsorbent for selective carbon dioxide capture and separation—a review. J. Clean. Prod. **242**, 118409 (2020). https://doi.org/10.1016/j.jclepro.2019.118409
4. A. Al Kindi, V.E. Sebastian, H. Al Hashemi, Non-intrusive corrosion monitoring technology, in *SmartPIMS and MatPIMS. AMPP Annual Conference + Expo*, (Nashville, TN, United States, 2025, April)
5. H. Al-Zawaidah, M. Kammerer, D.M. Mitrano, K. Waldschläger, Opportunities and challenges of a cap-and-trade system for plastics. Environ. Sci. Technol. **59**(4), 1887–1896 (2025). https://doi.org/10.1021/acs.est.4c04931
6. W. Ampomah, R.S. Balch, R.B. Grigg, R. Will, Z. Dai, M.D. White, Paper Presented at the SPE Improved Oil Recovery Conference, in *Farnsworth field CO_2-EOR project: Performance case history*, (Tulsa, Oklahoma, USA, 2016). https://doi.org/10.2118/179528-MS
7. D. Appah, V. Aimikhe, W. Okologume, Assessment of gas leak detection techniques in natural gas infrastructure: a review, in *SPE Nigeria Annual International Conference and Exhibition*, (Lagos, Nigeria, 2021). https://doi.org/10.2118/208236-MS
8. M.M.A. Awan, F.U.D. Kirmani, Evaluating the potential of depleted oil reservoirs for CO_2 sequestration through simulation modeling. Pet. Res. **10**(2), 425–431 (2025)
9. O. Babarinde, B. Schwartz, J. Meng, S. Kim, J.M. Segura, R.A. Schultz, H. Soroush, An overview of geological carbon sequestration and its geomechanical aspects. Geol. Soc., London, Special Publications, **528**(1), 61–72 (2022) https://doi.org/10.1144/SP528-2022-5
10. L.T. Bach, S.J. Gill, R.E.M. Rickaby, S. Gore, P. Renforth, CO_2 removal with enhanced weathering and ocean alkalinity enhancement: potential risks and CO-benefits for marine pelagic ecosystems. Front. Clim. **1**, 7 (2019). https://doi.org/10.3389/fclim.2019.00007
11. S. Bachu, D. Bonijoly, J. Bradshaw, R. Burruss, S. Holloway, N.P. Christensen, CO_2 storage capacity estimation: methodology and gaps. Int. J. Greenh. Gas Con. **1**(4), 430–443 (2007). https://doi.org/10.1016/S1750-5836(07)00086-2
12. S. Bachu, W.D. Gunter, E.H. Perkins, Aquifer disposal of CO_2: hydrodynamic and mineral trapping. Energy Convers. Manag. **35**(4), 269–279 (1994)

E. Fathi and F. Belyadi, *Subsurface CO_2 Sequestration*,
https://doi.org/10.1007/978-3-032-28296-5

13. R. Balch, R. Esser, N. Liu, *Monitoring CO_2 at an Enhanced Oil Recovery and Carbon Capture and Storage Project, Farnsworth Unit, Texas in CO_2 Summit II: Technologies and Opportunities* (Tri-State Generation & Transmission Association Inc. Frank Zhu, UOP/Honeywell Eds, ECI Symposium Series, Holly Krutka, 2016) https://dc.engconfintl.org/co2_summit2/30
14. R. Balch, B. McPherson, Integrating enhanced oil recovery and carbon capture and storage projects: a case study at Farnsworth field, Texas, in *Paper Presented at the SPE Western Regional Meeting*, (Anchorage, Alaska, USA, 2016). https://doi.org/10.2118/180408-MS
15. R. Balch, B. McPherson, M. Cather, A. Meyer, T. Xiao, W. Ampomah, R. Esser, M. Appold, M. White, N. Moodie, J. Heath, A. Luhman, G. El-kaseeh, K. McCormack, S. Lee, K. Hnottavange-Telleen, L. Huang, Z. Wu, R. Grigg, R. Will, *Southwest Regional Partnership on Carbon Sequestration: Phase III (Final Scientific/Technical Report)* (2025). https://doi.org/10.2172/2529468
16. B. Bane, Scientists unveil least costly carbon capture system to date. Pacific Northwest National Laboratory. (2023). https://www.pnnl.gov/news
17. J. Bartels, T. Vogt, V. Bertsch, Development of a strategic site selection framework for the green hydrogen economy: the example of Germany. Energy Convers. Manag. **342**, 120086 (2025). https://doi.org/10.1016/j.enconman.2025.120086
18. H. Belyadi, E. Fathi, F. Belyadi, Introduction to unconventional reservoirs, in *Hydraulic Fracturing in Unconventional Reservoirs*, (Elsevier, 2017), pp. 1–12
19. H. Belyadi, E. Fathi, F. Belyadi, *Hydraulic Fracturing in Unconventional Reservoirs: Theories, Operations, and Economic Analysis* (Gulf Professional Publishing, 2019)
20. W.M. Black, A review of drill-stem testing techniques and analysis. J. Pet. Technol. **8**, 21–30 (1956). https://doi.org/10.2118/589-G
21. H. Boer, Governing ecosystem carbon. Glob. Environ. Polit. **13**(4), 123–143 (2013). https://doi.org/10.1162/GLEP_a_00200
22. S. Bond, *Ship Design for CO_2 Transportation* (Global Gas Development Science Day, London, England, 2024)
23. I.C. Bourg, L.E. Beckingham, D.J. DePaolo, The nanoscale basis of CO_2 trapping for geologic storage. Environ. Sci. Technol. **49**(17), 10265–10284 (2015). https://doi.org/10.1021/acs.est.5b03003
24. D. Breeling, S. Long, Guest Post: How 'enhanced Weathering' Could Slow Climate Change and Boost Crop Yields. Carbon Brief (2018). https://www.carbonbrief.org/guest-post-how-enhanced-weathering-could-slow-climate-change-and-boost-crop-yields/
25. L. Bretschger, E. Grieg, Carbon taxes, CO_2 emissions, and the economy: the effects of fuel taxation in the UK. Energy Policy **195**, 114359 (2024). https://doi.org/10.1016/j.enpol.2024.114359
26. K. Bromberg Gedan, B.R. Silliman, M.D. Bertness, Centuries of human-driven change in salt marsh ecosystems. Annu. Rev. Mar. Sci. **1**(1), 117–141 (2009). https://doi.org/10.1146/annurev.marine.010908.163930
27. G. Brown, S. Fasham, P. Tomlinson, R. Crook, *A case study on the implementation of an acoustic automatic leak detection sonar (ALDS) in GoM and its application to current and future Brazilian fields* (OTC Brasil, Rio de Janeiro, Brazil, 2015). https://doi.org/10.4043/26270-MS
28. S. Budinis et al., *Direct Air Capture: A Key Technology for Net Zero* (International Energy Agency, 2023)
29. B.J.P. Buhre, L.K. Elliott, C.D. Sheng, R.P. Gupta, T.F. Wall, Oxy-fuel combustion technology for coal-fired power generation. Prog. Energy Combust. Sci. **31**(4), 283–307 (2005). https://doi.org/10.1016/j.pecs.2005.07.001
30. J. Burger, J. Nöhl, J. Seiler, P. Gabrielli, P. Oeuvray, V. Becattini, A. Reyes-Lúa, L. Riboldi, G. Sansavini, A. Bardow, Environmental impacts of carbon capture, transport, and storage supply chains: status and the way forward. Int. J. Greenh. Gas Con. **132**, 104039 (2024). https://doi.org/10.1016/j.ijggc.2023.104039
31. A. Busch, S. Alles, Y. Gensterblum, D. Prinz, D.N. Dewhurst, M.D. Raven, H. Stanjek, B.M. Krooss, Carbon dioxide storage potential of shales. Int. J. Greenh. Gas Control. **2**3 2008297–308 (2008). 1750-5836. https://doi.org/10.1016/j.ijggc.2008.03.003

32. T.A. Buscheck, K. Mansoor, X. Yang, H.M. Wainwright, S.A. Carroll, Downhole pressure and chemical monitoring for CO_2 and brine leak detection in aquifers above a CO_2 storage reservoir. Int. J. Greenh. Gas Con. **91**, 102812 (2019). https://doi.org/10.1016/j.ijggc.2019.102812
33. C. Callas, S.D. Saltzer, J.S. Davis, S.S. Hashemi, A.R. Kovscek, E.R. Okoroafor, G. Wen, M.D. Zoback, S.M. Benson, Criteria and workflow for selecting depleted hydrocarbon reservoirs for carbon storage. Appl. Energy **324**, 119668 (2022). https://doi.org/10.1016/j.apenergy.2022.119668
34. D.A. Camerini, J.P. von der Weid, C.S. Camerini, C.E. Maia, Proceedings of the 2004 international pipeline conference. 2004 international pipeline conference, volumes 1, 2, and 3, in *Leak detector pig for oil pipelines*, (Calgary, Alberta, Canada, ASME, 2004), pp. 1065–1068. https://doi.org/10.1115/IPC2004-0095
35. I. Campello Gómez, C. Gutiérrez, Desorption strategies in CO_2 capture technologies: novel approaches and future perspectives. J. Environ. Chem. Eng. **13**(2), 116109 (2025). https://doi.org/10.1016/j.jece.2025.116109
36. X. Cao, Q. Li, L. Xu, Y. Tan, A review of in situ carbon mineralization in basalt. J. Rock Mech. Geotech. Eng. **16**(4), 1467–1485 (2024)
37. T.R. Carr, B.J. Carney, B. Panetta, Evaluating direct deep-use geothermal potential in the Appalachian Basin. Emerg. Energy: Build. Futur. Soc. Explor. (2024b)
38. A. Cavanagh, R.S. Haszeldine, The Sleipner storage site: Capillary flow modeling of a layered CO_2 plume requires fractured shale barriers within the Utsira formation. Int. J. Greenh. Gas Control. **21**, 101–112 (2014). https://doi.org/10.1016/j.ijggc.2013.11.017
39. R.A. Chadwick, B.P. Marchant, G.A. Williams, CO_2 storage monitoring: leakage detection and measurement in subsurface volumes from 3D seismic data at Sleipner. Energy Procedia, **63**, 4224–4239. 1876-6102. (2014). https://doi.org/10.1016/j.egypro.2014.11.458. (https://www.sciencedirect.com/science/article/pii/S1876610214022735)
40. C. Chairunissa, D.K. Amanu, G. Astari, E. Indrayana, SPE/IATMI Asia Pacific Oil & Gas Conference and Exhibition, Virtual, in *Kedung Keris full well stream pipeline fiber optic leak detection system*, (2021). https://doi.org/10.2118/205776-MS
41. H. Chalmers, J. Gibbins, Carbon capture and storage: More energy or less carbon? J. Renew. Sustain. Energy **2**(3), 031006 (2010). https://doi.org/10.1063/1.3446897
42. S. Chen, S. Chen, Y. Zhang, H. Chai, L. Qin, Y. Gong, An investigation of the role of N-methyl-diethanolamine in non-aqueous solution for CO_2 capture process using ^{13}C NMR spectroscopy. Int. J. Greenh. Gas Con. **39**, 166–173 (2015). https://doi.org/10.1016/j.ijggc.2015.05.011
43. T. Chen, L. Huang, Seismicity characterization around the Farnsworth field site for combined large-scale CO_2 storage and EOR. Int. J. Greenh. Gas Con. **33**, 63–68 (2015). https://doi.org/10.1016/j.ijggc.2014.11.024
44. Y. Chen, S. Chen, D. Li, X. Jiang, Density-driven convection for CO_2 solubility trapping in saline aquifers: Modeling and influencing factors. Geotechnics **3**(1), 70–103 (2023)
45. S.-H. Cho, S.K. Park, T.S. Kim, et al., Performance analyses of oxy-fuel power generation systems including CO_2 capture: Comparison of two cycles using different recirculation fluids. J. Mech. Sci. Technol. **24**, 1947–1954 (2010). https://doi.org/10.1007/s12206-010-0623-x
46. S.-H. Cho, M. Soh, B.C. English, T.E. Yu, C.N. Boyer, Targeting payments for forest carbon sequestration given ecological and economic objectives. Forest Policy Econ. **100**, 214–226 (2019). https://doi.org/10.1016/j.forpol.2018.12.003
47. H. Chopra, C.H. Stanfield, Y. Teng, P.K. Thallapally, R. Kahsar, K. Khandelwal, et al., Scrubbing the sky: reimagining the air as a resource. ACS Sustain. Resour. Manag. **40** (2026)
48. D.E. Clark, E.H. Oelkers, I. Gunnarsson, B. Sigfússon, S.Ó. Snæbjörnsdóttir, E.S. Aradóttir, S.R. Gíslason, CarbFix2: CO_2 and H_2S mineralization during 3.5 years of continuous injection into basaltic rocks at more than 250°C. Geochim. Cosmochim. Acta **279**, 45–66 (2020)
49. P.L.C. Clarkson. Liquid CO_2 carriage by sea: an introduction. (2022). https://www.clarksons.com/home/news-and-insights/2022/liquid-co2-carriage-by-sea-an-introduction/

50. M.L. Couëslan, V. Smith, G. El-Kaseeh, J. Gilbert, N. Preece, L. Zhang, J. Gulati, Development and implementation of a seismic characterization and CO_2 monitoring program for the Illinois Basin–Decatur project. Greenh. Gases: Sci. Technol. **4**(5), 626–644 (2014). https://doi.org/10.1002/ghg.1452
51. K. Coyne, Crushing it: can ground-up rock help save the earth. **CSA News, 68**(4), 4–11 (2023). https://doi.org/10.1002/csan.20986
52. S. Crooks, D. Herr, J. Tamelander, D. Laffoley, J. Vandever, *Mitigating Climate Change through Restoration and Management of Coastal Wetlands and near-Shore Marine Ecosystems: Challenges and opportunities (Environment Department Papers No. 121)* (World Bank, 2011)
53. T.M. Daley, D. Vasco, J. Ajo-Franklin, L. Dobeck, L. Spangler, M. Leonti, Final Scientific/Technical Report for project "Geomechanical Monitoring for CO_2 Hub Storage: Production and Injection at Kevin Dome (2017). https://doi.org/10.2172/1366437
54. J. Davison, Performance and costs of power plants with capture and storage of CO_2. Energy **32**(7), 1163–1176 (2007). https://doi.org/10.1016/j.energy.2006.07.039
55. S. Davoodi, M. Al-Shargabi, D.A. Wood, V.S. Rukavishnikov, K.M. Minaev, Review of technological progress in carbon dioxide capture, storage, and utilization. Gas Sci. Eng. **117**, 205070 (2023). https://doi.org/10.1016/j.jgsce.2023.205070
56. S.G. de Ávila, M.A. Logli, J.R. Matos, Kinetic study of the thermal decomposition of monoethanolamine (MEA), diethanolamine (DEA), triethanolamine (TEA) and methyldiethanolamine (MDEA). Int. J. Greenh. Gas Con. **42**, 666–671 (2015). https://doi.org/10.1016/j.ijggc.2015.10.001
57. G. de Koeijer, Y. Enge, K. Sanden, O.F. Graff, O. Falk-Pedersen, T. Amundsen, S. Overå, CO_2 technology Centre Mongstad—design, functionality and emissions of the amine plant. Energy Procedia **4**, 1207–1213 (2011). https://doi.org/10.1016/j.egypro.2011.01.175
58. D. Dempsey, S. Kelkar, R. Pawar, Passive injection: a strategy for mitigating reservoir pressurization, induced seismicity and brine migration in geologic CO_2 storage. Int. J. Greenh. Gas Con. **28**, 96–113 (2014). https://doi.org/10.1016/j.ijggc.2014.06.002
59. H. Deng, C. Yu, Q. Jiang, X. Shi, X. Zhou, X. Chen, Accelerated CO_2 mineralization in basalt via reaction process control: mineralization effect, mineral evolution, and reservoir property implications. Gas Sci. Eng. **140**, 205648 (2025)
60. M. Desservettaz, M. Pikridas, I. Stavroulas, A. Bougiatioti, E. Liakakou, N. Hatzianastassiou, J. Sciare, N. Mihalopoulos, E. Bourtsoukidis, Emission of volatile organic compounds from residential biomass burning and their rapid chemical transformations. Sci. Total Environ. **903**, 166592 (2023). https://doi.org/10.1016/j.scitotenv.2023.166592
61. J. Dolson, Z. He, B.W. Horn, Search and Discovery, in *Advances and perspectives on stratigraphic trap exploration: making the subtle trap obvious*, (2018), p. 60054
62. Drax Group. Drax sets ambition to become world's first carbon negative company. (2019)
63. Y. Du, T. Gao, G.T. Rochelle, A.S. Bhown, Zero- and negative-emissions fossil-fired power plants using CO_2 capture by conventional aqueous amines. Int. J. Greenh. Gas Con. **111**, 103473 (2021). https://doi.org/10.1016/j.ijggc.2021.103473
64. C.M. Duarte, W.C. Dennison, R.J.W. Orth, T.J.B. Carruthers, The charisma of coastal ecosystems: addressing the imbalance. Estuar. Coasts **31**(2), 233–238 (2008). https://doi.org/10.1007/s12237-008-9038-7
65. C.M. Duarte, I.J. Losada, I.E. Hendriks, I. Mazarrasa, N. Marbà, The role of coastal plant communities for climate change mitigation and adaptation. Nat. Clim. Chang. **3**(11), 961–968 (2013). https://doi.org/10.1038/nclimate1970
66. H. Ducklow, D. Steinberg, K. Buesseler, Upper ocean carbon export and the biological pump. Oceanography **14**(4), 50–58 (2001). https://doi.org/10.5670/oceanog.2001.06
67. B. Dziejarski, R. Krzyżyńska, K. Andersson, Current status of carbon capture, utilization, and storage technologies in the global economy: a survey of technical assessment. Fuel **342**, 127814 (2023). https://doi.org/10.1016/j.fuel.2023.127814

68. P.A. Eigbe, O.O. Ajayi, O.T. Olakoyejo, O.L. Fadipe, S. Efe, A.O. Adelaja, A general review of CO_2 sequestration in underground geological formations and assessment of depleted hydrocarbon reservoirs in the Niger Delta. Appl. Energy **350**, 121723 (2023). https://doi.org/10.1016/j.apenergy.2023.121723
69. B. Eisler, G.A. Lanan, Fiber optic leak detection systems for subsea pipelines, in *Offshore Technology Conference*, (Houston, TX, United States, 2012). https://doi.org/10.4043/23070-MS
70. W. El Sherbeny, I. Nuic, G. Hasan, A. Abdesslam, T. Hassane, Magnetic flux leakage (MFL) technology provides the industry's most precise pipe integrity and corrosion evaluation, accurately characterizing casing and tubing strength, in *Technology Overview and Case History. SPE North Africa Technical Conference and Exhibition*, (Cairo, Egypt, 2015). https://doi.org/10.2118/175871-MS
71. W.J. Elliot, H. Rhee, Impacts of forest biomass operations on forest hydrologic and soil erosion processes. Trees, For. People **7**, 100186 (2022)
72. F. Engel, A. Kather, Improvements on the liquefaction of a pipeline CO2 stream for ship transport. Int. J. Greenh. Gas Con. **72**, 214–221 (2018)
73. EPA 430-F-21-004 October 2021. U.S. Cement Industry Carbon Intensities (2019)
74. D. Erik, *Craft, An Economic History of Weather Forecasting* (University of Richmond ECONOMIC HISTORY ASSOCIATION, 2023)
75. I. Ershaghi, *Solved Problems in Well Testing* (Springer International Publishing, 2023)
76. R.A. Esposito, J.C. Pashin, P.M. Walsh, Citronelle dome: A giant opportunity for multizone carbon storage and enhanced oil recovery in the Mississippi interior Salt Basin of Alabama. Environ. Geosci. **15**(2), 53–62 (2008)
77. European Centre for Medium-Range Weather Forecasts, Copernicus Atmosphere Monitoring Service, in *Global Fire Assimilation System (GFAS) v1.2 [Dataset]*, (2023) https://ads.atmosphere.copernicus.eu/
78. J. Ewald, T. Sterner, E. Sterner, Understanding the resistance to carbon taxes: drivers and barriers among the general public and fuel-tax protesters. Resour. Energy Econ. **70**, 101331 (2022). https://doi.org/10.1016/j.reseneeco.2022.101331
79. S. Fahad, S.B. Chavan, A.R. Chichaghare, A.R. Uthappa, M. Kumar, V. Kakade, A. Pradhan, D. Jinger, G. Rawale, D.K. Yadav, et al., Agroforestry systems for soil health improvement and maintenance. Sustainability **14**(22), 14877 (2022). https://doi.org/10.3390/su142214877
80. R. Fahy, Quantification & reduction of methane (GHG) emissions through aerial intelligence (ADIPEC, Abu Dhabi, UAE, 2023). https://doi.org/10.2118/216027-MS
81. E. Fathi, M.F. Adenan, N. Moryan, F. Belyadi, H. Belyadi, High-precision single-leak detection and localization in single-phase liquid pipelines using the negative pressure wave technique: an application in a real-field case study. SPE J. (2023). https://doi.org/10.2118/217455-PA
82. E. Fathi, I.Y. Akkutlu, Multi-component gas transport and adsorption effects during CO_2 injection and enhanced shale gas recovery. Int. J. Coal Geol. **123**, 52–61 (2014)
83. E. Fathi, D. Arcentales, F. Belyadi, Impacts of different operation conditions and geological formation characteristics on CO_2 sequestration in Citronelle dome, Alabama. Energies **16**(7) (2023b). https://doi.org/10.3390/en16073191
84. E. Fathi, F. Belyadi, M.F. Adenan, C. Pacheco, Application of fiber optics for completion design optimization: a methodological approach and key findings. Fuels **5**(1), 33–52 (2024b)
85. E. Fathi, T.R. Carr, M.F. Adenan, B. Panetta, A. Kumar, B.J. Carney, High-quality fracture network mapping using high frequency logging while drilling (LWD) data: MSEEL case study. Mach. Learn. Appl. **10**, 100421 (2022)
86. E. Fathi, G.B. Liwuitekong Dongho, B. Heidari, T.R. Carr, F. Belyadi, I. Bilgesu, Optimizing deep geothermal drilling for energy sustainability in the Appalachian Basin. Sustainability **16**(18), 8053 (2024a). https://doi.org/10.3390/su16188053
87. N. Fernández-Miranda, M.A. Lopez-Anton, T. Torre-Santos, M. Díaz-Somoano, M.R. Martínez-Tarazona, Impact of oxy-fuel conditions on elemental mercury re-emission in wet flue gas desulfurization systems. Environ. Sci. Technol. **50**(13), 7247–7253 (2016). https://doi.org/10.1021/acs.est.6b00210

88. P. Ferreira-Aparicio, M.J. Benito, J.L. Sanz, New trends in reforming technologies: from hydrogen industrial plants to multifuel microreformers. Catal. Rev. **47**(4), 491–588 (2005). https://doi.org/10.1080/01614940500364958
89. R.J. Finley, An overview of the Illinois Basin-Decatur project. Greenh. Gases: Sci. Technol. **4**(5), 571–579 (2014). https://doi.org/10.1002/ghg.1423
90. C.E. Fox, *CO_2 EOR Carbon Balance at SACROC [PowerPoint Slides]* (Kinder Morgan CO2 Company, Houston, TX, United States, 2009)
91. Q. Gao, Y. Cheng, E. Fathi, S. Ameri, Analysis of stress-field variations expected on subsurface faults and discontinuities in the vicinity of hydraulic fracturing. SPE Reserv. Eval. Eng. **19**(1), 54–69 (2016)
92. K.B. Gedan, B.R. Silliman, M.D. Bertness, Centuries of human-driven change in salt marsh ecosystems. Annu. Rev. Mar. Sci. **1**, 117–141 (2009). https://doi.org/10.1146/annurev.marine.010908.163930
93. R. Geyer, J.R. Jambeck, K.L. Law, Production, use, and fate of all plastics ever made. Sci. Adv. **3**(7), e1700782 (2017). https://doi.org/10.1126/sciadv.1700782
94. Global Centre for Maritime Decarbonization. (2024) Landmark study on offloading onboard captured carbon dioxide. https://www.gcformd.org/landmark-study-on-offloading-onboard-captured-carbon-dioxide/
95. M. Godec, G. Koperna, J. Gale, CO_2-ECBM: a review of its status and global potential. Energy Procedia **63**, 5858–5869 (2014). https://doi.org/10.1016/j.egypro.2014.11.619
96. A. Goodman, A. Hakala, G. Bromhal, D. Deel, T. Rodosta, S. Frailey, M. Small, D. Allen, V. Romanov, J. Fazio, N. Huerta, D. McIntyre, B. Kutchko, G. Guthrie, U.S. DOE methodology for the development of geologic storage potential for carbon dioxide at the national and regional scale. Int. J. Greenh. Gas Con. **5**(4), 952–965 (2011). https://doi.org/10.1016/j.ijggc.2011.03.010
97. T. Grant, A. Guinan, C. Shih, S. Lin, D. Vikara, D. Morgan, D. Remson, Comparative analysis of transport and storage options from a CO_2 source perspective. Int. J. Greenh. Gas Con. **72**, 175–191 (2018). https://doi.org/10.1016/j.ijggc.2018.06.012
98. S.E. Greenberg, Scaling up of deep saline storage in Illinois: From the Illinois Basin-Decatur project to the Illinois industrial CCS project, in *CCS Technical Workshop 2016*, (Tokyo, Japan, 2016)
99. N. Gruber, D. Clement, B.R. Carter, R.A. Feely, S. van Heuven, M. Hoppema, M. Ishii, R.M. Key, A. Kozyr, S.K. Lauvset, C.L. Monaco, J.T. Mathis, A. Murata, A. Olsen, F.F. Perez, C.L. Sabine, T. Tanhua, R. Wanninkhof, The oceanic sink for anthropogenic CO_2 from 1994 to 2007. Science **363**(6432), 1193–1199 (2019). https://doi.org/10.1126/science.aau5153
100. D. Guan, P. Gao, Z. Jiang, Q. Fan, Q. Li, Y. Zhou, L. Zhang, J. Zhao, Y. Song, L. Yang, Spatial evolution of CO_2 storage in depleted natural gas hydrate reservoirs and its synergistic efficiency analysis. Appl. Energy **376**(Part A), 124247 (2024)
101. Z.C. Guo, Z.X. Fu, Current situation of energy consumption and measures taken for energy saving in the iron and steel industry in China. Energy **35**(11), 4356–4360 (2010). https://doi.org/10.1016/j.energy.2009.04.008
102. Hamling, J.A., Gorecki, C.D., Klapperich, R.J., Saini, D., Steadman, E.N. Overview of the Bell Creek combined CO_2 storage and CO_2 enhanced oil recovery project, Energy Procedia, Volume 37, 2013, Pages 6402–6411, ISSN 1876-6102. https://doi.org/10.1016/j.egypro.2013.06.570. (https://www.sciencedirect.com/science/article/pii/S1876610213008138)
103. A. Hamza, I.A. Hussein, M.J. Al-Marri, M. Mahmoud, R. Shawabkeh, S. Aparicio, CO_2 enhanced gas recovery and sequestration in depleted gas reservoirs: a review. J. Pet. Sci. Eng. **196**, 107685 (2021)
104. M.C. Hänsel, M. Franks, M. Kalkuhl, O. Edenhofer, Optimal carbon taxation and horizontal equity: a welfare-theoretic approach with application to German household data. J. Environ. Econ. Manag. **116**, 102730 (2022). https://doi.org/10.1016/j.jeem.2022.102730
105. E. Hanson, C. Nwakile, V.O. Hammed, Carbon capture, utilization, and storage (CCUS) technologies: evaluating the effectiveness of advanced CCUS solutions for reducing CO_2 emissions. Results Surf. Interfaces **18**, 100381 (2025). https://doi.org/10.1016/j.rsurfi.2025.100381

106. J.A. Harju, M.D. Jensen, E.N. Steadman, J.A. Sorensen, E.M. O'Leary, 2006 IEEE cement industry technical conference, in *The plains CO_2 reduction (PCOR) partnership—identifying CO_2 sequestration opportunities for the cement industry in the central interior of North America*, (2006), pp. 1–7. https://doi.org/10.1109/CITCON.2006.1635716
107. M.M. Hasan, F. Faruque, F. Boukouvala, E.L. First, C.A. Floudas, Nationwide, regional, and statewide CO_2 capture, utilization, and sequestration supply chain network optimization. Ind. Eng. Chem. Res. **53**(18), 7489–7506 (2014). https://doi.org/10.1021/ie402931c
108. M.M. Hasan, R. Haque, M.I. Jahirul, M.G. Rasul, Pyrolysis of plastic waste for sustainable energy recovery: technological advancements and environmental impacts. Energy Convers. Manag. **326**, 119511 (2025). https://doi.org/10.1016/j.enconman.2025.119511
109. T. He, Z. Liu, H. Son, T. Gundersen, W. Lin, Comparative analysis of cryogenic distillation and chemical absorption for carbon capture in integrated natural gas liquefaction processes. J. Clean. Prod. **383**, 135264 (2023). https://doi.org/10.1016/j.jclepro.2022.135264
110. J.E. Heath, S.J. Bauer, S.T. Broome, T.A. Dewers, M.A. Rodriguez, Petrologic and petrophysical evaluation of the Dallas Center structure, Iowa, for compressed air energy storage in the mount Simon sandstone (SAND2013-7819). Sandia National Laboratories (2013)
111. N. Hernández-Hernández, L.T. Bach, M.F. Montero, J. Taucher, I. Baños, W. Guan, M. Espósito, A. Ludwig, E.P. Achterberg, U. Riebesell, J. Arístegui, High CO_2 under nutrient fertilization increases primary production and biomass in subtropical phytoplankton communities: a mesocosm approach. Front. Mar. Sci. **5**, 213 (2018). https://doi.org/10.3389/fmars.2018.00213
112. B. Hill, S. Hovorka, S. Melzer, Geologic carbon storage through enhanced oil recovery. Energy Procedia **37**, 6808–6830 (2013). https://doi.org/10.1016/j.egypro.2013.06.614
113. K. Hnottavange-Telleen, E. Chabora, R.J. Finley, S.E. Greenberg, S. Marsteller, Risk management in a large-scale CO_2 geosequestration pilot project, Illinois, USA. Energy Procedia **4**, 4044–4051 (2011). https://doi.org/10.1016/j.egypro.2011.02.346
114. M.T. Ho, A. Bustamante, D.E. Wiley, Comparison of CO_2 capture economics for iron and steel mills. Int. J. Greenh. Gas Con. **19**, 145–159 (2013). https://doi.org/10.1016/j.ijggc.2013.08.003
115. A. Holm, D. Jennejohn, L. Blodgett, *Geothermal Energy and Greenhouse Gas Emissions* (Geothermal Energy Association (GEA), 2012)
116. W.Y. Hong, A techno-economic review on carbon capture, utilisation and storage systems for achieving a net-zero CO_2 emissions future. Carbon Capture Sci. Technol. **3**, 100044 (2022). https://doi.org/10.1016/j.ccst.2022.100044
117. P.P. Hosseininoosheri, S.A. Hosseini, V. Nuñez-López, L.W. Lake, Impact of field development strategies on CO_2 trapping mechanisms in a CO_2-EOR field: A case study in the Permian Basin (SACROC unit). Int. J. Greenh. Gas Con. **72**, 92–104 (2018). https://doi.org/10.1016/j.ijggc.2018.03.002
118. R. Hou, C. Fong, B.D. Freeman, M.R. Hill, Z. Xie, Current status and advances in membrane technology for carbon capture. Sep. Purif. Technol. **300**, 121863 (2022). https://doi.org/10.1016/j.seppur.2022.121863
119. K.Z. House, A.C. Baclig, M. Ranjan, E.A. van Nierop, J. Wilcox, H.J. Herzog, Economic and energetic analysis of capturing CO_2 from ambient air. Proc. Natl. Acad. Sci. **108**(51), 20428–20433 (2011). https://doi.org/10.1073/pnas.1012253108
120. S.D. Hovorka, T.A. Meckel, R.H. Treviño, Monitoring a large-volume injection at Cranfield, Mississippi—Project design and recommendations. Int. J. Greenh. Gas Con. **18**, 345–360 (2013). https://doi.org/10.1016/j.ijggc.2013.03.021
121. S.D. Hovorka, T.A. Meckel, R.H. Trevino, J. Lu, J.-P. Nicot, J.-W. Choi, D. Freeman, P. Cook, T.M. Daley, J.B. Ajo-Franklin, B.M. Freifeld, C. Doughty, C.R. Carrigan, D. LaBrecque, Y.K. Kharaka, J.J. Thordsen, T.J. Phelps, C. Yang, K. Romanak, T. Zhang, R.M. Holt, J.S. Lindler, R.J. Butsch, Monitoring a large volume CO_2 injection: year two results from SECARB project at Denbury's Cranfield, Mississippi, USA. Energy Procedia **4**, 3478–3485 (2011). https://doi.org/10.1016/j.egypro.2011.02.274

122. J. Hua, K. Wang, J. Lin, Y. Qian, Carbon tax vs. carbon cap-and-trade: Implementation of carbon border tax in cross-regional production. Int. J. Prod. Econ. **274**, 109317 (2024). https://doi.org/10.1016/j.ijpe.2024.109317
123. S. Hughes, A. Zoelle, *Cost of Capturing CO_2 from Industrial Sources (Report No. DOE/NETL-2022/3300)* (National Energy Technology Laboratory, 2022)
124. F. Huq, S.B. Haderlein, O.A. Cirpka, M. Nowak, P. Blum, P. Grathwohl, Flow-through experiments on water–rock interactions in a sandstone caused by CO_2 injection at pressures and temperatures mimicking reservoir conditions. Appl. Geochem. **58**, 136–146 (2015). https://doi.org/10.1016/j.apgeochem.2015.04.006
125. M. Hussain, T. Zhang, M. Chaudhry, I. Jamil, S. Kausar, I. Hussain, Review of prediction of stress corrosion cracking in gas pipelines using machine learning. Machines **12**(1), 42 (2024). https://doi.org/10.3390/machines12010042
126. IEA, *A Vision for Clean Cooking Access for all*, ed. by Licence: CC BY 4.0, (IEA, Paris. https://www.iea.org/reports/a-vision-for-clean-cooking-access-for-all, 2023)
127. IEAGHG, The status and challenges of CO_2 shipping infrastructures (2020). https://ieaghg-publications.s3.eu-north-1.amazonaws.com/Technical+Reports/2020-10+The+Status+and+Challenges+of+CO2+Shipping+Infrastructures.pdf
128. G. Indelicato, P.E. Lapenna, R. Concetti, M. Caputo, M. Valorani, G. Magnotti, F. Creta, Numerical investigation of high pressure CO-diluted combustion using a flamelet-based approach. Combust. Sci. Technol. **192**(11), 2028–2049 (2020). https://doi.org/10.1080/00102202.2020.1811243
129. A. Ingelson, A. Kleffner, N. Nielson, Long-term liability for carbon capture and storage in depleted North American oil and gas reservoirs-a comparative analysis. Energy LJ **31**, 431 (2010)
130. I. Iyegbekedo, E. Fathi, T.R. Carr, F. Belyadi, Machine learning-driven quantification of CO_2 plume dynamics at Illinois Basin Decatur project sites using microseismic data. Energies **17**(17), 4421 (2024)
131. I.B. Iyegbekedo, S.K. Yerravally, E. Fathi, N. Garapati, Sub-surface analysis for geothermal hot water system for Morgantown, West Virginia, in *Proceedings of the 50th Workshop on Geothermal Reservoir Engineering*, (Stanford University, 2025)
132. M.O. Jeffries, J.E. Overland, D.K. Perovich, The Arctic shifts to a new normal. Phys. Today **66**, 35–40 (2013)
133. W. Jia, B. McPherson, SACROC CMG model, 2018-01-10. https://edx.netl.doe.gov/dataset/sacroc-cmg-model. https://doi.org/10.18141/1465116
134. F. Johnsson, Perspectives on CO_2 capture and storage. Greenh. Gases: Sci. Technol. **1**(2), 119–133 (2011). https://doi.org/10.1002/ghg.9
135. M.W. Jones, G.P. Peters, T. Gasser, R.M. Andrew, C. Schwingshackl, J. Gütschow, R.A. Houghton, P. Friedlingstein, J. Pongratz, C. Le Quéré, National contributions to climate change due to historical emissions of carbon dioxide, methane and nitrous oxide. Earth System Science Data **15**(12), 5301–5369 (2023). https://doi.org/10.5194/essd-15-5301-2023
136. S.M. Kang, E. Fathi, R.J. Ambrose, I.Y. Akkutlu, R.F. Sigal, CO_2 storage capacity of organic-rich shales. SPE J. **16**(4), 842–855 (2011)
137. A. Karayil, A. Elseragy, A.M. Aliyu, An assessment of CO_2 capture technologies towards global carbon net neutrality. Energies **17**(6), 1460 (2024). https://doi.org/10.3390/en17061460
138. P. Kelemen, S.M. Benson, H. Pilorgé, P. Psarras, J. Wilcox, An overview of the status and challenges of CO_2 storage in minerals and geological formations. Frontiers in Climate **1**, 9 (2019). https://doi.org/10.3389/fclim.2019.00009
139. S. Khandoozi, R. Hazlett, M. Fustic, A critical review of CO_2 mineral trapping in sedimentary reservoirs—from theory to application: pertinent parameters, acceleration methods and evaluation workflow. Earth Sci. Rev. **244**, 104515 (2023)
140. Y.K. Kharaka, J.J. Thordsen, E. Kakouros, G. Ambats, W.N. Herkelrath, S.R. Beers, J.T. Birkholzer, J.A. Apps, N.F. Spycher, L. Zheng, R.C. Trautz, H.W. Rauch, K.S. Gullickson, Changes in the chemistry of shallow groundwater related to the 2008 injection of CO_2 at the

ZERT field site, Bozeman, Montana. Environ. Earth Sci. **60**, 273–284 (2010). https://doi.org/10.1007/s12665-009-0401-1
141. K. Kim, D. Kim, Y. Na, Y. Song, J. Wang, A review of carbon mineralization mechanism during geological CO_2 storage. Heliyon **9**(12), e23135 (2023). https://doi.org/10.1016/j.heliyon.2023.e23135
142. J. Koornneef, A. Ramirez, T. van Harmelen, A. van Horssen, W. Turkenburg, A. Faaij, The impact of CO_2 capture in the power and heat sector on the emission of SO_2, NOx, particulate matter, volatile organic compounds and NH_3 in the European Union. Atmos. Environ. **44**(11), 1369–1385 (2010). https://doi.org/10.1016/j.atmosenv.2010.01.022
143. G. Koperna, A. Oudinot, G. McColpin, J. Heath, N. Liu, A. Wells, G. Young, 2009 International Coalbes & Shale Gas Symposium, in *CO_2-ECBM/storage activities at the San Juan Basin's Pump Canyon test site*, (2009)
144. G.J. Koperna, D. Riestenberg, SPE International Conference on CO_2 Capture, Storage, and Utilization, in *Carbon dioxide enhanced coalbed methane and storage: Is there promise*, (San Diego, CA, United States, 2009). https://doi.org/10.2118/126627-MS
145. N.V.S. Korlapati, F. Khan, Q. Noor, S. Mirza, S. Vaddiraju, Review and analysis of pipeline leak detection methods. J. Pipeline Sci. Eng. **2**(4), 100074 (2022). https://doi.org/10.1016/j.jpse.2022.100074
146. H. Kumar, S. Ravikumar, 2010 international conference on recent advances in space technology services and climate change, in *An approach of CO_2 capture technology for mitigating global warming and climate change—an overview*, (2010), pp. 364–371. https://doi.org/10.1109/RSTSCC.2010.5712870
147. E.J. Kutsienyo, M.S. Appold, M.E. Cather, Investigation of the effect of injected CO_2 on the morrow B sandstone through laboratory batch reaction experiments: Implications for CO_2 sequestration in the Farnsworth unit, northern Texas, USA. Energies **16**(12), 4611 (2023). https://doi.org/10.3390/en16124611
148. R. Lahann, J. Rupp, C. Medina, An evaluation of the seal capacity and CO_2 retention properties of the Eau Claire formation (Cambrian). Environ. Geosci. **21**(2), 83–106 (2014). https://doi.org/10.1306/eg.04281313010
149. P. Larkin, W. Leiss, J. Arvai, M. Dusseault, M. Fall, R. Gracie, A. Heyes, D. Krewski, An integrated uncertainty and contingency planning and management framework for carbon capture and storage: a Canadian perspective. Int. J. Uncertain. Conting. Plan. Manag. **22**(1), 21–45 (2019). https://doi.org/10.1504/IJRAM.2019.103336
150. R.G. Larkin, SPE Improved Oil Recovery Symposium, in *Hydrodynamic trapping of CO_2 geosequestered in saline aquifers*, (Tulsa, OK, United States, 2010). https://doi.org/10.2118/128205-MS
151. E. Larson, C. Greig, J. Jenkins, E. Mayfield, A. Pascale, C. Zhang, J. Drossman, R. Williams, S. Pacala, R. Socolow, E. Baik, R. Birdsey, R. Duke, R. Jones, B. Haley, E. Leslie, K. Paustian, A. Swan, *October 29* (Potential pathways, infrastructure, and impacts. Princeton University, Net-zero America, 2021) https://netzeroamerica.princeton.edu/
152. S. Learn, Y. Cheng, R. Dolan, S. Shahidi, Evaluation of internal leak detection techniques, in *PSIG Annual Meeting*, (New Orleans, LA, United States, 2015)
153. S. Lee, H.S. Kim, J. Park, B.M. Kang, C.-H. Cho, H. Lim, W. Won, Scenario-based techno-economic analysis of steam methane reforming process for hydrogen production. Appl. Sci. **11**, 6021 (2021). https://doi.org/10.3390/app11136021
154. H. Lei, S. Yang, L. Zu, Z. Wang, Y. Li, Oil recovery performance and CO_2 storage potential of CO_2 water-alternating-gas injection after continuous CO_2 injection in a multilayer formation. Energy Fuel **30**(11), 8922–8931 (2016). https://doi.org/10.1021/acs.energyfuels.6b01307
155. D.Y.C. Leung, G. Caramanna, M.M. Maroto-Valer, An overview of current status of carbon dioxide capture and storage technologies. Renew. Sust. Energ. Rev. **39**, 426–443 (2014). https://doi.org/10.1016/j.rser.2014.07.093
156. A. Levasseur, Recovery act: Oxy-combustion technology development for industrial-scale boiler applications. Task 4—testing in Alstom's 15 MWth boiler simulation facility (report no. DOE-NETL-40763). U.S. Department of Energy (2014). https://doi.org/10.2172/1159325

157. E. Lewis, J. Moore, R. McDowell, K. Carter, P. Dinterman, W. Harrison, C. Medina, R. Ortt, B. Slater, M. Solis, T. Sparks, The Midwest regional carbon sequestration partnership petroleum fields database: compilation, utilization, and support for CCUS activities. Environ. Geosci. **28**(1), 25–32 (2021). https://doi.org/10.1306/eg.12172020008
158. N. Li, W. Feng, J. Yu, F. Chen, Q. Zhang, S. Zhu, Y. Hu, Y. Li, Recent advances in geological storage: Trapping mechanisms, storage sites, projects, and application of machine learning. Energy Fuel **37**(14), 10087–10111 (2023). https://doi.org/10.1021/acs.energyfuels.3c01433
159. Q. Li, X. Huang, N. Li, T. Qi, R. Wang, L. Wang, S. An, Energy-efficient biphasic solvents for industrial CO_2 capture: absorption mechanism and stability characteristics. Energy **293**, 130710 (2024). https://doi.org/10.1016/j.energy.2024.130710
160. Q. Li, R. Song, H. Shi, J. Ma, X. Liu, X. Li, U-tube based near-surface environmental monitoring in the Shenhua carbon dioxide capture and storage (CCS) project. Environ. Sci. Pollut. Res. **25**(22), 21405–21414 (2018). https://doi.org/10.1007/s11356-018-1252-7
161. S. Li, D. Tasnady, Biochar for soil carbon sequestration: Current knowledge, mechanisms, and future perspectives. **9**, 67 (2023). https://doi.org/10.3390/c9030067
162. X. Li, B. Jiang, Y. Bai, Manufacturers' decision making on emission reduction based on blockchain technology—Analysis under the government's carbon cap-and-trade policy. J. Clean. Prod. **499**, 145224 (2025). https://doi.org/10.1016/j.jclepro.2025.145224
163. Library of Congress, U.S. withdrawal from the Paris Agreement: Process and potential effects (Report No. R48504), (2025). https://www.congress.gov/crs-product/R48504
164. J.T. Litynski, S.M. Klara, H.G. McIlvried, R.D. Srivastava, The United States Department of Energy's regional carbon sequestration partnerships program: a collaborative approach to carbon management. Environ. Int. **32**(1), 128–144 (2006). https://doi.org/10.1016/j.envint.2005.05.019
165. J.T. Litynski, S. Plasynski, H.G. McIlvried, C. Mahoney, R.D. Srivastava, The United States department of energy's regional carbon sequestration partnerships program validation phase. Environ. Int. **34**(1), 127–138 (2008). https://doi.org/10.1016/j.envint.2007.07.005
166. J.T. Litynski, T. Rodosta, D. Vikara, R. Srivastava, U.S. DOE'S R&D program to develop infrastructure for carbon storage: overview of the regional carbon sequestration partnerships and other R&D field projects. Energy Procedia **37**, 6527–6543 (2013). https://doi.org/10.1016/j.egypro.2013.06.584
167. G. Liu, C. Ehlig–Economides, Consistent model for injection and falloff pressure match of diagnostic fracture injection tests (DFITs). SPE Drill & Compl **34**, 345–355 (2019). https://doi.org/10.2118/187191-PA
168. J. Liu, J. Hou, Q. Fan, H. Chen, China's national ETS: global and local lessons, Energy Rep., **8**(Supplement 6), 428–437 (2022). ISSN 2352-4847. https://doi.org/10.1016/j.egyr.2022.03.097
169. L.R. López, P. Dessì, A. Cabrera-Codony, L. Rocha-Melogno, B. Kraakman, V. Naddeo, M.D. Balaguer, S. Puig, CO_2 in indoor environments: From environmental and health risk to potential renewable carbon source. Sci. Total Environ. **856**(2), 159088 (2023). https://doi.org/10.1016/j.scitotenv.2022.159088
170. J.E. Lovich, Golden Eagle mortality at a wind-energy facility near Palm Springs, California. U.S. Geological Survey, Southwest Biological Science Center, Western Birds **46**, 76–80 (2015)
171. Y. Ma, S. Yi, M. Wang, Biomimetic mineralization for carbon capture and sequestration. Carbon Capture Sci. Technol. **13**, 100257 (2024)
172. I.M. Madeira, H.R. Barreto Orlande, M.J. Colaço, Leak detection of pipelines, in *Handbook of Pipeline Engineering*, (Springer International Publishing, Cham, 2024), pp. 1–23
173. M. Mallet, home, environmental, human Healthon posted on November 28, 2024, in *Toxic blade time bomb*, (2024)
174. D. Manzo, R. Thai, H.L. Thu, G.K. Venayagamoorthy Fuel cell technology review: types, economy, applications, and vehicle-to-grid scheme, Sustain. Energy Technol. Assess. **75**, 104229 (2025) ISSN 2213-1388. https://doi.org/10.1016/j.seta.2025.104229
175. O. Massarweh, M. Al-khuzaei, M. Al-Shafi, Y. Bicer, A.S. Abushaikha, Blue hydrogen production from natural gas reservoirs: a review of application and feasibility. J. CO_2 Util. **70**, 102438 (2023). https://doi.org/10.1016/j.jcou.2023.102438

176. J.M. Matter, M. Stute, S.Ó. Snæbjörnsdóttir, E.H. Oelkers, S.R. Gíslason, E.S. Aradóttir, B. Sigfusson, I. Gunnarsson, H. Sigurdardóttir, E. Gunnlaugsson, G. Axelsson, H.A. Alfredsson, D. Wolff-Boenisch, K. Mesfin, D. Fernandez de la Reguera, J. Taya, J. Hall, K. Dideriksen, W.S. Broecker, Rapid carbon mineralization for permanent disposal of anthropogenic carbon dioxide emissions. Science **352**(6291), 1312–1314 (2016)
177. H. McLaughlin, A.A. Littlefield, M. Menefee, A. Kinzer, T. Hull, B.K. Sovacool, M.D. Bazilian, J. Kim, S. Griffiths, Carbon capture utilization and storage in review: Sociotechnical implications for a carbon reliant world. Renew. Sust. Energ. Rev. **177**, 113215 (2023). https://doi.org/10.1016/j.rser.2023.113215
178. C.J. Mcowen, L.V. Weatherdon, J.V. Bochove, E. Sullivan, S. Blyth, C. Zockler, D. Stanwell-Smith, N. Kingston, C.S. Martin, M. Spalding, S. Fletcher, A global map of saltmarshes. Biodivers. Data J. **5**, e11764 (2017). https://doi.org/10.3897/BDJ.5.e11764
179. B.J. McPherson, Looking ahead: Research agenda for the study of carbon sequestration. Geophys. Monogr. Ser. **183**, 305–316 (2009). https://doi.org/10.1029/2008GM000792
180. B.J. McPherson, R.S. Balch, Lessons learned from the southwest regional partnership on carbon sequestration: a comprehensive, 22-year duration field demonstration program, in *Paper Presented at the SPE Europe Energy Conference and Exhibition*, (Turin, Italy, 2024). https://doi.org/10.2118/220068-MSr
181. W.N. Meier et al., Arctic Sea ice in transformation: A review of recent observed changes and impacts on biology and human activity. Rev. Geophys. **52**, 185–217 (2014). https://doi.org/10.1002/2013RG000431
182. R. Metcalfe, K. Thatcher, G. Towler, A. Paulley, J. Eng, Sub-surface uncertainty and contingency planning for the endurance CO_2 store of the White rose project, UK. Energy Procedia **114**, 4313–4320 (2017). https://doi.org/10.1016/j.egypro.2017.03.1578
183. E. Meuleman, Validation of Transformational CO_2 Capture Solvent Technology with Revolutionary Stability (Apollo), in *U.S. Department of Energy National Energy Technology Laboratory Carbon Management Project Review Meeting, United States*, (2022)
184. H. Mikulčić, I.R. Skov, D.F. Dominković, S.R.W. Alwi, Z.A. Manan, R. Tan, N. Duić, S.N.H. Mohamad, X. Wang, Flexible carbon capture and utilization technologies in future energy systems and the utilization pathways of captured CO_2. Renew. Sust. Energ. Rev. **114**, 109338 (2019). https://doi.org/10.1016/j.rser.2019.109338
185. M. Mizanuzzaman, M.H. Kabir, Challenges and threats of rising sea level: a scenario assessment on the coastal areas in Bangladesh. Open Access Libr. J. **12**, 1–1 (2025). https://doi.org/10.4236/oalib.1112814
186. S. Moazzem, M.G. Rasul, M.M.K. Khan, A review on technologies for reducing CO_2 emission from coal-fired power plants, in *Clean Energy for Sustainable Development*, ed. by M.M.K. Khan, (2012), pp. 27–46). IntechOpen. https://doi.org/10.5772/31876
187. M.M. Mohd Pauzi, N. Azmi, K.K. Lau, Emerging solvent regeneration technologies for CO_2 capture through offshore natural gas purification processes. Sustainability **14**(7), 4350 (2022). https://doi.org/10.3390/su14074350
188. M. Montero-Hidalgo, F. Tuya, F. Otero-Ferrer, R. Haroun, F. Santos-Martín, Mapping and assessing seagrass meadows changes and blue carbon under past, current, and future scenarios. Sci. Total Environ. **872**, 162244 (2023). https://doi.org/10.1016/j.scitotenv.2023.162244
189. J. Morales Pedraza, *Status and Perspective in the Use of Coal for Electricity Generation in the North America Region* (Elsevier, 2019)
190. A. Morgan, R. Grigg, W. Ampomah, A gate-to-gate life cycle assessment for the CO2-EOR operations at Farnsworth unit (FWU). Energies **14**(9), 2499 (2021). https://doi.org/10.3390/en14092499
191. D. Morgan, A. Sheriff, M. Mark-Moser, G. Liu, T. Grant, G. Creason, D. Vikara, L. Cunha, Modeling the cost of onshore CO_2 pipeline transport and onshore CO_2 saline storage, in *SPE/AAPG/SEG CCUS Conference*, (TX, United States, Houston, 2024)
192. E. Motaei, T. Ganat, K. Azhar, H. Al-Hadrami, Harnessing deep transient testing for reservoir characterization and CO_2 emission reduction in challenging geological settings. Pet. Res. **9**(3), 380–392 (2024). https://doi.org/10.1016/j.ptlrs.2024.01.014

193. M.T. Mota-Martinez, J.P. Hallett, N. Mac Dowell, Solvent selection and design for CO_2 capture—how we might have been missing the point. Sustain. Energy Fuels **1**(10), 2078–2090 (2017). https://doi.org/10.1039/C7SE00404D
194. A. Muscat, E.M. de Olde, I.J.M. de Boer, R. Ripoll-Bosch, The battle for biomass: a systematic review of food-feed-fuel competition. Glob. Food Sec. **25**, 100330 (2020). https://doi.org/10.1016/j.gfs.2019.100330
195. NOAA National Centers for Environmental Information, Climate at a glance: Global rankings (2025). https://www.ncei.noaa.gov/access/monitoring/climate-at-a-glance/global/rankings
196. NOAA National Centers for Environmental Information, 2021 was world's 6th-warmest year on record (2025). https://www.noaa.gov/news/2021-was-worlds-6th-warmest-year-on-record
197. V. Núñez-López, E. Moskal, Potential of CO_2-EOR for near-term decarbonization. Front. Clim. **1**, 5 (2019). https://doi.org/10.3389/fclim.2019.00005
198. K. Orchard, CO_2 transport: the role of CO_2 shipping, in *UK CCS Research Centre Workshop*, (2019)
199. Organisation for Economic Co-operation and Development, Global plastics outlook (2022). https://stats.oecd.org/viewhtml.aspx?datasetcode=PLASTIC_WASTE_5&lang=en
200. R.J. Orth, J.S. Lefcheck, K.S. McGlathery, L. Aoki, M.L. Luckenbach, K.A. Moore, M.P.J. Oreska, R. Snyder, D.J. Wilcox, B. Lusk, Restoration of seagrass habitat leads to rapid recovery of coastal ecosystem services. Sci. Adv. **6**(41), eabc6434 (2020). https://doi.org/10.1126/sciadv.abc6434
201. K.A. Oyewole, O.B. Okedere, K.O. Rabiu, K.O. Alawode, S. Oyelami, Carbon dioxide emission, mitigation and storage technologies pathways. Sustainable. Environment **9**(1) (2023). https://doi.org/10.1080/27658511.2023.2188760
202. M. Ozaki, T. Ohsumi, CCS from multiple sources to offshore storage site complex via ship transport. Energy Procedia **4**, 2992–2999 (2011). https://doi.org/10.1016/j.egypro.2011.02.157
203. Y. Pan, R.A. Birdsey, J. Fang, R. Houghton, P.E. Kauppi, W.A. Kurz, O.L. Phillips, A. Shvidenko, S.L. Lewis, J.G. Canadell, P. Ciais, R.B. Jackson, S.W. Pacala, A.D. McGuire, S. Piao, A. Rautiainen, S. Sitch, D. Hayes, A large and persistent carbon sink in the world's forests. Science **333**(6045), 988–993 (2011). https://doi.org/10.1126/science.1201609
204. P.W. Parfomak, Carbon dioxide pipelines: safety issues (CRS report IN11944). Congressional Research Service (2022). https://crsreports.congress.gov/product/pdf/IN/IN11944
205. S. Parker, Riviera, in *New Wave of LCO2 Ships Emerges, but Ports Remain a Bottleneck*, (2024) https://www.rivieramm.com/news-content-hub/news-content-hub/new-wave-of-lco2-ships-emerges-but-ports-remain-a-bottleneck-80312
206. C.L. Parkinson, A 40-y record reveals gradual Antarctic Sea ice increases followed by decreases at rates far exceeding the rates seen in the Arctic. Proc. Natl. Acad. Sci. USA **116**(29), 14414–14423 (2019). https://doi.org/10.1073/pnas.1906556116
207. W. Parry, Live Science, in *Could Fertilizing the Oceans Reduce Global Warming*, (2012) https://www.livescience.com/21684-geoengineering-ocean-iron-fertilization.html
208. M. Patterson, P. Singh, H. Cho, The current state of the industrial energy assessment and its impacts on the manufacturing industry. Energy Rep. **8**, 7297–7311 (2022). https://doi.org/10.1016/j.egyr.2022.05.242
209. P.A.E. Pogge von Strandmann, K.W. Burton, S.O. Snaebjornsdottir, B. Sigfusson, E.S. Aradottir, I. Gunnarsson, H.A. Alfredsson, K.G. Mesfin, E.H. Oelkers, S.R. Gislason, Rapid CO_2 mineralisation into calcite at the CarbFix storage site quantified using calcium isotopes. Nat. Commun. **10**(1), 1983 (2019)
210. C.K. Preston, *The Carbon Capture Project at Air Products' Port Arthur Hydrogen Production Facility* (IEAGHG, 2018)
211. T.D. Rathnaweera, W. Wu, Y. Ji, R.P. Gamage, Understanding injection-induced seismicity in enhanced geothermal systems: From the coupled thermo-hydro-mechanical-chemical process to anthropogenic earthquake prediction. Earth Sci. Rev. **205**, 103182 (2020). https://doi.org/10.1016/j.earscirev.2020.103182

212. J. Reynolds, A. Kam, PSIG Annual Meeting, in *An evaluation of negative pressure wave leak detection: challenges, limitations and use cases*, (London, United Kingdom, 2019)
213. S. Rezaei, A. Liu, P. Hovington, Emerging technologies in post-combustion carbon dioxide capture & removal. Catal. Today **423**, 114286 (2023). https://doi.org/10.1016/j.cattod.2023.114286
214. H. Ritchie, How much of global greenhouse gas emissions come from plastics? Our World in Data (2023). https://ourworldindata.org/ghg-emissions-plastics
215. J.J. Roberts, S.M.V. Gilfillan, L. Stalker, M. Naylor, Geochemical tracers for monitoring offshore CO_2 stores. Int. J. Greenh. Gas Con. **65**, 218–234 (2017). https://doi.org/10.1016/j.ijggc.2017.07.021
216. J. Robinson, E.E. Popova, A. Yool, M. Srokosz, R.S. Lampitt, J.R. Blundell, How deep is deep enough? Ocean iron fertilization and carbon sequestration in the Southern Ocean. Geophys. Res. Lett. **41**(7), 2489–2495 (2014). https://doi.org/10.1002/2013GL058799
217. T. Rodosta, W. Aljoe, G. Bromhal, D. Damiani, U.S. DOE regional carbon sequestration partnership initiative: new insights and lessons learned. Energy Procedia **114**, 5580–5592 (2017). https://doi.org/10.1016/j.egypro.2017.03.1698
218. T. Rodosta, J. Litynski, S. Plasynski, L. Spangler, et al., US Department of Energy's regional carbon sequestration partnership initiative: update on validation and development phases. Energy Procedia **4**, 3457–3464 (2011). https://doi.org/10.1016/j.egypro.2011.02.271
219. J. Rogelj, P.M. Forster, E. Kriegler, et al., Estimating and tracking the remaining carbon budget for stringent climate targets. Nature **571**, 335–342 (2019). https://doi.org/10.1038/s41586-019-1368-z
220. N. Routley, Mapped: The 1.2 billion people without access to electricity. Visual Capitalist (2019). https://www.visualcapitalist.com/1-2-billion-people-without-access-to-electricity/
221. M. Ruth, S.A. Gabriel, K.L. Palmer, D. Burtraw, A. Paul, Y. Chen, B.F. Hobbs, D. Irani, J. Michael, K.M. Ross, R. Conklin, J. Miller, Economic and energy impacts from participation in the regional greenhouse gas initiative: a case study of the state of Maryland, Energy Policy **36**, 6, 2008, 2279–2289, ISSN 0301-4215. https://doi.org/10.1016/j.enpol.2008.03.012
222. D. Saebea, P. Ruengrit, A. Arpornwichanop, Y. Patcharavorachot, Gasification of plastic waste for synthesis gas production. Energy Rep. **6**(Suppl. 1), 202–207 (2020). https://doi.org/10.1016/j.egyr.2019.08.043
223. D. Saini, Engineering aspects of geologic CO_2 storage. Spring (2017). https://doi.org/10.1007/978-3-319-56074-8
224. A. Samir, F.H. Ashour, A.A.A. Hakim, et al., Recent advances in biodegradable polymers for sustainable applications. NPJ Mater. Degrad. **6**(1), 68 (2022). https://doi.org/10.1038/s41529-022-00277-7
225. M.A. Sattar, M.M. Garcia, R. Banasiak, L.M. Portela, L. Babout, Electrical resistance tomography for control applications: quantitative study of the gas-liquid distribution inside a cyclone. Sensors **20**, 6069 (2020). https://doi.org/10.3390/s20216069
226. Schaper A., (2021). A survey of active U.S. carbon capture projects. https://oaktrust.library.tamu.edu/server/api/core/bitstreams/8ef77e1b-5c49-45fc-924f-4088eff1be21/content
227. C. Scheurer, R.E. Hincapie, E. Neubauer, A. Metz, D. Ness, Sorption of nanomaterials to sandstone rock. Nano **12**, 200 (2022). https://doi.org/10.3390/nano12020200
228. R.A. Schultz, S. Williams-Stroud, B. Horváth, J. Wickens, et al., Underground energy-related product storage and sequestration: Site characterization, risk analysis and monitoring. Geol. Soc., London, Special Publications, **528**, 37–59 (2022). https://doi.org/10.1144/SP528-2022-66
229. M. Schwander, Carbon capture and storage: an industry viewpoint. Sustain. Technol., Syst. Policies **2012**(1) (2012). https://doi.org/10.5339/stsp.2012.ccs.10
230. S. Shackley, C. McLachlan, C. Gough, The public perception of carbon dioxide captures and storage in the UK: Results from focus groups and a survey. Clim. Pol. **4**(4), 377–398 (2004). https://doi.org/10.1080/14693062.2004.9685538
231. C.-H. Shei, J.C.-E. Liu, I.-Y.L. Hsieh, Distributional effects of carbon pricing: An analysis of income-based versus expenditure-based approaches. J. Clean. Prod. **465**, 141446 (2024). https://doi.org/10.1016/j.jclepro.2024.141446

232. F. Sheikh, Abu Dhabi international petroleum exhibition and conference, in *Commercialization of Al Reyadah—World's 1st Carbon Capture CCUS Project from Iron and Steel Industry for Enhanced Oil Recovery CO_2-EOR*, (Abu Dhabi, United Arab Emirates, 2021). https://doi.org/10.2118/207676-MS
233. M. Shen, L. Tong, S. Yin, C. Liu, L. Wang, W. Feng, Y. Ding, Cryogenic technology progress for CO_2 capture under carbon neutrality goals: A review. Sep. Purif. Technol. **299**, 121734 (2022). https://doi.org/10.1016/j.seppur.2022.121734
234. SIGTTO, *SIGTTO Information Paper No.15:A Listing of Design Guidelines for Liquefied Gas Terminals (Referencing Ports and Jetties)* (Witherby & Co. Ltd., Livingston, UK, 1997)
235. A. Singh, K. Stéphenne, Shell Cansolv CO_2 capture technology: achievement from first commercial plant. Energy Procedia **63**, 1678–1685 (2014). https://doi.org/10.1016/j.egypro.2014.11.177
236. A.K. Sleiti, W. Al-Ammari, S. Ahmed, J. Kapat, Direct-fired oxy-combustion supercritical-CO_2 power cycle with novel preheating configurations—Thermodynamic and exergoeconomic analyses. Energy **226**, 120441 (2021). https://doi.org/10.1016/j.energy.2021.120441
237. S. Snæbjörnsdóttir, S.R. Gislason, CO_2 storage potential of basaltic rocks offshore Iceland. Energy Procedia **86**, 371–380 (2016). https://doi.org/10.1016/j.egypro.2016.01.038
238. S.Ó. Snæbjörnsdóttir, E.H. Oelkers, K. Mesfin, E.S. Aradóttir, K. Dideriksen, I. Gunnarsson, E. Gunnlaugsson, J.M. Matter, M. Stute, S.R. Gislason, The chemistry and saturation states of subsurface fluids during the in-situ mineralisation of CO_2 and H_2S at the CarbFix site in SW-Iceland. Int. J. Greenhouse Gas Control **58**, 87–102 (2017). https://doi.org/10.1016/j.ijggc.2017.01.007
239. S.Ó. Snæbjörnsdóttir, F. Wiese, T. Fridriksson, H. Ármansson, G.M. Einarsson, S.R. Gislason, CO_2 storage potential of basaltic rocks in Iceland and the oceanic ridges. Energy Procedia **63**, 4585–4600 (2014)
240. M.R. Soltanian, M.A. Amooie, D.R. Cole, D.E. Graham, S.A. Hosseini, S. Hovorka, S.M. Pfiffner, T.J. Phelps, J. Moortgat, Simulating the Cranfield geological carbon sequestration project with high-resolution static models and an accurate equation of state. Int. J. Greenhouse Gas Control **54**(1), 282–296 (2016). https://doi.org/10.1016/j.ijggc.2016.10.002
241. X.Y.D. Soo, J.J.C. Lee, W.-Y. Wu, L. Tao, C. Wang, Q. Zhu, J. Bu, Advancements in CO_2 capture by absorption and adsorption: a comprehensive review. J. CO_2 Util. **81**, 102727 (2024). https://doi.org/10.1016/j.jcou.2024.102727
242. L.H. Spangler, A. Cihan, B. Winkelman, et al., Big Sky Regional Carbon Sequestration Partnership (Phase III Final Scientific/Technical Report) (2020)
243. H. Spidle, Smart methane emission detection system development. Oil & Natural Gas 2020 Integrated Review (2020)
244. Splash247.com, *Shipping of carbon dioxide readies to go mainstream* (Splash Extra, 2023) https://splash247.com/shipping-of-carbon-dioxide-readies-to-go-mainstream/
245. R. Stanger, T. Wall, R. Spörl, M. Paneru, S. Grathwohl, M. Weidmann, G. Scheffknecht, D. McDonald, K. Myöhänen, J. Ritvanen, S. Rahiala, T. Hyppänen, J. Mletzko, A. Kather, S. Santos, Oxyfuel combustion for CO_2 capture in power plants. Int. J. Greenhouse Gas Control **40**, 55–125 (2015). https://doi.org/10.1016/j.ijggc.2015.06.010
246. R.M. Stephens, *Quest CCS Project RAM Study—Final Report* (Shell Global Solution, 2010) https://open.alberta.ca/dataset/46ddba1a-7b86-4d7c-b8b6-8fe33a60fada/resource/a2b53125-0fbf-4eda-a314-005fe2931dd8/download/ramfinalreport.pdf
247. S. Suk, S. Lee, Y.S. Jeong, The Korean emissions trading scheme: Business perspectives on the early years of operations. Clim. Pol. **18**(6), 715–728 (2018). https://doi.org/10.1080/14693062.2017.1346499
248. S.H. Tak, S.K. Park, T.S. Kim, et al., Performance analyses of oxy-fuel power generation systems including CO_2 capture: Comparison of two cycles using different recirculation fluids. J. Mech. Sci. Technol. **24**, 1947–1954 (2010). https://doi.org/10.1007/s12206-010-0623-x
249. The White House Office of Domestic Climate Policy, Methane emissions reduction action plan: Critical and commonsense steps to cut pollution and consumer costs, while boosting good-paying jobs and American competitiveness (2021)

250. S. Themann, H.M. Schmidt, D. Esser, Measurement, monitoring, and verification of CO_2 storage: An Integrated Approach, in *SPE International Conference on CO_2 Capture, Storage, and Utilization*, (San Diego, CA, United States, 2009). https://doi.org/10.2118/129127-MS
251. C. Thiberville, Y. Wang, P. Waltrich, W. Williams, S.I. Kam, Modeling of smart pigging for pipeline leak detection. SPE Prod. Oper. **35**, 610–627 (2020). https://doi.org/10.2118/198648-PA
252. K.W. Thoning, P.P. Tans, W.D. Komhyr, Atmospheric carbon dioxide at Mauna Loa observatory: 2. Analysis of the NOAA GMCC data, 1974–1985. J. Geophys. Res. **94**(D6), 8549–8565 (1989). https://doi.org/10.1029/JD094iD06p08549
253. K. Thuerbach, N. Johnson, Analyzing California carbon allowance prices: a case for the rate-adjusted Price floor. Commodity Insights Digest (2024) bayes-cid.com
254. I. Tiseo, *Cement Industry Emissions Worldwide—Statistics & Facts* (Statista, Energy & Environment, Emissions, 2025)
255. S. Topham, S. Bazzanella, S. Luhr, L. Zhao, A. Otto, D. Stolten, Carbon dioxide, in *Ullmann's Encyclopedia of Industrial Chemistry*, (Wiley-VCH Verlag GmbH & Co. KGaA, 2014), p. 43
256. R. Trautz, D. Thomas, D. Miller, M. Robertson, G. Koperna, D. Riestenberg, Geophysical monitoring using active seismic techniques at the Citronelle Alabama CO_2 storage demonstration site, Int. J. Greenh. Gas Con. **99**, 103084 (2020). ISSN 1750-5836. https://doi.org/10.1016/j.ijggc.2020.103084
257. United Nations Climate Change. (2024). From Billions to Trillions: Setting a New Goal on Climate Finance
258. United Nations Framework Convention on Climate Change, Paris agreement, (2015). https://unfccc.int/sites/default/files/english_paris_agreement.pdf
259. S. Valluri, V. Claremboux, S. Kawatra, Opportunities and challenges in CO_2 utilization. J. Environ. Sci. **113**, 322–344 (2022). https://doi.org/10.1016/j.jes.2021.05.043
260. B. Van der Bruggen, Microfiltration, ultrafiltration, nanofiltration, reverse osmosis, and forward osmosis, in *Fundamental Modelling of Membrane Systems*, ed. by P. Luis, (Elsevier, 2018), pp. 25–70. https://doi.org/10.1016/B978-0-12-813483-2.00002-2
261. D.P. Van Vuuren et al., The representative concentration pathways: An overview. Clim. **109**(1–2), 5–31 (2011)
262. K. Varshney, M. Pedersen Zari, N. Bakshi, Carbon sequestration and habitat provisioning through building-integrated vegetation: A global survey of experts. Buildings **12**, 1458 (2022). https://doi.org/10.3390/buildings12091458
263. Vattenfall. (2008). Vattenfall inaugurates world's first pilot for a coal-fired power plant with CO_2 capture [Press release]. https://group.vattenfall.com/press-and-media/press-releases/2008/vattenfall-inaugurates-worlds-first-pilot-for-a-coal-fired-power-plant-with-co2-capture
264. J. von Rothkirch, O. Ejderyan, Anticipating the social fit of CCS projects by looking at place factors. Int. J. Greenhouse Gas Control **110**, 103399 (2021). https://doi.org/10.1016/j.ijggc.2021.103399
265. K. Walker, A.P. Garcia, J. Nunn, G. Lyman, SPE Energy Transition Symposium, in *CO_2 leak detection and conformance verification using borehole gravity*, (Houston, TX, United States, 2023). https://doi.org/10.2118/215734-MS
266. J.E. Walsh, Melting ice: What is happening to Arctic Sea ice, and what does it mean for us. Oceanography **26**, 171–181 (2013)
267. H. Wang, Z. Kou, Z. Ji, S. Wang, Y. Li, Z. Jiao, M. Johnson, J.F. McLaughlin, Investigation of enhanced CO_2 storage in deep saline aquifers by WAG and brine extraction in the Minnelusa sandstone, Wyoming. Energy **265**, 126379 (2023). https://doi.org/10.1016/j.energy.2022.126379
268. Warwick, P. D., Attanasi, E. D., Blondes, M. S., Brennan, S. T., Buursink, M. L., Cahan, S. M., Doolan, C. A., Freeman, P. A., Karacan, C. Ö., Lohr, C. D., Merrill, M. D., Olea, R. A., Shelton, J. L., Slucher, E. R., & Varela, B. A. (2022). National assessment of carbon dioxide enhanced oil recovery and associated carbon dioxide retention resources—Results (U.S. Geological Survey circular 1489). https://doi.org/10.3133/cir1489

269. P.D. Warwick, E.D. Attanasi, R.A. Olea, M.S. Blondes, P.A. Freeman, S.T. Brennan, M.D. Merrill, M.K. Verma, C.Ö. Karacan, J.L. Shelton, C.D. Lohr, H. Jahediesfanjani, J.N. Roueché, A probabilistic assessment methodology for carbon dioxide enhanced oil recovery and associated carbon dioxide retention (U.S. Geological Survey scientific investigations report 2019–5115). (2019). https://doi.org/10.3133/sir20195115
270. M. Waycott, C.M. Duarte, T.J.B. Carruthers, R.J. Orth, W.C. Dennison, S. Olyarnik, A. Calladine, J.W. Fourqurean, K.L. Heck, A.R. Hughes, G.A. Kendrick, W.J. Kenworthy, F.T. Short, S.L. Williams, Accelerating loss of seagrasses across the globe threatens coastal ecosystems. Proc. Natl. Acad. Sci. **106**(30), 12377–12381 (2009). https://doi.org/10.1073/pnas.0905620106
271. B. Wei, B. Wang, X. Li, M. Aishan, Y. Ju, *CO_2 Storage in Depleted Oil and Gas Reservoirs: A Review* (Advances in Geo-Energy Research, 2023)
272. J. Wei, Y. Wang, Y. Liu, Installation and pricing decision in a two-channel supply chain under carbon cap-and-trade mechanism. Sustain. Futur. **10**, 100876 (2025). https://doi.org/10.1016/j.sftr.2025.100876
273. N. Wei, X. Li, Y. Wang, R.T. Dahowski, C.L. Davidson, G.S. Bromhal, A preliminary sub-basin scale evaluation framework of site suitability for onshore aquifer-based CO_2 storage in China. Int. J. Greenhouse Gas Control **12**, 231–246 (2013). https://doi.org/10.1016/j.ijggc.2012.10.012
274. M.H. Weik, *Fiber Optics Standard Dictionary*, 3rd edn. (Springer, 1997). https://doi.org/10.1007/978-1-4615-6149-4
275. F.J. Wentz, M. Schabel, Effects of satellite orbital decay on MSU lower tropospheric temperature trends. J. Clim. **11**(1), 1–9 (1998). https://doi.org/10.1175/1520-0442(1998)011<0001:EOSODO>2.0.CO;2
276. B. Wetenhall, H. Aghanjani, H. Chalmers, S.D. Benson, M.-C. Ferrari, J. Li, J.M. Race, P. Singh, J. Davison, Impact of CO_2 impurity on CO_2 compression, liquefaction, and transportation. Energy Procedia **63**, 2764–2778 (2017)
277. L. Whitmarsh, D. Xenias, C.R. Jones, Framing effects on public support for carbon capture and storage. Palgrave Commun. **5**(1), 17 (2019). https://doi.org/10.1057/s41599-019-0217-x
278. Å. Widén, B. Malm Renöfält, E. Degerman, D. Wisaeus, R. Jansson, Environmental flow scenarios for a regulated river system: Projecting catchment-wide ecosystem benefits and consequences for hydroelectric production. Water Resour. Res. **58**, e2021WR030297 (2022). https://doi.org/10.1029/2021WR030297
279. D.J. Wuebbles, D.W. Fahey, K.A. Hibbard, D.J. Dokken, B.C. Stewart, Maycock, T. K. (Eds.)., *Climate Science Special Report: Fourth National Climate Assessment* (Volume I. U.S, Global Change Research Program, 2017). https://doi.org/10.7930/J0J964J6(USCGRP)
280. R. Wüstenhagen, M. Wolsink, M.J. Bürer, Social acceptance of renewable energy innovation: An introduction to the concept. Energy Policy **35**(5), 2683–2691 (2007). https://doi.org/10.1016/j.enpol.2006.12.001
281. P. Xing, J. McLennan, J. Moore, In-situ stress measurements at the Utah frontier observatory for research in geothermal energy (FORGE) site. Energies **13**(21), 5842 (2020). https://doi.org/10.3390/en13215842
282. C. Yang, K. Romanak, S. Hovorka, R.M. Holt, J. Lindner, R. Trevino, Near-surface monitoring of large-volume CO_2 injection at Cranfield: Early field test of SECARB phase III. SPE J. **18**(3), 486–494 (2013). https://doi.org/10.2118/163075-PA
283. X. Yang, T. Buscheck, K. Mansoor, Z. Wang, K. Gao, L. Huang, D. Appriou, S. Carroll, Assessment of geophysical monitoring methods for detection of brine and CO_2 leakage in drinking water aquifers. Int. J. Greenhouse Gas Control **90**, 102803 (2019). https://doi.org/10.1016/j.ijggc.2019.102803
284. X. Yang, R.N. Lassen, K.H. Jensen, M.C. Looms, Monitoring CO_2 migration in a shallow sand aquifer using 3D crosshole electrical resistivity tomography. Int. J. Greenhouse Gas Control **42**, 534–544 (2015). https://doi.org/10.1016/j.ijggc.2015.09.005
285. H. Yin, B. Yu, Z. Zhang, L. Jia, J. Xue, X. Chen, S. Wang, J. Han, Z. Xing, Evaluating the impact of different afforestation timescales on the soil quality in a typical watershed valley

using the criteria importance through intercriteria correlation method and gray target model. Front. For. Glob. Chang. **5**, 997949 (2022). https://doi.org/10.3389/ffgc.2022.997949
286. J. You, W. Ampomah, A. Morgan, Q. Sun, X. Huang, A comprehensive techno-eco-assessment of CO_2 enhanced oil recovery projects using a machine-learning assisted workflow. Int. J. Greenhouse Gas Control **111**, 103480 (2021). https://doi.org/10.1016/j.ijggc.2021.103480
287. M.R. Zakaria, M.A.A. Farid, Y. Andou, I. Ramli, M.A. Hassan, Production of biochar and activated carbon from oil palm biomass: Current status, prospects, and challenges. Ind. Crop. Prod. **199**, 116767 (2023). https://doi.org/10.1016/j.indcrop.2023.116767
288. O.A.M. Zamani, D. Knez, Well integrity in salt cavern hydrogen storage. Energies **17**(14), 3586 (2024)
289. Y. Zapata, M.R. Kristensen, N. Huerta, C. Brown, C.S. Kabir, Z. Reza, CO_2 geological storage: Critical insights on plume dynamics and storage efficiency during long-term injection and post-injection periods. J. Nat. Gas Sci. Eng. **83**, 103542 (2020). https://doi.org/10.1016/j.jngse.2020.103542
290. H. Zentou, B. Hoque, M.A. Abdalla, A.F. Saber, O.Y. Abdelaziz, M. Aliyu, A.M. Alkhedhair, A.J. Alabduly, M.M. Abdelnaby, Recent advances and challenges in solid sorbents for CO_2 capture. Carbon Capture Sci. Technol. **15**, 100386 (2025). https://doi.org/10.1016/j.ccst.2025.100386
291. Zero Emissions Platform & Carbon Capture & Storage Association. (2024). Achieving a European market for CO_2 transport by ship
292. J. Zhang, Pore pressure prediction from well logs: Methods, modifications, and new approaches. Earth-Sci. Rev. **108**(1–2), 50–63 (2011)
293. X. Zhao, H. Deng, W. Wang, et al., Impact of naturally leaking carbon dioxide on soil properties and ecosystems in the Qinghai-Tibet plateau. Sci. Rep. **7**, 3001 (2017). https://doi.org/10.1038/s41598-017-02500-x
294. X. Zhao, K.H. Smith, M.A. Simioni, W. Tao, S.E. Kentish, W. Fei, G.W. Stevens, Comparison of several packings for CO_2 chemical absorption in a packed column. Int. J. Greenhouse Gas Control **5**(5), 1163–1169 (2011). https://doi.org/10.1016/j.ijggc.2011.07.006
295. B. Zheng, J. Xu, Carbon capture and storage development trends from a techno-paradigm perspective. Energies **7**(8), 5221–5250 (2014). https://doi.org/10.3390/en7085221

MIX
Papier aus verantwortungsvollen Quellen
Paper from responsible sources
FSC® C105338

If you have any concerns about our products,
you can contact us on
ProductSafety@springernature.com

In case Publisher is established outside the EU,
the EU authorized representative is:
Springer Nature Customer Service Center GmbH
Europaplatz 3, 69115 Heidelberg, Germany

Printed by Libri Plureos GmbH
in Hamburg, Germany